# Inflammatory Bowel Disease: Translation from Basic Research to Clinical Practice

FALK SYMPOSIUM 140

# Inflammatory Bowel Disease: Translation from Basic Research to Clinical Practice

Edited by

**J.-F. Colombel**
*Hôpital Claude Huriez*
*CHRU Lille*
*Lille*
*France*

**J. Schölmerich**
*Innere Medizin I*
*Klinikum der Universität Regensburg*
*Regensburg*
*Germany*

**C. Gasché**
*Universitätskliniken Wien*
*Klinik für Innere Medizin IV*
*Wien*
*Austria*

**B. Vucelic**
*University Hospital Rebro*
*Division of Gastroenterology*
*Zagreb*
*Croatia*

*Proceedings of the Falk Symposium 140 held in Dubrovnik, Croatia,*
*May 7–8, 2004*

 Springer

Library of Congress Cataloging-in-Publication Data is available.

ISBN 1-4020-2847-4

---

Published by Springer,
PO Box 17, 3300 AA Dordrecht, The Netherlands

Sold and distributed in North, Central and South America
by Springer,
101 Philip Drive, Norwell, MA 02061 USA

In all other countries, sold and distributed
by Springer,
PO Box 322, 3300 AH Dordrecht, The Netherlands

*Printed on acid-free paper*

Printed and bound in Great Britain by MPG Books Limited, Bodmin, Cornwall.

# Contents

CONTENTS

CONTENTS

# CONTENTS

# List of principal contributors

**T Ahmad**
Gastroenterology Unit
University of Oxford
Gibson Laboratories
Radcliffe Infirmary
Woodstock Road
Oxford, OX2 6HE
United Kingdom

**S Bar-Meir**
Chaim Sheba Medical Center
Department of Gastroenterology
2 Sheba Road
IL-52621 Tel Hashomer
Israel

**CR Boland**
Baylor University Medical Center
Division of Gastroenterology
3500 Gaston Avenue
Dallas, TX 75246
USA

**E Butruk**
Institute of Oncology
Department of Gastroenterology
ul. Roentgena 5
PL-02 781 Warsaw
Poland

**R Caprilli**
Dipartimento di Scienze Cliniche
I Cattedra di Gastroenterologia
Università di Roma 'La Sapienza'
Policlinico Umberto I
Viale del Policlinico 155
I-00161 Roma
Italy

**J-F Colombel**
Hôpital Claude Huriez
CHRU Lille
Gastroenterology and Hepatology
1, Place de Verdun
F-59037 Lille
France

**J Cosnes**
Service de Gastroenterologie et
   Nutrition
Hôpital Saint-Antoine
184 rue du Faubourg Saint-Antoine
F-75571 Paris cedex 12
France

**S Cukovic-Cavka**
Division of Gastroenterology
Department of Internal Medicine
University Hospital Rebro
Kispaticeva 12
HR-10000 Zagreb
Croatia

**G D'Haens**
Imelda GI Clinical Research Center
Imeldalaan 9
B-2820 Bonheiden
Belgium

**MC Dubinsky**
UCLA School of Medicine
Cedars-Sinai Medical Center
Pediatric IBD Center, Suite 1165W
8635 West Third Street
Los Angeles, CA 90048
USA

**LJ Egan**
Mayo Clinic
Division of Gastroenterology and
  Hepatology
200 First Street SW
Rochester, MN 55905
USA

**RN Fedorak**
University of Alberta
Division of Gastroenterology
Suite 205 College Plaza
8215-112 Street
Edmonton, AB T6G 2C8
Canada

**D Franchimont**
Université Libre de Bruxelles
Hôpital Erasme
Department of Gastroenterology
Route de Lennik 808
B-1070 Bruxelles
Belgium

**C Gasché**
Universitätskliniken Wien
Klinik für Innere Medizin IV
Währinger Gürtel 18-20
A-1090 Wien
Austria

**M Gassull**
Hospital Universitari Germans
Trias i Pujol
Servicio de Gastroenterologia
Carretera del Canyet s/n
E-08916 Badalona
Spain

**MB Grisham**
Department of Molecular and
  Cellular Physiology
LSU Health Sciences Center
1501 Kings Highway
Shreveport, LA 71130-3932
USA

**B Hamm**
Department of Radiology
Charité – Universitätsmedizin Berlin
Campus Charité Mitte
Humboldt Universität zu Berlin
Schumannstrasse 20-21
D-10117 Berlin
Germany

**SB Hanauer**
University of Chicago
Department of Medicine,
  Gastroenterology Section
5841 S. Maryland Avenue, MC 4076
Chicago, IL 60637
USA

**DP Jewell**
University of Oxford
Radcliffe University
Gastroenterology Unit
Nuffield Department of Medicine
Woodstock Road
Oxford, OX2 6HE
United Kingdom

**MA Kamm**
St Mark's Hospital
Gastroenterology Unit
Watford Road
Harrow, HA1 3UJ
United Kingdom

**S Kolacek**
Children's Hospital
Pediatric Gastroenterology and
  Nutrition
Klaiceva 16
HR-10000 Zagreb
Croatia

**Z Krastev**
Clinic of Gastroenterology
Department of Internal Medicine
Medical Faculty of Sofia
University Hospital St. Ivan Rilsky
15, Blvd Ivan Geshov
BG-1431 Sofia
Bulgaria

**R Löfberg**
Karolinska Institute
HMQ Sophia Hospital
IBD Unit
S-141 86 Huddinge
Sweden

**M Lukáš**
Charles University
General Faculty Hospital
IV Medical Department
U nemocnice 2
CZ-12000 Praha 2
Czech Republic

**S Markovic**
University Medical Center, Llubljana
Department of Gastroenterology
Japljeva 2
SLO-1000 Ljubljana
Slovenia

**F Michelassi**
Department of Surgery
Weill Medical College of Cornell
  University
New York Presbyterian Hospital-
  Weill Cornell Medical Center
525 East 68th Street
New York, NY 10021
USA

**P Michetti**
CHUV
Departement de Médecine Interne
Division de la Gastroentérologie et
  Hépatologie
Rue du Bugnon 46
CH-1011 Lausanne
Switzerland

**W Miehsler**
Medical University of Vienna
Department of Internal Medicine IV
Division of Gastroenterology and
  Hepatology
Währinger Gürtel 18-20
A-1090 Wien
Austria

**MF Neurath**
Department of Medicine I
University of Mainz
Langenbeckstrasse 1
D-55101 Mainz
Germany

**CA O'Morain**
Adelaide and Meath Hospitals
Department of Gastroenterology
Trinity College
IRL-Dublin 24
Ireland

**DJ Philpott**
Institut Pasteur
Groupe d'Immunité Innée et
  Signalisation
28 rue du Dr Roux
F-75724 Paris cedex 15
France

**DK Podolsky**
Gastrointestinal Unit and Center for
  the Study of Inflammatory Bowel
  Disease
Harvard Medical School and
  Massachusetts General Hospital
55 Fruit Street
Boston, MA 02114-2696
USA

**JJ Powell**
St Thomas' Hospital
The Rayne Institute
Gastrointestinal and Immunology
  Department
London SE1 7EH
United Kingdom

**D Rachmilewitz**
Shaare Zedek Medical Center
Department of Medicine
Beit Vagan, PO Box 3235
IL-91 031 Jerusalem
Israel

**GAW Rook**
Centre for Infectious Diseases and
  International Health
Windeyer Institute of Medical
  Sciences
Royal Free and Univ. College
  Medical School
46 Cleveland Street
London, W1P 6DB
United Kingdom

**WJ Sandborn**
Mayo Clinic
Division of Gastroenterology and
  Hepatology
200 First Street SW
Rochester, MN 55905
USA

**J Schölmerich**
Innere Medizin I
Klinikum der Universität Regensburg
D-93042 Regensburg
Germany

**M Schwab**
Dr Margarete Fischer-Bosch
  Institute of Clinical Pharmacology
Auerbachstrasse 112
D-70376 Stuttgart
Germany

**EF Stange**
Department of Internal Medicine I
Robert-Bosch-Krankenhaus
Auerbachstrasse 110
D-70376 Stuttgart
Germany

**RW Stockbrügger**
Academisch Ziekenhuis Maastricht
Afd. Interne Geneeskunde
Werkgroep Gastroent/Hepatology
Postbus 5800
NL-6202 AZ Maastricht
The Netherlands

**H Tilg**
Department of Medicine
University Hospital Innsbruck
Anichstrasse 35
A-6020 Innsbruck
Austria

**A Timmer**
Innere Medizin I
Klinikum der Universitat Regensburg
D-93042 Regensburg
Germany

**Z Tulassay**
Semmelweiss University
Medical School
II Department of Medicine
Szentkiraly u. 46
H-1088 Budapest
Hungary

**K Turetschek**
Department of Radiology
Medical University of Vienna
Währinger Gürtel 18-20
A-1090 Wien
Austria

**GNJ Tytgat**
Department of Gastroenterology
Academic Medical Centre
Meibergdreef 9
NL-1105 AZ Amsterdam
The Netherlands

**S Vermeire**
Internal Medicine - Division of
  Gastroenterology
University Hospital Gasthuisberg
Herestraat 49
B-3000 Leuven
Belgium

**B Vucelic**
University Hospital Rebro
Division of Gastroenterology
Department of Medicine
Kispaticeva 12
HR-10000 Zagreb
Croatia

# Preface

The amount of information on the pathogenesis of IBD is rapidly growing. This is reflected by a continuous increase in the number of papers presented at numerous international GI meetings and published in various relevant journals. To make things more difficult for practicing physicians, there is also a large number of clinical trials being published which require periodical critical reviews and recommendations. Faced with this issues, the scientific committee of the Falk symposium No. 140 decided to introduce a novel format of the meeting that is also reflected in the title of the Symposium "Translation from basic research to clinical practice". Sessions were designed in a way that they started with the information from basic sciences on different aspects of these complex diseases and further lead to its clinical implications. Special attention was payed to the mechanisms of action of established drugs. The last two sessions were clinically oriented and focused on the most difficult aspects of both Crohn's disease and ulcerative colitis. We expected the chosen format of the Symposium to enable participants to enjoy the lectures given by leading experts in the field and at the same time have the benefit of receiving an up-to-date information on the clinical application of the new knowledge. The outstanding faculty assembled from all over the world enabled the realisation of our plans in the best possible way. Combined with the beauty of historical Dubrovnik, the meeting was exciting indeed.

The editors are indebted to Dr. Dr. Herbert Falk and the Falk Foundation, Freiburg for their support and, in cooperation with BTI O-Tours Zagreb, for the exceptional organisation of the meeting.

*Jean-Frederic Colombel*
*Christoph Gasche*
*Juergen Schoelmerich*
*Boris Vucelic*

# Section I
# Genetics

**Chair: D.P. JEWELL and Z. TULASSAY**

# 1
# Molecular classification of inflammatory bowel disease

T. AHMAD

## INTRODUCTION

Diagnosis and classification of disease is central to the practice of medicine, permitting the identification of subgroups that differ in aetiology, natural history, prognosis and response to therapy. Inflammatory bowel disease (IBD) has traditionally been categorized as either ulcerative colitis (UC) or Crohn's disease (CD) on the basis of clinical, radiological and histological criteria. However, over the past 40 years the heterogeneous nature of these diseases has become increasingly apparent. These clinical observations have suggested the existence of phenotypic subtypes based on features such as disease location, behaviour, natural history, and response to treatment. Recent advances in our understanding of the genetics of IBD, in particular the identification of CARD15, have provided the opportunity to explore the genetic basis for this clinical heterogeneity. The rapidly emerging genetic data are beginning to challenge traditional clinical classifications, promising a new taxonomy of disease based on molecular rather than clinical definitions. Such a classification is vital in understanding the molecular mechanisms and complex environmental interactions specific to disease subgroups. In time this may permit accurate diagnosis and prediction of disease course, complications and response to treatment. This presentation reviews recent progress towards a molecular classification of IBD.

## THE CLINICAL CLASSIFICATION OF THE IBD

Early attempts to classify patients with CD were made in the 1970s based on the anatomical location of disease. Data from subsequent studies validated this approach by demonstrating that disease location has important implications for both medical[1,2] and surgical therapy[3–5]. In 1988 additional distinct subgroups of CD were described, defined by disease behaviour[6]. Perforating disease was shown to be distinct from non-perforating disease, a finding that has been confirmed in subsequent studies[7]. Moreover, recurrent postoperative

3

disease was shown to follow the same behaviour pattern as the primary disease[6]. Data from these observational studies led to the development of a formal consensus classification of CD, with the objective of standardizing the description of study populations in clinical trials, and to aid the correlation of putative aetiological factors with particular clinical phenotypes. The first such formal attempt was produced in 1992[8] and subsequently refined in 1998 to produce the now widely used Vienna classification[9]. This categorizes CD phenotype according to three clinical variables: age at diagnosis [below 40 years (A1), equal to or above 40 years (A2)]; location [terminal ileum (L1), colon (L2), ileocolonic (L3), upper gastrointestinal (L4)]; and behaviour [non-stricturing non-penetrating (B1), stricturing (B2), penetrating (B3)]. Data from recent serological studies support this clinical classification of CD: anti-*Saccharomyces cerevisiae* antibodies (ASCA) appear to define a group of patients with fibrostenosing small bowel disease whilst anti-neutrophil cyto-plasmic antibodies (ANCA) define a group with left-sided colitis similar to UC[10,11].

In contrast to CD, there is considerably less evidence for the existence of discrete clinical subgroups in UC. Extent of inflammation is commonly used to classify disease as this predicts the clinical severity, need for surgery and development of cancer. However, it is clear that this is not a stable character-istic with disease extension and regression occurring in more than two-thirds of patients with time. Other clinical characteristics commonly used to classify UC include the pattern of relapse and remission, the response to drugs (e.g. steroid responsive, refractory and resistant) and the need for surgery.

## EPIDEMIOLOGICAL EVIDENCE FOR A GENETIC CONTRIBUTION TO THE DISEASE HETEROGENEITY OF IBD

From the mid-1990s evidence for a possible genetic basis for these subgroups emerged from epidemiological studies in multiply affected families[12–17]. The evidence is particularly convincing for CD, which has been more extensively studied than UC. In a study from Oxford 82% of affected siblings were concordant for disease type (CD or UC), 76% for disease extent and 84% for extra-intestinal manifestations. Concordance was greater in siblings than in parent–child pairs[13]. A similar study from Johns Hopkins of 60 multiply-affected CD families identified concordance rates of 86% for disease site and 82% for disease behaviour[14]. Similar findings have been reported from centres in France[15] and Belgium[16]. Of note, concordance rates in the French study increased with the number of affected relatives within the family[15]. Less data on familial concordance are available for UC. In a study from Oxford concor-dance rates for disease extent were 53% in 17 parent–child pairs and 69% in 35 sibling pairs[13]. These figures were lower than for CD–CD pairs, consistent with other evidence suggesting a smaller contribution of genetics to disease susceptibility in UC compared to CD.

# TOWARDS A MOLECULAR CLASSIFICATION OF IBD

## A genetic model of IBD

A wealth of data now supports a genetic model of IBD in which an individual's clinical manifestations of IBD result from the interaction of one or more of a number of genetic variants, with environmental factors. This model provides the framework for a future molecular classification of IBD. It proposes a hierarchy of genetic influence, with disease susceptibility being determined by a limited number of susceptibility genes either acting alone or in combination (epistasis); while disease phenotype is determined by the interaction of both susceptibility and modifying genes. Different patterns of gene variation may lead to similar or identical clinical phenotypes. Such genetic heterogeneity, which may occur within homogeneous populations, is likely to be most marked between different ethnic groups.

## Susceptibility genes

The development in the early 1990s of detailed microsatellite marker maps[18], together with advances in molecular and computational technologies, facilitated simultaneous linkage analyses at multiple markers distributed across the genome. This technique was employed by Hugot and colleagues in 1996 to carry out the first genome-wide screen for CD[19]. Since then, subsequent IBD genome scans from seven independent groups have been published. Putative loci have been identified on almost every chromosome, suggesting that overall genetic susceptibility to IBD is conferred by genetic variation at a number of loci. Importantly only seven of these, designated IBD1–7, meet strict criteria for significant linkage[20] and have been replicated in independent studies (Figure 1). These loci include regions on chromosomes 16q (IBD1), 12q (IBD2), 6p (IBD3), 14q (IBD4), 5q (IBD5), 19p (IBD6) and 1p (IBD7). Some loci have been shown to be specific to either UC (e.g. IBD2) or CD (e.g. IBD1), whilst others confer common susceptibility to IBD[21].

Typically, genome-wide scans can only map susceptibility genes to relatively large loci, which may contain several hundred genes. Narrowing down these loci may be achieved using both linkage and association strategies with an increased marker density. In 2001 this approach was successfully used for the first time in any complex disease to identify the first CD susceptibility gene, CARD15, within the IBD1 locus[22]. More recently similar mapping strategies have been used to identify putative IBD genes on chromosomes 5q (IBD5)[23] and 10q[23].

### Phenotype modifier genes

It is proposed that modifier genes have no effect on disease susceptibility, but in the presence of susceptibility genes act to modify disease phenotype, including disease progression, complications and response to treatment. Such modifiers may also act to influence disease penetrance[24], explaining the apparent differences in IBD inheritance patterns between families.

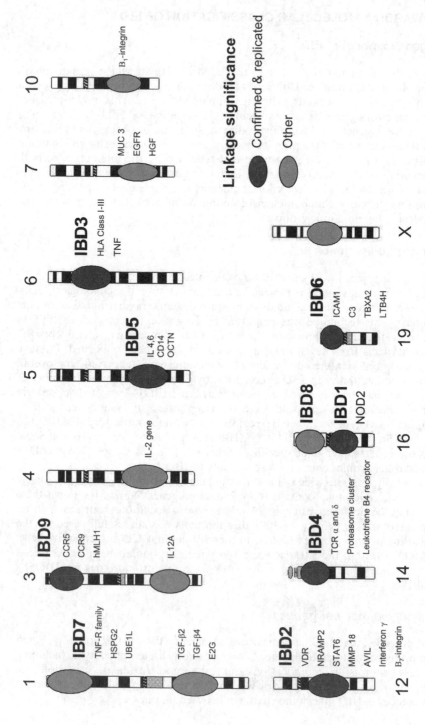

**Figure 1** IBD linkage areas. *Significance levels defined by Lander and Kruglyak (Modified from ref. 21)

# NOD2/CARD15

Association between three variants of NOD2/CARD15 and susceptibility to CD (and not UC), first described in 2001[22,25,26], has been replicated in a number of cohorts[21]. These variants comprise a frameshift mutation (SNP13, Leu1007finsC), which leads to a truncation of the protein transcript and two other common non-conservative polymorphisms (SNP8, Arg702Trp; SNP12, Gly908Arg). In addition, a further 27 rare, potentially disease-causing mutations have been identified in a large European cohort of patients[27]. Consistent across most North American and European cohorts, carriage of a single NOD2 variant is associated with a 2–3-fold risk of disease. This relative risk rises to more than 20 in individuals carrying two variant alleles, supporting a recessive model for CD[28,29]. In these populations the population-attributable risk of the three common mutations for CD has been estimated to be 20–30%. However, the attributable risk of these polymorphisms varies with ethnicity. The association appears to be not as strong among the Finns[30], the Irish[31] and the Scots[32]. The frequency of CARD15 mutations is also lower in African-American compared to Caucasian cohorts[33] and in the Japanese[34,35], Korean[36], and Chinese[37] populations CARD15 variants are not found. CD is common among Jews, particularly of Ashkenazi origin. However, the attributable and relative risks of CARD15 allele variants in this population are considerably less than those observed in other high-risk Caucasian groups, reflecting, in part, the higher prevalence of CARD15 variants (in particular the 908Arg allele) in the healthy Jewish population[38,39].

## CARD15 – association with location and behaviour of CD

More than 10 studies have examined the phenotypic expression of the NOD2/CARD15 variants. Phenotype definitions, methods of classification and duration of follow-up have varied widely. Nonetheless all but one of these studies have reported association with either ileal[40–44], or fibrostenosing disease[27,38,45], or both[46]. Dissecting out whether the primary association is with ileal rather than stenosing disease is difficult, as these clinical variables are not independent. These associations are present in Jewish[38] and non-Jewish populations, and are apparent with all three of the NOD2/CARD15 variants, although in two studies the relative risk was greatest with the frameshift mutation[38,40]. Interestingly the association also appears to be present in patients with ileocolonic disease, suggesting that the NOD2/CARD15 variants are associated with ileal disease irrespective of the presence of colonic disease. This unprecedented concordance between centres is particularly significant considering the difficulties of phenotype definition discussed above. However, it is important to note that not all patients with ileal disease have NOD2/CARD15 mutations, suggesting that mutations in other genes (or other NOD2 mutations) might produce a similar phenotype. The variant alleles do not appear to determine severity of ileal or fibrostenosing disease, as judged by the time to first surgery[47] or surgical recurrence[40].

It is not yet clear precisely how CARD15 variants cause ileal or fibrostenosing disease, but a number of hypotheses have been proposed. One potential explanation for this association is the prominent expression of CARD15 in Paneth cells[48,49]. Paneth cells are specialized intestinal epithelial cells involved in host defence against enteric infection. They respond to bacterial products by secreting antimicrobial peptides. The greatest number of Paneth cells is present in the terminal ileum, and they are generally absent from the colon and rectum, except in IBD when they may also occur in the right colon. Inherited CARD15 mutations may therefore cause ileal disease by abrogating the appropriate responses of Paneth cells. An alternative or complementary explanation is that ileal involvement follows selective colonization of the ileum by bacterial species which are normally suppressed by intact CARD15 function[50].

## CARD 15 – association with familial disease?

Several epidemiological studies have reported a higher incidence of ileal or ileocolonic disease in familial cases of CD[15,51,52]. Seven studies analysed association between family history and NOD2/CARD15 status, of which three reported significantly higher frequencies of NOD2/CARD15 polymorphism in patients with familial disease[39,41,46]. Of note in the study from New York, comprising 481 exclusively Jewish patients, association with the NOD2 variants was present only in patients with familial disease[39].

## CARD15 – association with early age at diagnosis

Five studies analysed association between age of onset and NOD2/CARD15 status, of which three reported association between early age of diagnosis and possession of two variant alleles[27,39,40].

## IBD5 and CD

The IBD5 linkage area on chromosome 5 (5q31–33) was originally identified in 1999 in a genome scan from the USA[53] and subsequently replicated in a Canadian cohort[54]. This locus appears specific to CD. Linkage evidence increased when analysis included families with one individual diagnosed with CD under age 16 (LOD score 3.9). Rioux et al. went on to fine-map this region by genotyping an increased number of microsatellite markers. Eleven known genes were sequenced for possible polymorphisms[54]. Of a total of 16 polymorphisms identified, only two demonstrated association, neither of which appeared to be an attractive candidate for disease. Further mapping of this area identified a single, highly conserved 250 kb haplotype associated with CD that spanned an important cytokine gene cluster[55]. The genotype relative risk for this haplotype was shown to be 2 for heterozygotes, rising to 6 for homozygotes. This association has been replicated in other studies, although the relative risks associated with possession of this haplotype were less than in the initial Canadian report[56–58].

A recent study claims to have identified the causative variants in the IBD5 locus. By re-sequencing the five genes lying within the boundaries of the IBD5 haplotype they identified 10 novel SNP including two putative SNP in the organic cationic transporter genes OCTN1 (sodium-dependent) and OCTN2

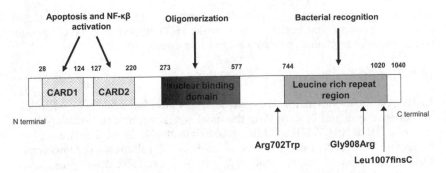

**Figure 2** Structure of the CARD15 gene and location of the Crohn's disease-associated variants. The numbers represent the amino-acid position. CARD, caspase recruitment domain; NBD, nucleotide-binding domain; LRR, leucine-rich repeat

(sodium-independent) which alter transcription and transporter functions. These genes are widely expressed but *in-situ* hybridization shows that they are particularly expressed in epithelial cells, CD68+ macrophages, and CD43+ T cells. A missense substitution in OCTN1 (L503F) and a G→C transversion in the OCTN2 promoter form a highly conserved haplotype associated with susceptibility to CD only. Possession of the risk haplotype confers a 3–4-fold risk of disease, comparable with possession of a single CARD15 variant allele. This risk rises to 7–10-fold in individuals possessing variants at both loci, suggesting these loci interact in an epistatic fashion to confer susceptibility to CD[23].

## IBD5 – association with location and behaviour of CD

In a genotype–phenotype study from Oxford the IBD5 risk haplotype described by Rioux et al. was shown to be specifically associated with perianal CD, with the greatest relative risk seen in individuals homozygous for the risk haplotype (RR 3–4)[58]. A subsequent large study failed to identify a genotype-phenotype association with this haplotype, although specific analysis for association with perianal disease was not carried out. Phenotypic analysis with the OCTN risk haplotype has not yet been reported[57].

## HLA complex and IBD

The HLA region (encompassed within IBD3) is a highly complex, gene-dense region located on chromosome 6p. It encodes the classical class I and II genes essential for normal lymphocyte function and, in addition, a further 200+ genes, many of which have immunoregulatory roles. While there can be no doubt that genes in the HLA region contribute to IBD susceptibility, the majority of reported associations in phenotypically unselected CD and UC cohorts have been inconsistent, even within defined ethnic groups, reflecting many of the methodological problems discussed earlier. More convincing associations with discrete phenotypic subgroups have emerged from recent studies, suggesting that perhaps genes in the HLA region may have a greater role in modifying phenotype than determining disease susceptibility

## HLA and UC

In UC the most convincing HLA associations have been with class II alleles and disease susceptibility. To date the most consistent associations have been between DRB1*0103, DRB1*1502, and DRB1*0401, all of which have been highlighted by a recent meta-analysis[59]. Within UC subgroups the uncommon DRB1*0103 is consistently associated with both extensive disease and severe disease requiring colectomy[60-64]. Among patients requiring colectomy carriage of DRB1*0103 is also associated with shorter time to surgery (9 months vs 5.2 years, $p = 0.002$)[63]. However, the low frequency of this allele, even in the UC population, suggests that this association is unlikely to be clinically useful in predicting disease course.

## HLA – susceptibility locus for colonic CD

Interestingly, DRB1*0103, which has been associated with both UC suscept-ibility and clinical subtype, has also been associated with CD[40,65,66]. In a recent Canadian study a subgroup analysis revealed that an association with DRB1*0103 was observed only in CD patients with pure colonic disease[66]. Although this study is underpowered, the observation of a shared association of DRB1*0103 with UC and colonic CD provides a tantalizing clue as to the potential molecular basis for definition of this subgroup. Furthermore, the existence of a shared UC and CD susceptibility allele on chromosome 6 may explain why both forms of IBD can coexist in a family at a frequency greater than expected by chance. Indeed, whereas NOD2/CARD15 is strongly associated with ileal CD, the HLA region is particularly associated with colonic CD. In addition to the association of colonic disease with the rare DRB1*0103 allele, we have recently shown an association with the common classical autoimmune haplotype (A1B8DR3) and colonic CD, with or without ileal disease, with a relative risk of 3[40]. While this requires replication, these data are consistent with reports from a Belgian group demonstrating association of colonic disease with the TNF promoter allele –308A, which occurs on this haplotype[67].

## HLA genes and the extraintestinal manifestations of IBD

Phenotype may also be defined by the presence of extraintestinal manifesta-tions of IBD, and although inevitably limited by inadequate sample size a number of genetic associations have been described. In studies comprising both UC and CD patients a migratory pauciarticular large joint arthritis was shown to be associated with HLA-DRB1*0103, B*27 and B*35, whereas a chronic, small joint, symmetrical arthritis was associated with HLA-B*44[68]. Uveitis has been associated with HLA-B*27 and DRB1*0103, and erythema nodosum with the TNF promoter SNP TNF-1031C[69].

## CURRENT CHALLENGES TO THE MOLECULAR CLASSIFICATION OF IBD

There are a number of obstacles that currently challenge the development of a molecular classification of IBD.

The first challenge is to narrow down the large linkage loci to areas encompassing a manageable number of candidate genes for further study. This task is made more difficult by the apparent clustering of multiple candidate genes to a limited number of genomic regions and by the complex and extensive patterns of linkage disequilibrium across these regions. Recent studies of the IBD1, IBD5 and Ch10 linkage regions have shown that this can now be successfully achieved with appropriately designed studies.

Demonstration that not all patients with ileal CD have NOD2/CARD15 variants suggests that the same phenotype may be caused by mutations in different genes. Such genetic heterogeneity may in part explain the failure to replicate genetic associations in different populations, and suggests that a future molecular classification will need to be population-specific.

The current clinical classification and definitions of IBD pose perhaps the greatest challenge to the development of a molecular classification. It is now clear that schemes such as the Vienna classification are insufficiently detailed to prevent ambiguous assignment of disease subgroup. Particularly problematic is the subclassification of disease behaviour into inflammatory (non-stricturing–non-penetrating), stricturing (stenosing), or fistulizing (penetrating) patterns. Not surprisingly, poor inter- and intraobserver variability results, even among experts[70]. This is likely to be a particular problem with multicentre studies.

Further difficulty is encountered in the classification of phenotypes which are not stable with time, such as CD behaviour. This was illustrated in a study by Cosnes, who found that, of more than 2000 CD patients initially without complications of stenosis or penetrating disease, only a small fraction remained free of complications after 20 years[71]. The stability of phenotype has also been examined in a Belgian population of patients with CD followed up to 25 years[72]. Although location of disease remained relatively stable there were striking changes in disease behaviour over time: within 10 years 46% of patients had changed disease behaviour from purely inflammatory disease to either stricturing (27.1%) or penetrating (29.4%). Similar problems are encountered when phenotyping UC patients, as disease may extend or regress unpredictably with time[73,74]. These observations should be kept in mind when interpreting data from studies that have evaluated disease behaviour at only a single point in time.

It is now clear that the expression of a specific genetic mutation is strongly influenced by an individual's genetic background. There are a growing number of examples from other human diseases[24] and from mouse models of IBD, that gene–gene interactions may affect disease penetrance, dominance and phenotype. Such genetic interaction adds further complexity to the identification of IBD genes. Recent studies in CD have provided evidence of epistatic interaction between CARD15 and CD14[75], IBD5[76], DLG5 113A[77] and OCTN 1 and 2[23].

**Table 1**  Future clinical applications of a molecular classification of IBD

---

Aetiology
   Facilitate understanding of molecular mechanisms specific to disease subgroups
   Enable identification of environmental factors specific to disease subgroups
   Provide insight into the aetiology of other disorders

Permit accurate early diagnosis based upon molecular mechanisms, rather than phenotype

Screening of 'at-risk' individuals
   Target prevention at individuals at risk

Predict disease course
   Need for surgery
   Postoperative recurrence
   Development of cancer
   Development of extraintestinal manifestations
   Development of osteoporosis

Therapeutics
   Drug development focused on specific molecularly defined disease subtypes
   Ensure disease homogeneity in pharmaceutical studies
   Optimal patient treatment tailored to specific disease subtype
   Prediction of adverse drug reactions

---

Finally the impact of environmental factors on a molecular classification of disease cannot be forgotten. Cigarette smoking is arguably the most important environmental factor identified in IBD. In CD, cigarette smoking increases disease susceptibility and exerts an effect on disease location, disease course, success of medical therapy and need for repeat surgery[78]. Smoking is a risk factor for ileal disease, independent of CARD15[79]. In contrast, in UC smoking cessation appears to increase the severity of disease, as reflected by the need for hospital admission and major medical therapy[80]. In discordant twin pairs with CD and UC, smokers tended to develop CD whilst non-smokers tended to develop UC, suggesting that, in individuals susceptible to IBD, smoking may interact with the genotype to influence the expression of disease type[81]. In UC patients, appendicectomy may influence overall susceptibility and modify the clinical course of disease. Finally studies in mice models of IBD suggest that the species of bacteria colonizing the gut may act to determine disease phenotype. Thus in the IL-10 knockout mouse colonization with *Escherichia coli* produces a caecal-dominant disease, whilst *E. faecalis* produces a distal colonic disease.

## CONCLUSIONS

For 30 years data from genetic studies in IBD have been disappointingly inconsistent. The recent identification of CARD15 represents a major land-mark in our understanding of IBD genetics. Unlike previously reported genetic associations the biological plausibility, strength of association, dose–response effect and consistency between independent studies, suggest a causal role for the CARD15 variants. Subsequent reports demonstrating the specific association with ileal or fibrostenosing CD have validated the approach of classifying patients into accurately defined clinical subgroups. Understanding the genetic basis for the complex clinical patterns of IBD phenotype represents the crucial first step in understanding the molecular mechanisms specific to subgroups of disease. The emerging genetic data suggest that IBD comprises a heterogeneous family of oligogenic inflammatory disorders in which the specific clinical manifestations of disease in any individual are determined by the interaction of a number of genetic variants and environmental factors.

Data from recent genetic studies are already being used in the laboratory to redefine disease on a molecular basis. This has immediate scientific application through the stratification of genetic studies, necessary given the extent of locus heterogeneity in IBD. Despite these advances, current knowledge of the genetic basis for IBD phenotype is of little clinical value in the diagnosis, classification or prediction of disease course. For the time being diagnosis and management will continue to be a clinical process. However, it is hoped that, as more is learnt about the genetic basis for individual subgroups, a molecular classification, with real clinical applications, will follow (Table 1).

## Acknowledgements

Tariq Ahmad is supported by grants from the National Association of Crohn's and Colitis (NACC, UK) and the Broad Foundation (USA).

## References

1. Farmer RG, Hawk WA, Turnbull RB Jr. Clinical patterns in Crohn's disease: a statistical study of 615 cases. Gastroenterology. 1975;68:627–35.
2. Franchimont DP, Louis E, Croes F, Belaiche J. Clinical pattern of corticosteroid dependent Crohn's disease. Eur J Gastroenterol Hepatol. 1998;10:821–5.
3. Farmer RG, Whelan G, Fazio VW. Long-term follow-up of patients with Crohn's disease. Relationship between the clinical pattern and prognosis. Gastroenterology. 1985;88:1818–25.
4. Whelan G, Farmer RG, Fazio VW, Goormastic M. Recurrence after surgery in Crohn's disease. Relationship to location of disease (clinical pattern) and surgical indication. Gastroenterology. 1985;88:1826–33.
5. Griffiths AM, Wesson DE, Shandling B, Corey M, Sherman PM. Factors influencing postoperative recurrence of Crohn's disease in childhood. Gut. 1991;32:491–5.
6. Greenstein AJ, Lachman P, Sachar DB et al. Perforating and non-perforating indications for repeated operations in Crohn's disease: evidence for two clinical forms. Gut. 1988;29: 588–92.
7. Aeberhard P, Berchtold W, Riedtmann HJ, Stadelmann G. Surgical recurrence of perforating and nonperforating Crohn's disease. A study of 101 surgically treated patients. Dis Colon Rectum. 1996;39:80–7.
8. Sachar DB, Andrews HA, Farmer RG et al. Proposed classification of patient subgroups in Crohn's disease. Gastroenterol Int. 1992;5:141–54.
9. Gasche C, Scholmerich J, Brynskov J et al. A simple classification of Crohn's disease: report of the Working Party for the World Congresses of Gastroenterology, Vienna 1998. Inflamm Bowel Dis. 2000;6:8–15.
10. Vasiliauskas EA, Kam LY, Karp LC, Gaiennie J, Yang H, Targan SR. Marker antibody expression stratifies Crohn's disease into immunologically homogeneous subgroups with distinct clinical characteristics. Gut. 2000;47:487–96.

11. Vasiliauskas EA, Plevy SE, Landers CJ et al. Perinuclear antineutrophil cytoplasmic antibodies in patients with Crohn's disease define a clinical subgroup. Gastroenterology. 1996;110:1810–19.

12. Polito JM 2nd, Childs B, Mellits ED, Tokayer AZ, Harris ML, Bayless TM. Crohn's disease: influence of age at diagnosis on site and clinical type of disease. Gastroenterology. 1996;111:580–6.

13. Satsangi J, Grootscholten C, Holt H, Jewell DP. Clinical patterns of familial inflammatory bowel disease. Gut. 1996;38:738–41.

14. Bayless TM, Tokayer AZ, Polito JM 2nd, Quaskey SA, Mellits ED, Harris ML. Crohn's disease: concordance for site and clinical type in affected family members – potential hereditary influences. Gastroenterology. 1996;111:573–9.

15. Colombel JF, Grandbastien B, Gower-Rousseau C et al. Clinical characteristics of Crohn's disease in 72 families. Gastroenterology. 1996;111:604–7.

16. Peeters M, Nevens H, Baert F et al. Familial aggregation in Crohn's disease: increased age-adjusted risk and concordance in clinical characteristics. Gastroenterology. 1996;111:597–603.

17. Annese V, Andreoli A, Astegiano M et al. Clinical features in familial cases of Crohn's disease and ulcerative colitis in Italy: a GISC study. Italian Study Group for the Disease of Colon and Rectum. Am J Gastroenterol. 2001;96:2939–45.

18. Weissenbach J, Gyapay G, Dib C et al. A second-generation linkage map of the human genome. Nature. 1992;359:794–801.

19. Hugot JP, Laurent-Puig P, Gower-Rousseau C et al. Mapping of a susceptibility locus for Crohn's disease on chromosome 16. Nature. 1996;379:821–3.

20. Lander E, Kruglyak L. Genetic dissection of complex traits: guidelines for interpreting and reporting linkage results. Nat Genet. 1995;11:241–7.

21. Ahmad T, Tamboli CP, Jewell D, Colombel JF. Clinical relevance of advances in genetics and immunogenetics of IBD. Gastroenterology. 2004 (In press).

22. Hugot JP, Chamaillard M, Zouali H et al. Association of NOD2 leucine-rich repeat variants with susceptibility to Crohn's disease. Nature. 2001;411:599–603.

23. Peltekova VD, Wintle RF, Rubin LA et al. Functional variants of OCTN cation transporter genes are associated with Crohn disease. Nat Genet. 2004;36:471–5.

24. Nadeau JH. Modifier genes in mice and humans. Nat Rev Genet. 2001;2:165–74.

25. Ogura Y, Bonen DK, Inohara N et al. A frameshift mutation in NOD2 associated with susceptibility to Crohn's disease. Nature. 2001;411:603–6.

26. Hampe J, Cuthbert A, Croucher PJ et al. Association between insertion mutation in NOD2 gene and Crohn's disease in German and British populations. Lancet. 2001;357:1925–8.

27. Lesage S, Zouali H, Cezard JP et al. CARD15/NOD2 mutational analysis and genotype–phenotype correlation in 612 patients with inflammatory bowel disease. Am J Hum Genet. 2002;70:845–57.

28. Orholm M, Iselius L, Sorensen TI, Munkholm P, Langholz E, Binder V. Investigation of inheritance of chronic inflammatory bowel diseases by complex segregation analysis. Br Med J. 1993;306:20–4.

29. Forabosco P, Collins A, Latiano A et al. Combined segregation and linkage analysis of inflammatory bowel disease in the IBD1 region using severity to characterise Crohn's disease and ulcerative colitis. On behalf of the GISC. Eur J Hum Genet. 2000;8:846–52.

30. Helio T, Halme L, Lappalainen L et al. CARD15/NOD2 gene variants are associated with familially occurring and complicated forms of Crohn's disaese. Gut. 2003;52:558–62.

31. Bairead E, Harmon DL, Curtis AM et al. Association of NOD2 with Crohn's disease in a homogeneous Irish population. Eur J Hum Genet. 2003;11:237–44.

32. Crichton D, Arnott IDR, Watts D et al. NOD2/CARD15 mutations in a Scottish Crohn's disease population. Gastroenterology. 2002;122:A298.

33. Bonen DK, Nicolae DL, Moran T et al. Racial differences in Nod2 variation: characterization of Nod2 in African-Americans with Crohn's disease. Gastroenterology. 2002;122:A29 (abstract).

34. Inoue N, Tamura K, Kinouchi Y et al. Lack of common Nod2 variants in Japanese patients with Crohn's disease. Gastroenterology. 2002;123:86–91.

35. Yamazaki K, Takazoe M, Tanaka T, Kazumori T, Nakamura Y. Absence of mutation in the NOD2/CARD15 gene among 483 Japanese patients with Crohn's disease. J Hum Genet. 2002;47:469–72.

36. Croucher PJ, Mascheretti S, Hampe J et al. Haplotype structure and association to Crohn's disease of CARD15 mutations in two ethnically divergent populations. Eur J Hum Genet. 2003;11:6–16.

37. Leong RW, Armuzzi A, Ahmad T et al. NOD2/CARD15 gene polymorphisms and Crohn's disease in the Chinese population. Aliment Pharmacol Ther. 2003;17:1465–70.

38. Abreu MT, Taylor KD, Lin YC et al. Mutations in NOD2 are associated with fibrostenosing disease in patients with Crohn's disease. Gastroenterology. 2002;123:679–88.

39. Zhou Z, Lin XY, Akolkar PN et al. Variation at NOD2/CARD15 in familial and sporadic cases of Crohn's disease in the Ahskenazi Jewish population. Am J Gastroenterol. 2002;97: 3095–101.

40. Ahmad T, Armuzzi A, Bunce M et al. The molecular classification of the clinical manifestations of Crohn's disease. Gastroenterology. 2002;122:854–66.

41. Cuthbert AP, Fisher SA, Mirza MM et al. The contribution of NOD2 gene mutations to the risk and site of disease in inflammatory bowel disease. Gastroenterology. 2002;122:867–74.

42. Murillo L, Crusius JB, van Bodegraven AA, Alizadeh BZ, Pena AS. CARD15 gene and the classification of Crohn's disease. Immunogenetics. 2002;54:59–61.

43. Hampe J, Grebe J, Nikolaus S et al. Association of NOD2 (CARD 15) genotype with clinical course of Crohn's disease: a cohort study. Lancet. 2002;359:1661–5.

44. Vermeire S, Wild G, Kocher K et al. CARD15 genetic variation in a Quebec population: prevalence, genotype–phenotype relationship, and haplotype structure. Am J Hum Genet. 2002;71:74–83.

45. Radlmayr M, Torok HP, Martin K, Folwaczny C. The c-insertion mutation of the NOD2 gene is associated with fistulizing and fibrostenotic phenotypes in Crohn's disease (letter). Gastroenterology. 2002;122:2091–2.

46. Vavassori P, Borgiani P, D'Apice MR et al. 3020insC mutation within the NOD2 gene in Crohn's disease: frequency and association with clinical pattern in an Italian population. Dig Liver Dis. 2002;34:153.

47. Brant SR, Picco MF, Achkar JP et al. Defining complex contributions of NOD2/CARD15 gene mutations, age at onset, and tobacco use on Crohn's disease phenotypes. Inflamm Bowel Dis. 2003;9:281–9.

48. Lala S, Ogura Y, Osborne C et al. Crohn's disease and the NOD2 gene : a role for Paneth cells. Gastroenterology. 2003;125:47–57.

49. Ogura Y, Lala S, Xin W et al. Expression of NOD2 in Paneth cells: a possible link to Crohn's ileitis. Gut. 2003;52:1591–7.

50. Darfeuille-Michaud A, Boudeau J, Bulois P et al. High prevalence of invasive *Escherichia coli* strains in ileal Crohn's disease. 2004;127:412–21.

51. Cottone M, Brignola C, Rosselli M et al. Relationship between site of disease and familial occurrence in Crohn's disease. Dig Dis Sci. 1997;42:129–32.

52. Halme L, Turunen U, Helio T et al. Familial and sporadic inflammatory bowel disease: comparison of clinical features and serological markers in a genetically homogeneous population. Scand J Gastroenterol. 2002;37:692–8.

53. Ma Y, Ohmen JD, Li Z et al. A genome-wide search identifies potential new susceptibility loci for Crohn's disease. Inflamm Bowel Dis. 1999;5:271–8.

54. Rioux JD, Silverberg MS, Daly MJ et al. Genomewide search in Canadian families with inflammatory bowel disease reveals two novel susceptibility loci. Am J Hum Genet. 2000; 66:1863–70.

55. Rioux JD, Daly MJ, Silverberg MS et al. Genetic variation in the 5q31 cytokine gene cluster confers susceptibility to Crohn disease. Nat Genet. 2001;29:223–8.

56. Negoro K, McGovern DPB, Kinouchi Y et al. Analysis of the IBD5 locus and potential gene–gene interactions in Crohn's disease. Gut. 2003;52:541–6.

57. Giallourakis C, Stoll M, Miller K et al. IBD5 is a general risk factor for inflammatory bowel disease: replication of association with Crohn's disease and identification of a novel association with ulcerative colitis. Am J Hum Genet. 2003;73:205–11.

58. Armuzzi A, Ahmad T, Ling KL et al. Genotype–phenotype analysis of the Crohn's disease susceptibility haplotype on chromosome 5q31. Gut. 2003;52:1133–9.

59. Stokkers PC, Reitsma PH, Tytgat GN, van Deventer SJ. HLA-DR and -DQ phenotypes in inflammatory bowel disease: a meta-analysis. Gut. 1999;45:395–401.

60. Satsangi J, Welsh KI, Bunce M et al. Contribution of genes of the major histocompatibility complex to susceptibility and disease phenotype in inflammatory bowel disease. Lancet. 1996;347:1212–17.

61. Roussomoustakaki M, Satsangi J, Welsh K et al. Genetic markers may predict disease behavior in patients with ulcerative colitis. Gastroenterology. 1997;112:1845–53.

62. Bouma G, Crusius JB, Garcia-Gonzalez MA et al. Genetic markers in clinically well defined patients with ulcerative colitis (UC). Clin Exp Immunol. 1999;115:294–300.

63. Ahmad T, Armuzzi A, Neville M et al. The contribution of human leucocyte antigen complex genes to disease phenotype in ulcerative colitis. Tissue Antigens. 2003;62:527–35.

64. Yamamoto-Furusho JK, Uscanga LF, Vargas-Alarcon G et al. Clinical and genetic heterogeneity in Mexican patients with ulcerative colitis. Hum Immunol. 2003;64:119–23.

65. Trachtenberg EA, Yang H, Hayes E et al. HLA class II haplotype associations with inflammatory bowel disease in Jewish (Ashkenazi) and non-Jewish Caucasian populations. Hum Immunol. 2000;61:326–33.

66. Silverberg MS, Mirea L, Bull SB et al. A population- and family-based study of Canadian families reveals association of HLADRB1*0103 with colonic involvement in inflammatory bowel disease. Inflamm Bowel Dis. 2003;9:1–9.

67. Louis E, Peeters M, Franchimont D et al. Tumour necrosis factor (TNF) gene polymorphism in Crohn's disease (CD): influence on disease behaviour? Clin Exp Immunol. 2000;119: 64–8.

68. Orchard TR, Thiyagaraja S, Welsh KI, Wordsworth BP, Hill Gaston JS, Jewell DP. Clinical phenotype is related to HLA genotype in the peripheral arthropathies of inflammatory bowel disease. Gastroenterology. 2000;118:274–8.

69. Orchard TR, Chua CN, Ahmad T, Cheng H, Welsh KI, Jewell DP. Uveitis and erythema nodosum in inflammatory bowel disease: clinical features and the role of HLA genes. Gastroenterology. 2002;123:714–18.

70. Steinhart AH, Girgrah N, McLeod RS. Reliability of a Crohn's disease clinical classification scheme based on disease behavior. Inflamm Bowel Dis. 1998;4:228–34.

71. Cosnes J, Cattan S, Blain A et al. Long-term evolution of disease behavior of Crohn's disease. Inflamm Bowel Dis. 2002;8:244–50.

72. Louis E, Collard A, Oger AF, Degroote E, Aboul Nasr El Yafi FA, Belaiche J. Behaviour of Crohn's disease according to the Vienna classification: changing pattern over the course of the disease. Gut. 2001;49:777–82.

73. Powell-Tuck J, Ritchie JK, Lennard-Jones JE. The prognosis of idiopathic proctitis. Scand J Gastroenterol. 1977;12:727–32.

74. Moum B, Ekbom A, Vatn MH, Elgjo K. Change in the extent of colonoscopic and histological involvement in ulcerative colitis over time. Am J Gastroenterol. 1999;94: 1564–9.

75. Klein W, Tromm A, Griga T et al. Interaction of polymorphisms in the CARD15 and CD14 genes in patients with Crohn disease. Scand J Gastroenterol. 2003;38:834–6.

76. Mirza MM, Fisher SA, King K et al. Genetic evidence for interaction of the 5q31 cytokine locus and the CARD15 gene in Crohn disease. Am J Hum Genet. 2003;72:1018–22.

77. Stoll M, Corneliussen B, Costello CM et al. Genetic variation in DLG5 is associated with inflammatory bowel disease. Nat Genet. 2004;36:476–80.

78. Rubin DT, Hanauer S. Smoking and inflammatory bowel disease. Eur J Gastroenterol Hepatol. 2000;12:855–62.

79. Brant SR, Picco MF, Achkar JP et al. Defining complex contributions of NOD2/CARD15 gene mutations, age at onset, and tobacco use on Crohn's disease phenotypes. Inflamm Bowel Dis. 2003;9:281–9.

80. Beaugerie L, Massot N, Carbonnel F, Cattan S, Gendre JP, Cosnes J. Impact of cessation of smoking on the course of ulcerative colitis. Am J Gastroenterol. 2001;96:2113–16.

81. Bridger S, Lee JCW, Bjarnason I, Lennard Jones J, MacPherson AJ. In siblings with similar genetic susceptibility for inflammatory bowel disease, smokers tend to develop Crohn's disease and non-smokers develop ulcerative colitis. Gut. 2002;51:21–5.

# 2
# NOD proteins and defence against microbial infection

D. J. PHILPOTT

## INTRODUCTION

Epithelial cells comprising mucosal surfaces constitute the first line of defence against microbial pathogens. These cells are in the unique position of being in constant contact with bacteria and bacterial products, yet these factors that are normally proinflammatory for other cell types do not induce epithelial cells to initiate a defensive response. In the context of the colon the refractory nature of epithelial cells to bacteria seems logical, since mounting an inflammatory response to the normal microbial flora would indeed be detrimental to the host. However, epithelial cells possess the ability to discriminate between pathogenic and non-pathogenic bacteria since infection of these cells by pathogenic bacteria induces proinflammatory responses[1]. Our findings suggest that a discriminatory system has evolved in these cells based on the inside versus outside presentation of bacterial products known collectively as 'pathogen-associated molecular patterns' or 'PAMP'. PAMP include microbial motifs such as lipopolysaccharide and peptidoglycan from bacteria and double-stranded RNA from viruses[2]. Epithelial cells are refractory to extracellular bacteria and PAMP since dedicated pattern-recognition receptors, e.g. Toll-like receptors (TLR), are either not expressed by these cells or expressed at very low levels[1,3]. However, PAMP presented inside the cell can initiate an inflammatory response. We have shown that a protein called NOD1 (also known as CARD4) plays a role in this intracellular detection system in epithelial cells. NOD1 is a cytosolic protein of approximately 90 kDa. Strikingly, this protein is similar to the plant disease resistance proteins 'N' from tobacco and 'RPS2' of *Arabidopsis*. Similar to these proteins, NOD1 has an C-terminal leucine-rich repeat (LRR) and a central nucleotide-binding site (NBS). NOD1 also possesses an N-terminal CARD domain (for 'caspase activating and recruitment domain') found also in the pro-apoptotic proteins, CED4 from *Caenorhabditis elegans* and APAF-1 in mammals[4,5]. Upon infection with *Shigella flexneri*, a Gram-negative bacterium that is capable of invading epithelial cells, NOD1 oligomerizes, which is an event shown previously to be necessary for NOD1 activation[6]. Different constructs of NOD1 were then tested for their ability to act as dominant-negative inhibitors of the activation of NFκB and c-Jun N-terminal kinase (JNK) following *S. flexneri* infection, as both of these signalling pathways are necessary for the induction of the proinflammatory gene programme. A NOD1 molecule lacking the CARD domain as well as the LRR domain of NOD1 were able to dose-dependently block the activation of both JNK and NFκB in infected cells. Finally, we could show that infection by *S. flexneri* initiates a signal transduction cascade within the infected cell, resulting in the oligomerization of NOD1, recruitment of the kinase RICK to the complex that then recruits the Ikappa kinase complex[6]. Activation of this complex by recruitment is the penultimate step in the activation of NFκB[7]. These results indicate that two distinct pathways exist for the detection of bacterial PAMP: one pathway that is involved in extracellular detection (through TLR) and one pathway involving NOD1 that detects bacterial factors brought inside infected cells. This work also represents the first demonstration of a physiological stimulator of the NOD1–RICK–IKK cascade leading to NFκB activation.

## NOD1: A SENSOR OF GRAM-NEGATIVE BACTERIAL INFECTION

Our work then focused on the characterization of NOD1 ligand. From our studies it was evident that the ligand was present in bacterial supernatants isolated from Gram-negative bacteria, including *S. flexneri* and *Escherichia coli*. Therefore, we screened a number of different potential PAMP for their ability to activate NFκB in a NOD1-dependent manner. Strikingly, we found that purified peptidoglycans (PGN) from Gram-negative bacteria were able to activate NOD1, leading to NFκB induction. PGN preparations from the Gram-positive bacteria, *Bacillus subtilis* and *Staphylococcus aureus*, were unable to stimulate this response. Through the analysis of Gram-negative bacterial PGN fractions separated by reverse-phase HPLC we were then able to determine that the minimal naturally occurring PGN fragment that stimulates NOD1 is a disaccharide of N-acetyl glucosamine-N-acetyl muramic acid linked to a tripeptide where the terminal amino acid is meso-diaminopimimic acid (meso-DAP)[8]. The presence of DAP in the PGN can be considered as a general signature of Gram-negative bacterial infection since most, but not all, Gram-positive organisms have lysine in this position in their PGN. What is also striking for NOD1 is that it is highly specific for the tripeptide structure since the presence of an additional amino acid to the DAP abrogates the sensing of this bacterial product by NOD1[8].

More recently, we and others have been able to narrow down the PGN structure sensed by NOD1 in more detail, thus defining the minimal motif. Markedly, NOD1 can sense just the terminal two amino acids within the naturally occurring fragment: d-glutamine-mesoDAP (referred to as iE-DAP by the group of Inohara)[9,10]. This minimal motif is a signature of bacterial infection since neither of these amino acids exists in mammals.

In the context of *in-vivo* infection, NOD1 appears to be a key molecule in the sensing of Gram-negative bacterial infection since isolated intestinal epithelial cells from mice deficient in NOD1 can no longer react to the intracellular presentation of Gram-negative bacterial products. Moreover, NOD1 appears to be the only functional bacterial sensor in intestinal epithelial cells, highlighting the key role of this PRR in host defence[8].

Structure/function analysis of NOD1 has identified that the LRR domain is probably the domain that interacts with the bacterial ligand, directly or indirectly. A NOD1 construct lacking the LRR domain no longer responds to the ligand[11]. Molecular modelling of the LRR domain suggests that this domain conforms to a horseshoe-like structure, similar to the LRR domains of internalin A from *Listeria monocytogenes* and the ribonuclease inhibitor[12]. Based on these structures, and information from the ligand interaction domains in these two proteins, we carried out site-directed mutagenesis of the LRR domain of NOD1 in the putative ligand interaction domain on the concave surface of the molecule. We discovered that a number of these mutations resulted in a molecule that could no longer sense the ligand. Current studies will aim to examine the binding properties of these mutant proteins *in vitro*, and elucidation of the crystal structure of this molecule.

## NOD2 AS A GENERAL SENSOR OF BACTERIAL INFECTION

NOD2 is a close homologue of NOD1 and was recently identified to be the first susceptibility gene involved in the chronic inflammatory bowel disease, Crohn's disease (CD)[13,14]. Mutations in NOD2 are also responsible for another human genetic disease called Blau syndrome, in which sufferers also present symptoms of inflammation[15]. The expression of profile of NOD2 is more or less restricted to monocytes/macrophages but expression can be up-regulated in other cell types upon treatment with proinflammatory stimuli such as interferon gamma or tumour necrosis factor alpha[16]. Our studies also led us to examine the bacterial ligand sensed by NOD2. Like NOD1, this protein also senses a PGN fragment; however, NOD2 is specific for muramyl dipeptide or MDP[17,18]. NOD2 sensing of MDP is stereospecific since only MDP with amino acids in the L, D form and not in the L, L form, can be sensed by NOD2. MDP is the minimal bioactive PGN fragment from both Gram-positive and Gram-negative bacteria, making NOD2 a general sensor of bacterial infection. Interestingly, MDP has been known for decades for its immunomodulatory action; it is a component of Freund's complete adjuvant. With the discovery of NOD2 as the host receptor for MDP, the immunostimulatory properties of this compound can now be investigated in more detail.

## ROLE OF NOD2 IN CD

The most common mutation in the *NOD2* gene that is associated with CD is an insertion mutation at position 3020 that leads to the deletion of the terminal LRR of the protein. When we compared the wild-type versus the mutant protein for the ability of these molecules to sense MDP, it was observed that the mutant NOD2 can no longer detect MDP to initiate NFκB activation[18]. Moreover, the group of Gabriel Nunez showed that peripheral blood mononuclear cells isolated from patients with CD could not respond to MDP in terms of NFκB activation and cytokine induction[17]. The implications of these findings therefore suggest that the defect in CD patients may be the inability to respond normally to bacterial products. An apparent paradox thus arises: how can the lack of bacterial detection and loss of subsequent activation of proinflammatory signals translate into the chronic and severe inflammation of the bowel that we see during CD?

There are four possible hypotheses that may be proposed to explain the role of NOD2 in the pathogenesis of CD. One possible explanation is that the intestinal mucosa is maintained at a certain level of inflammation by NOD2, and possibly NOD1, in order to keep in check any possible bacterial breach of the epithelial barrier. If bacteria or their products enter within the submucosa, NOD2 induces a local inflammatory response that suppresses any possible infection. In CD it can be hypothesized that, without this sentinel function of NOD2, bacteria and/or their products that gain access to the submucosa overstimulate the system, aggravating defence systems that would not normally be active. This effect would manifest itself as the severe and chronic inflammation that is observed in the intestinal tissues of CD patients[19,20].

Secondly, NOD2 may function directly as an antibacterial factor. Hisamatsu et al. showed that the number of viable internalized *Salmonella typhimurium* in Caco-2 intestinal epithelial cells stably transfected with a NOD2 expression plasmid was lower than in vector-transfected cells, suggesting a direct anti-bacterial activity for NOD2[21]. Furthermore, cells expressing the frame-shift mutant of NOD2 associated with CD could not control bacterial infection as well as cells transfected with the wild-type construct. Although the mechanism by which NOD2 can be directly antibacterial was not investigated, it can be hypothesized that NOD proteins may regulate the expression or release of antimicrobial factors, including reactive oxygen intermediates, nitric oxide, trefoil peptides or antimicrobial peptides such as defensins. Consistent with this latter point, NOD2 is highly expressed in Paneth cells[22]. These specialized cells of the ileum, which are located at the base of the intestinal crypts, secrete defensins and other antimicrobial peptides in response to bacterial products, including MDP[23]. Using a monoclonal antibody directed against NOD2, protein expression was observed in the cytoplasm of these cells and also closely associated with the intracellular granules containing antimicrobial factors[22]. These findings suggest that NOD2 may play a role in the regulation of Paneth cell responses to intestinal bacteria. Accordingly, mutations in NOD2 may contribute to bacterial proliferation and thereby set up a chronic inflammatory state at the level of the intestinal mucosa. Furthermore, these findings are also consistent with the fact that Paneth cell distribution is more concentrated in the ileum and disease associated to this portion of the intestine is selectively associated with mutations in NOD2.

Besides NFκB induction, NOD1 and NOD2 have been described to also participate in the activation of an apoptotic response[5]. Accordingly, another hypothesis for the role of NOD2 in the pathogenesis of CD might involve dysregulation of apoptosis by mutated forms of NOD2. The genes encoding NOD1 and NOD2 were first cloned based on the homology of their CARD domains with that of APAF-1, an activator of caspase 9. Subsequently, it was shown that NOD1[24] and NOD2[25] interact with the CARD domain of caspase 9 and, although not activating apoptosis directly, these molecules can potentiate the apoptotic response when co-expressed with this caspase. Although, in these studies, apoptosis was examined in the context of overexpression of either NOD1 or NOD2 with caspase 9, it is possible that in the appropriate cell type, or perhaps with an essential co-signal, these molecules may directly induce an apoptotic response that may be crucial in the regulation of the immune response. So far, the possibility that the mutations in NOD2 associated with CD have lost their ability to regulate caspase 9-dependent apoptosis has not been examined. If this hypothesis is verified, however, the loss of apoptotic activity by mutated NOD2 could manifest at two different levels. On one hand, epithelial cells with an altered ability to undergo apoptosis could favour the development of necrosis, resulting in the release of 'danger signals' and promoting the induction of inflammation. On the other hand, restricted apoptosis of certain immune cells could also contribute to loss of tolerance and aggravation of the inflammatory reaction seen during CD. In fact, 'escape' from apoptosis could account for the prolonged survival of lymphocytes that have been observed to infiltrate CD-associated mucosa[26–28]. Although the exact role in the pathology of CD is not known, antibody therapy directed against tumour necrosis factor[29] or the interleukin 6 receptor[30] increases the sensitivity of these lymphocytes to undergo apoptosis, and controls intestinal inflammation both in patients and in different experimental models of colitis. Therefore, it will be interesting to examine if NOD2 mutations are linked to this resistance to apoptosis in lymphocytes and the subsequent dysregulation of inflammatory signals.

Finally, another possibility is that NOD2 negatively influences the activation of other signal transduction pathways, parallel to the idea of the 'guard hypothesis' for certain R proteins in plants[31]. According to this hypothesis, NOD2 would 'guard' additional signalling proteins involved in proinflammatory pathways, and thus prevent the activation of defence responses. In support of this idea, Chen et al. recently showed that NOD2 interacts with TAK1, a MAP kinase kinase kinase that is an essential component of the signalling pathways downstream of the receptors for many inflammatory cytokines and LPS[32]. Mutated NOD2 could no longer interact with TAK1 and suppress proinflammatory signalling via this molecule. Although preliminary, these findings may suggest that mutated NOD2 in CD patients can no longer carry out an inhibitory function, and thereby exacerbates inflammation driven by other proinflammatory pathways that may use TAK1 as a mediator. In support of this hypothesis, Wantanabe et al. recently showed that intact NOD2 is required for 'down-regulating' the response to Toll-like receptor 2 ligands; NOD2 deficiency increased TLR2-mediated activation of NFκB and enhanced Th1 responses[33].

## CONCLUSION

The discovery of the association between NOD2 mutations and CD represents a key piece of evidence that will help to resolve the mystery behind the pathogenesis of this disease. For the moment the link between NOD2 and disease development, however, remains obscure, and even paradoxical. Indeed, recent evidence has shown that the loss of responsiveness of this innate immune receptor to bacterial ligands somehow results in overactivation of the immune system and the development of inflammatory lesions. Nevertheless, the precise determination of the peptidoglycan ligand of NOD2 now concentrates interest into understanding the interaction between bacteria and the innate immune response, and how these interactions are relayed to the adaptive immune response. Moreover, further characterization of the knockout mouse of NOD2, and application of this mouse in inflammatory bowel disease models, will also aid in understanding the pathogenesis of this disease[34].

## Acknowledgements

I thank the members of my laboratory, and that of Philippe Sansonetti, for their contributions and helpful discussions.

## References

1. Philpott DJ, Girardin SE, Sansonetti PJ. Innate immune responses of epithelial cells following infection with bacterial pathogens. Curr Opin Immunol. 2001;13:410–6.
2. Takeda K, Kaisho T, Akira S. Toll-like receptors. Annu Rev Immunol. 2003;21:335–76.
3. Backhed F, Hornef M. Toll-like receptor 4-mediated signaling by epithelial surfaces: necessity or threat? Microbes Infect. 2003;5:951–9.
4. Chamaillard M, Girardin SE, Viala J, Philpott DJ. Nods, Nalps and Naip: intracellular regulators of bacterial-induced inflammation. Cell Microbiol. 2003;5:581–92.
5. Inohara N, Nunez G. NODs: intracellular proteins involved in inflammation and apoptosis. Nat Rev Immunol. 2003;3:371–82.
6. Girardin SE, Tournebize R, Mavris M et al. CARD4/Nod1 mediates NF-kappaB and JNK activation by invasive *Shigella flexneri*. EMBO Reports. 2001;2:736–42.
7. Inohara N, Koseki T, Lin J et al. An induced proximity model for NF-kappa B activation in the Nod1/RICK and RIP signaling pathways. J Biol Chem. 2000;275:27823–31.
8. Girardin SE, Boneca IG, Carneiro LA et al. Nod1 detects a unique muropeptide from gram-negative bacterial peptidoglycan. Science. 2003;300:1584–7.
9. Girardin SE, Travassos LH, Herve M et al. Peptidoglycan molecular requirements allowing detection by Nod1 and Nod2. J Biol Chem. 2003;278:41702–8.
10. Chamaillard M, Hashimoto M, Horie Y et al. An essential role for NOD1 in host recognition of bacterial peptidoglycan containing diaminopimelic acid. Nat Immunol. 2003;4:702–7.
11. Inohara N, Ogura Y, Chen FF, Muto A, Nunez G. Human Nod1 confers responsiveness to bacterial lipopolysaccharides. J Biol Chem. 2001;276:2551–4.
12. Kobe B, Kajava AV. The leucine-rich repeat as a protein recognition motif. Curr Opin Struct Biol. 2001;11:725–32.
13. Hugot JP, Chamaillard M, Zouali H et al. Association of NOD2 leucine-rich repeat variants with susceptibility to Crohn's disease. Nature. 2001;411:599–603.
14. Ogura Y, Bonen DK, Inohara N et al. A frameshift mutation in NOD2 associated with susceptibility to Crohn's disease. Nature. 2001;411:603–6.
15. Miceli-Richard C, Lesage S, Rybojad M et al. CARD15 mutations in Blau syndrome. Nat Genet. 2001;29:19–20.

16. Gutierrez O, Pipaon C, Inohara N et al. Induction of Nod2 in myelomonocytic and intestinal epithelial cells via nuclear factor-kappa B activation. J Biol Chem. 2002;277: 41701–5.
17. Inohara N, Ogura Y, Fontalba A et al. Host recognition of bacterial muramyl dipeptide mediated through NOD2. Implications for Crohn's disease. J Biol Chem. 2003;278:5509–12.
18. Girardin SE, Boneca IG, Viala J et al. Nod2 is a general sensor of peptidoglycan through muramyl dipeptide (MDP) detection. J Biol Chem. 2003;278:8869–72.
19. Philpott DJ, Viala J. Towards an understanding of the role of NOD2/CARD15 in the pathogenesis of Crohn's disease. Best Pract Res Clin Gastroenterol. 2004;18:555–68.
20. Inohara N, Nunez G. The NOD: a signaling module that regulates apoptosis and host defense against pathogens. Oncogene. 2001;20:6473–81.
21. Hisamatsu T, Suzuki M, Reinecker HC, Nadeau WJ, McCormick BA, Podolsky DK. CARD15/NOD2 functions as an antibacterial factor in human intestinal epithelial cells. Gastroenterology. 2003;124:993–1000.
22. Ogura Y, Lala S, Xin W et al. Expression of NOD2 in Paneth cells: a possible link to Crohn's ileitis. Gut. 2003;52:1591–7.
23. Ayabe T, Satchell DP, Pesendorfer P et al. Activation of Paneth cell alpha-defensins in mouse small intestine. J Biol Chem. 2002;277:5219–28.
24. Inohara N, Koseki T, del Peso L et al. Nod1, an Apaf-1-like activator of caspase-9 and nuclear factor-kappaB. J Biol Chem. 1999;274:14560–7.
25. Ogura Y, Inohara N, Benito A, Chen FF, Yamaoka S, Nunez G. Nod2, a Nod1/Apaf-1 family member that is restricted to monocytes and activates NF-kappaB. J Biol Chem. 2001;276:4812–8.
26. Boirivant M, Marini M, Di Felice G et al. Lamina propria T cells in Crohn's disease and other gastrointestinal inflammation show defective CD2 pathway-induced apoptosis. Gastroenterology. 1999;116:557–65.
27. Ina K, Itoh J, Fukushima K et al. Resistance of Crohn's disease T cells to multiple apoptotic signals is associated with a Bcl-2/Bax mucosal imbalance. J Immunol. 1999; 163:1081–90.
28. Meresse B, Dubucquoi S, Tourvieille B, Desreumaux P, Colombel JF, Dessaint JP. CD28+ intraepithelial lymphocytes with long telomeres are recruited within the inflamed ileal mucosa in Crohn disease. Human Immunol. 2001;62:694–700.
29. ten Hove T, van Montfrans C, Peppelenbosch MP, van Deventer SJ. Infliximab treatment induces apoptosis of lamina propria T lymphocytes in Crohn's disease. Gut. 2002;50:206–11.
30. Atreya R, Mudter J, Finotto S et al. Blockade of interleukin 6 trans signaling suppresses T-cell resistance against apoptosis in chronic intestinal inflammation: evidence in Crohn disease and experimental colitis in vivo. Nature Med. 2000;6:583–8.
31. Dangl JL, Jones JD. Plant pathogens and integrated defence responses to infection. Nature. 2001;411:826–33.
32. Chen CM, Gong Y, Zhang M, Chen JJ. Reciprocal cross-talk between Nod2 and TAK1 signaling pathways. J Biol Chem. 2004;279:25876–82.
33. Watanabe T, Kitani A, Murray PJ, Strober W. NOD2 is a negative regulator of Toll-like receptor 2-mediated T helper type 1 responses. Nat Immunol. 2004 (In press).
34. Pauleau AL, Murray PJ. Role of nod2 in the response of macrophages to toll-like receptor agonists. Mol Cell Biol. 2003;23:7531–9.

# 3
# Pharmacogenetics of inflammatory bowel disease therapy

## M. SCHWAB and E. SCHAEFFELER

## INTRODUCTION

The response of individual patients to the same drug given in the same dose varies considerably for several substances. Many patients will expect the desired drug effect, some may suffer from well-known adverse drug reactions, others may experience no effects, and very rarely a patient will die from severe side-effects. It is currently very difficult for physicians to prescribe the optimal drug in the optimal dose for each patient, since prediction of a patient's response to a specific drug is rarely possible.

From a clinical point of view, excessive numbers of patients suffer from adverse drug reactions. In the USA such reactions are estimated to be from the fourth to sixth leading cause of deaths in hospital patients[1]. In addition, the cost burden is tremendous. A large percentage of health budgets are spent on inefficient drug therapy. The application of pharmacogenetic knowledge offers the opportunity of individually tailored, safe and efficient drug treatment.

## HISTORICAL BACKGROUND

Pharmacogenetics is actually not a novel field in clinical pharmacology. Indeed, the term 'pharmacogenetics' was coined by the German geneticist Friedrich Vogel no less than 40 years ago[2]. Important inherited differences in drug response first emerged during the 1950s. The observation of prolonged postoperative muscle relaxation after administration of suxamethonium was attributed to an atypical variant of plasma cholinesterase (butyrylcholinesterase)[3]. At the same time it was shown that the haemolysis observed in a significant fraction of patients treated with antimalarial drugs such as primaquine or chloroquine cosegregated with an anomaly of erythrocyte glucose 6-phosphate dehydrogenase activity – which was also shown to be under genetic control. Studies likewise disclosed that the pharmacokinetics and side-effects of isoniazid – such as peripheral neuropathy – are influenced by inherited variants of arylamine N-acetyltransferase (NAT2)[4,5]. The field of pharmacogenetics

24

experienced an enormous increase in attention in 1977 when two independent research groups in London and Bonn observed markedly increased side-effects in some volunteers given the antihypertensive drug debrisoquine[6] and the antiarrhythmic sparteine[7]. Both drugs are catalysed by the cytochrome P450 2D6 (CYP2D6) isozyme and it was shown that side-effects resulted from reduced oxidative metabolism, which in turn was shown to be under mono-genic control. Phenotyping larger numbers of individuals – by administration of probe drugs such as debrisoquine and sparteine and measurement of the ratio of parent to metabolite in the urine – disclosed that 5–10% of the white population exhibit the poor metabolizer ('PM') phenotype[8].

The introduction of routine genotyping methods, including the development of high-throughput techniques and the availability of sequence data from the human genome project, has led to an exponential increase in research in this field during the past few years. Thus, it has been proposed that, by the year 2010, personalized medicines based on pharmacogenetics should be part of routine clinical practice[9].

## INDIVIDUAL VARIATION IN DRUG RESPONSE

To understand why individual patients respond differently to drugs it is necessary to consider processes which are involved in absorption, transport and metabolism of a drug, as well as various drug targets (Figure 1). The effect of a drug will depend first on its systemic concentration and its concentration at the drug target. Drug concentration at its target will in many cases represent a mere function of the systemic concentration. Active transport processes via various membrane transporter, however, may influence local target concentra-tions.

The systemic concentration of a drug depends on several pharmacokinetic factors, commonly referred to by the acronym ADME (drug absorption, distribution, metabolism and elimination). It has long been known that a wide variety of individual factors may influence the pharmacokinetics of a drug and must therefore be taken into account when determining dosage for a given patient. In addition to physiological/pathophysiological and environmental factors it has become increasingly clear that hereditary variances in drug-metabolizing enzymes can exert considerable influence on drug concentrations.

The second crucial determinant of an observed drug effect is the response of the drug target to a given drug concentration. Hereditary variants have been discovered and characterized in receptors, drug transporters, ion channels, lipoproteins, coagulation factors and many other factors involved in immune response, cell development, cell cycle control and other functions, that significantly influence the manifestation and course of diseases. Many of these polymorphic structures are at the same time targets of specific drugs and can thus potentially influence the effect that a specific drug concentration will exert at the drug target.

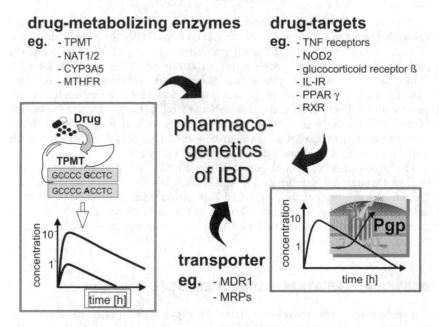

**Figure 1** Current research areas in pharmacogenetics of IBD. Pharmacogenetics currently comprises the study of polymorphic drug-metabolizing enzymes and drug transporters (pharmacokinetics) and drug targets such as drug receptors (pharmacodynamics)

## DRUG-METABOLIZING ENZYMES

A large number of enzymes play a role in drug metabolism. In brief, drug metabolism takes place in two phases. In phase I metabolism the xenobiotic is subject to oxidation, reduction or hydrolysis, leading to modifications in the functional chemical groups of the drug. Phase II of xenobiotic metabolism comprises conjugation reactions with small endogenous molecules, resulting in increased solubility in water and facilitating elimination of the drug. The reactions involved in xenobiotic metabolism lead in most cases to inactivation of the drug. There are also examples, however, of inactive prodrugs that are metabolized to the functionally active drug by drug-metabolizing enzymes (bioactivation). During the past 10 years polymorphisms have been described in genes coding for most drug-metabolizing enzymes[10].

### Thiopurine S-methyltransferase (TPMT)

One of the best-characterized polymorphic enzymes of phase II drug metabolism is thiopurine S-methyltransferase (TPMT), which is involved in the catabolic inactivation of thiopurine drugs such as azathioprine (AZA), 6-mercaptopurine (6-MP), and 6-thioguanine. For over 30 years AZA and 6-MP have been used as immunomodulatory agents in the treatment of inflammatory bowel disease (IBD)[11]. Potential risk factors for drug inefficacy as well as drug toxicity include pathophysiological, environmental and disease factors. Additionally, in the case of thiopurines of even greater importance in the determination of individual risk are inherited factors that affect pharmacokinetics. At the original recommended standard dose of up to 2.5 mg/kg per day, AZA has to be stopped in about 15% of both adult and paediatric patients due to severe haematotoxicity (e.g. pancytopenia) or gastrointestinal side-effects such as hepatotoxicity, pancreatitis, and gastrointestinal disturbances[12,13]. TPMT enzyme activity in humans is controlled by a common genetic polymorphism at the TPMT locus and approximately 1 in 200 individuals has very low TPMT activity, 11% have intermediate activity and 89% show normal/high activity. Inactivating mutations which are responsible for TPMT deficiency in a homozygous or compound heterozygous manner have been described and 20 variant TPMT alleles (TPMT*2-*18) are known so far[14]. The ethnic distribution of TPMT alleles, however, varies significantly among Caucasians, Asians, Africans and African-Americans[15]. TPMT*3A was found to be the most prevalent deficient allele in Caucasians, whereas the TPMT*3C allele is more frequently identified in African and Southeast Asian populations. All other inactive alleles are very rare in the populations studied and were described only in single cases[14]. Numerous clinical studies have shown that TPMT-deficient patients, including IBD patients, are at high risk for severe and sometimes fatal haematotoxicity due to the accumulation of cytotoxic metabolites after treatment with standard doses of thiopurines[16–18]. In addition, patients who are heterozygous carriers of one inactive TPMT allele also have an increased risk for thiopurine-related toxicity[19]. Therefore, in both cases, genotype-guided dose adjustment has been required (Figure 2)[20].

On the other hand it has been demonstrated that AZA-related gastrointestinal side-effects (e.g. hepatotoxicity or pancreatitis) cannot be explained by the TPMT polymorphism and other mechanisms for AZA-induced gastrointestinal toxicity may be responsible especially in the case of AZA-related hepatotoxicity[18]. Prospective determination of erythrocyte TPMT activity is advocated as a routine safety measure prior to therapy in order to avoid drug toxicity; however, the determination of the constitutive TPMT enzyme activity is time-consuming and has a number of serious limitations. If a deficient or heterozygous patient has received transfusions with red blood cells (RBC) from a homozygous wild-type individual, TPMT activity cannot be reliably determined within 30–60 days after transfusion[17]. Moreover, thiopurine administration itself alters TPMT activity in RBC with an increase of enzyme activity of approximately 20–30% compared to baseline, especially in heterozygous individuals[21]. Finally, some other clinically important drugs (e.g. sulphasalazine, olsalazine) have been identified as being partly potent inhibitors of TPMT in vitro and in vivo with the clinical consequence of an increased risk for developing leukopenia as demonstrated in IBD patients[22]. Therefore to avoid misclassification genotyping has been proposed as a reliable method to identify patients who are at a high risk for developing toxicity. However, evaluation of each diagnostic test needs estimation of sensitivity and specificity, as well as the positive and negative predictive value. Since in a recent large-scale study in our

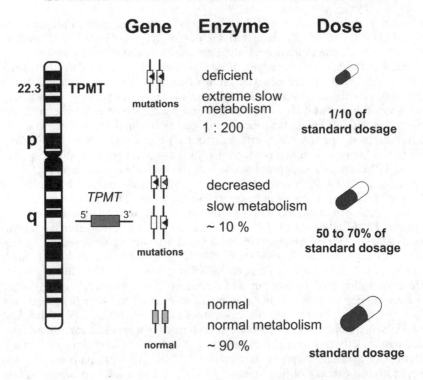

**Figure 2** Clinical consequences of *TPMT* polymorphism in treatment of IBD patients. Dose adjustment using approximately one-tenth of the standard dosage of thiopurines is necessary in the case of TPMT deficiency to avoid severe hematotoxicit[39]. Additionally, heterozygous individuals are at higher risk of developing adverse side-effects under standard dosage of thiopurines

laboratory all these parameters show values higher than 90%, genetic testing for *TPMT* is worthy of adoption into clinical practice[14]. Taking all *TPMT* variants into consideration the overall concordance rate between *TPMT* genetics and phenotypes was 98.4% in this study. Strikingly, testing for very rare mutant alleles does not significantly enhance the reliability of *TPMT* genotyping for intermediate methylators (86% vs. 89%), whereas all TPMT-deficient subjects were homozygous or compound heterozygous carriers for common *TPMT* alleles. Nevertheless, using automated genomic approaches such as DNA-microarrays or Maldi-TOF technology, genotyping will allow a rapid, time-and-cost saving screening for all known *TPMT*-inactivating muta-tions with maximal certainty for prediction. This may increase the routine clinical use of testing for TPMT to prospectively optimize thiopurine therapy.

Given the widespread use of thiopurine agents in IBD treatment *TPMT* genotyping offers the possibility of patient-tailored drug dosing and a sub-stantial reduction of haematopoietic toxicity (Figure 2).

# DRUG TRANSPORT

## MDR1/P-glycoprotein (P-gp)

In recent years it has become increasingly clear that drug transporters play a crucial role in drug absorption, distribution and elimination. The best-studied drug transporter is the *MDR1* gene product, the multidrug transporter P-glycoprotein (P-gp). This drug transporter was first described in tumour cells, where it contributes to multidrug resistance (MDR) against anticancer agents[23]. P-gp is a member of the large ATP-binding cassette superfamily of transport proteins. It is an integral membrane protein and is expressed in humans in the epithelial cells of the lower gastrointestinal tract, in the brush border of proximal renal tubule cells, in the liver at the biliary face of hepatocytes, in the apical membrane of pancreatic ductuli, on the luminal surface of capillary endothelial cells in the brain and testes, in the placenta, in the adreneal cortex and in some haematopoietic cells. In brief, P-gp acts as an efflux pump. The function of the protein and the distribution and polarity of P-gp expression suggest that the physiological function of this transporter is to provide a barrier against the entry of toxic compounds into the body, certain compartments of the body and cells, and to remove xenobiotics from circulation[24]. Experiments using the *mdr1*$^{-/-}$ knockout mouse model have provided further insights into the role of P-gp in drug therapy. It has been shown that the lack of this transporter gene has effects on the bioavailability, pharmacokinetics, distribution as well as drug effects and side-effects of important therapeutic drugs such as vinblastine, dexamethasone, digoxin and cyclosporine A[25]. In humans, clinical studies have shown that P-gp limits oral bioavailability of many of its substrates, including the immunosuppressant drugs cyclosporine A and tacrolimus, and HIV protease inhibitors[25,26]. Furthermore, the co-administration of P-gp inhibitors such as verapamil or PSC833 has been shown to increase the bioavailability of drugs whose absorption is normally limited by P-gp activity[27]. P-gp drug substrates commonly used in IBD therapy are corticosteroids such as prednisone and budesonide, as well as cyclosporine A and tacrolimus (FK506). Recently, we tested the hypothesis that budesonide and prednisone are substrates of P-gp, thereby possibly contributing to variable therapeutic effects. Using polarized, basal to apical transport of [$^3$H]budesonide and [$^3$H]prednisone in monolayers of L-MDR1 cells and Caco-2 cells, both of which express P-gp in their apical membrane, budesonide and prednisone were identified as substrates of the intestinal drug efflux pump, P-gp[28]. Therefore, drug secretion via P-gp into gut lumen might play a more important role in pharmacokinetics and pharmacodynamics of these corticosteroids than currently appreciated in gastroenterological practice.

## MDR1 polymorphisms

In the first systematic mutation screening, carried out by Hoffmeyer et al.[29], 15 mutations were found. So far 28 SNP have been identified, of which mutations in exon 21 (G2677T) and exon 26 (C3435T) are of particular interest because they are associated with differences in expression and/or function, as recently reviewed by Schwab et al.[30] and Marzolini et al.[31]. A two-fold reduction in intestinal (duodenal) P-gp expression was observed in subjects who were homozygous for the 3435T allele in exon 26 (Figure 3[29]). Based on this lower P-gp expression one would predict that the rate and extent of bioavailability of a P-gp substrate should be different from the wild-type genotype. In agreement with this assumption differences in pharmacokinetic parameters for the P-gp substrate digoxin were observed between genotypes in several studies[31]. The molecular mechanism responsible for the reduced P-gp expression associated with the exon 26 SNP has not been elucidated so far. However, the existence of a linkage disequilibrium with a second coding SNP in the *MDR1* gene has been studied extensively[30,31]. Although not complete such a linkage disequilibrium has been observed for the exon 21 SNP at the position 2677, suggesting that different haplotypes may be more predictive in relationship to treatment outcome and host susceptibility to diseases such as IBD[23]. Significant ethnic differences exist in the frequency of the various *MDR1* SNP[30,32].

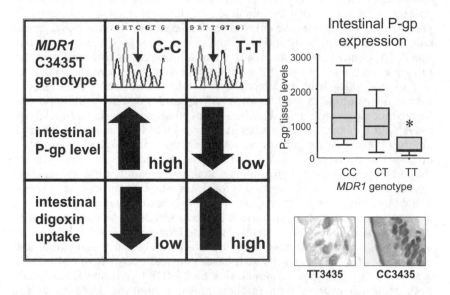

**Figure 3** Summary of the functional consequences of the *MDR1* polymorphism C3435T. The transition from C to T was associated with decreased intestinal expression of P-gp, and increased digoxin bioavailability resulting in a higher digoxin uptake[29]

In Crohn's disease poor reponse to corticosteroids has been related to increased expression of the intestinal drug efflux pump, P-gp[33]. This finding is important since steroid dependency and steroid resistance have been reported in 36% and 20% of patients with Crohn's disease[34] In parallel, the *MDR1* polymorphism at exon 26 (C3435T) was identified as significant factor for steroid weaning in paediatric heart transplant patients[35]. One might therefore

30

speculate that failure of standard therapy in Crohn's disease with budesonide or prednisone in particular might also be related to high expression of P-gp in intestinal mucosa and lymphocytes. This is supported by a recent observation in target cells of drugs in Crohn's disease. The MDR1 pump inhibitor PSC-833 significantly increased intracellular human T lymphocyte levels of cortisol[27].

In consequence, comprehensive studies elucidating the effect of *MDR1* polymorphims on the pharmacokinetics and pharmacodynamics of P-gp substrates (steroids, calacineurin inhibitors) used in IBD patients are necessary.

In this context the contribution of further tragets as another potential cause of steroid resistance in IBD, for example mutations of the glucocorticoid receptor gene resulting in altered ratios of glucocorticoid receptor α to its splice variant, glucocorticoid receptor β, should be taken into account[36].

## CLINICAL APPLICATION OF PHARMACOGENETICS

### Current dose recommendations are not appropriate for all patients

Traditionally, clinicians treating patients suffering from renal insufficiency use pocket dosing tables for the administration of drugs eliminated by the kidneys. From these tables the required dose adjustment can be read according to the patient's creatinine clearance. We anticipate that in future the dose of certain drugs may also take into account the patient's genotype to individually maximize therapeutic benefit and minimize risk.

In the case of TPMT, genotype-based dose recommendations for thiopurines appear to be helpful in treatment of IBD patients and have become clinical practice[37]. However, most important is the fact that the published recommendations are derived from retrospective, often small-scale, studies. Unfortunately, prospective clinical trials including pharmacoeconomic evaluations are presently lacking, also in the case of TPMT.

## FUTURE PROSPECTS FOR DRUG THERAPY

Despite a number of promising examples, the use of genotype information in drug prescribing is still in its infancy. Knowledge in this field, nevertheless, is rapidly growing. With greater understanding of the role of genetic factors in disease, and with growing evidence that diseases formerly considered homogeneous may actually comprise several distinct disease entities and require different treatment approaches, it is anticipated that progress in this field will enable more tailored therapy. Thus in addition to suggesting adjustments in drug dose, knowledge of the individual genotype may assist in subclassifying of patients[38].

It may be possible to identify genotype–response relationships at an early stage, so that the drug is marketed with appropriate recommendations. There is a need, therefore, for well-conducted studies post-licensing of a drug and as part of post-marketing surveillence to detect associations, as well as establish and test the validity and usefulness of the pharmacogenetic approach.

A 'genetic passport' has been suggested, which would carry a genetic fingerprint and indicate to a physician which drugs are safe and which should be avoided in a particular patient. There has been, and continues to be, a widespread discussion on the social and legal consequences of the collateral information produced by such a practice, especially in the context of predicting susceptibility to hitherto incurable diseases.

## References

1. Lazarou J, Pomeranz BH, Corey PN. Incidence of adverse drug reactions in hospitalized patients: a meta-analysis of prospective studies. J Am Med Assoc.1998;279:1200–5.
2. Vogel F. Moderne Probleme der Humangenetik. Ergebn Inn Med Kinderheilk. 1959;12:52–125.
3. Kalow W. Familial incidence of low pseudocholinesterase level. Lancet. 1956;2:576–7.
4. Bönicke R, Losboa BP. Über die Erbbedingtheit der intraindividuellen Konstanz der Isoniazidausscheidung beim Menschen (Untersuchungen an eineiigen Zwillingen). Naturwissenschaften. 1957;44:314.
5. Grant DM, Goodfellow GH, Sugamori K, Durette K. Pharmacogenetics of the human arylamine N-acetyltransferases. Pharmacology. 2000;61:204–11.
6. Mahgoub A, Idle JR, Dring LG, Lancaster R, Smith RL. Polymorphic hydroxylation of debrisoquine in man. Lancet. 1977;2:584–6.
7. Eichelbaum M, Spannbrucker N, Dengler HJ. N-oxidation of sparteine in man and its interindividual differences. Naunyn Schmiedebergs Arch Pharmacol. 1975;287(Suppl.): R94.
8. Griese EU, Zanger UM, Brudermanns U et al. Assessment of the predictive power of genotypes for the *in-vivo* catalytic function of CYP2D6 in a German population. Pharmacogenetics.1998;8:15–26.
9. Liggett SB. Pharmacogenetic applications of the human genome project. Nat Med. 2001;7: 281–3.
10. Evans WE, McLeod HL. Pharmacogenomics – drug disposition, drug targets, and side effects. N Engl J Med. 2003;348:538–49.
11. Schwab M, Klotz U. Pharmacokinetic considerations in the treatment of inflammatory bowel disease. Clin Pharmacokinet. 2001;40:723–51.
12. Present DH, Meltzer SJ, Krumholz MP, Wolke A, Korelitz BI. 6-Mercaptopurine in the management of inflammatory bowel disease: short- and long-term toxicity. Ann Intern Med. 1989;111:641–9.
13. Connell WR, Kamm MA, Ritchie JK, Lennard-Jones JE. Bone marrow toxicity caused by azathioprine in inflammatory bowel disease: 27 years of experience. Gut.1993;34:1081–5.
14. Schaeffeler E, Fischer C, Brockmeier D et al. Comprehensive analysis of thiopurine S-methyltransferase phenotype–genotype correlation in a large population of German-Caucasians and identification of novel TPMT variants. Pharmacogenetics. 2004;14:407–17.
15. McLeod HL, Krynetski EY, Relling MV, Evans WE. Genetic polymorphism of thiopurine methyltransferase and its clinical relevance for childhood acute lymphoblastic leukemia. Leukemia. 2000;14:567–72.
16. Colombel JF, Ferrari N, Debuysere H et al. Genotypic analysis of thiopurine S-methyltransferase in patients with Crohn's disease and severe myelosuppression during azathioprine therapy. Gastroenterology. 2000;118:1025–30.
17. Schwab M, Schaeffeler E, Marx C, Zanger U, Aulitzky W, Eichelbaum M. Shortcoming in the diagnosis of TPMT deficiency in a patient with Crohn's disease using phenotyping only. Gastroenterology. 2001;121:498–9.
18. Schwab M, Schäffeler E, Marx C et al. Azathioprine therapy and adverse drug reactions in patients with inflammatory bowel disease: impact of the thiopurine S-methyltransferase polymorphism. Pharmacogenetics. 2002;12:429–36.
19. Relling MV, Hancock ML, Rivera GK et al. Mercaptopurine therapy intolerance and heterozygosity at the thiopurine S-methyltransferase gene locus. J Natl Cancer Inst. 1999; 91:2001–8.

20. Lennard L. TPMT in the treatment of Crohn's disease with azathioprine. Gut. 2002;51: 143–6.
21. McLeod HL, Relling MV, Liu Q, Pui CH, Evans WE. Polymorphic thiopurine methyltransferase in erythrocytes is indicative of activity in leukemic blasts from children with acute lymphoblastic leukemia. Blood. 1995;85:1897–902.
22. Lowry PW, Franklin CL, Weaver AL et al. Leucopenia resulting from a drug interaction between azathioprine or 6-mercaptopurine and mesalamine, sulphasalazine, or balsalazide. Gut. 2001;49:656–4.
23. Ambudkar SV, Kimchi-Sarfaty C, Sauna ZE, Gottesman MM. P-glycoprotein: from genomics to mechanism. Oncogene. 2003;22:7468–85.
24. Ambudkar SV, Dey S, Hrycyna CA, Ramachandra M, Pastan I, Gottesman MM. Biochemical, cellular, and pharmacological aspects of the multidrug transporter. Annu Rev Pharmacol Toxicol. 1999;39:361–98.
25. Lin JH, Yamazaki M. Role of P-glycoprotein in pharmacokinetics: clinical implications. Clin Pharmacokinet. 2003;42:59–98.
26. Kimura Y, Matsuo M, Takahashi K et al. ATP hydrolysis-dependent multidrug efflux transporter: MDR1/P-glycoprotein. Curr Drug Metab. 2004;5:1–10.
27. Farrell RJ, Menconi MJ, Keates AC et al. P-Glycoprotein-170 inhibition significantly reduces cortisol and cyclosporin efflux from human intestinal epithelial cells and T lymphocytes. Aliment Pharmacol Ther. 2002;16:1021–31.
28. Dilger K, Schwab M, Fromm MF. Identification of budesonide and prednisone as substrates of the intestinal drug efflux pump P-glycoprotein. Inflamm Bowel Dis. 2004;10: 578–83.
29. Hoffmeyer S, Burk O, von Richter O et al. Functional polymorphisms of the human multidrug-resistance gene: multiple sequence variations and correlation of one allele with P-glycoprotein expression and activity *in vivo*. Proc Natl Acad Sci USA. 2000;97:3473–8.
30. Schwab M, Eichelbaum M, Fromm MF. Genetic polymorphisms of the human *MDR1* drug transporter. Annu Rev Pharmacol Toxicol. 2003;43:285–307.
31. Marzolini C, Paus E, Buclin T, Kim RB. Polymorphisms in human MDR1 (P-glycoprotein): recent advances and clinical relevance. Clin Pharmacol Ther. 2004;75:13–33.
32. Schaeffeler E, Eichelbaum M, Brinkmann U et al. Frequency of C3435T polymorphism of *MDR1* gene in African people. Lancet. 2001;358:383–4.
33. Farrell RJ, Murphy A, Long A et al. High multidrug resistance (P-glycoprotein 170) expression in inflammatory bowel disease patients who fail medical therapy. Gastroenterology. 2000;118:279-88.
34. Munkholm P, Langholz E, Davidsen M et al. Frequency of glucocorticoid resistance and dependency in Crohn's disease. Gut. 1994;35:360–2.
35. Zheng H, Webber S, Zeevi A et al. The MDR1 polymorphisms at exons 21 and 26 predict steroid weaning in pediatric heart transplant patients. Hum Immunol. 2002;63:765–70.
36. Farrell RJ, Kelleher D. Glucocorticoid resistance in inflammatory bowel disease. J Endocrinol. 2003;178:339–46.
37. Ho GT, Lees C, Satsangi J. Pharmacogenetics and inflammatory bowel disease: progress and prospects. Inflamm Bowel Dis. 2004;10:148–58.
38. Evans WE, Relling MV. Moving towards individualized medicine with pharmacogenomics. Nature. 2004;429:464–8.
39. Kaskas BA, Louis E, Hindorf U et al. Safe treatment of thiopurine S-methyltransferase deficient Crohn's disease patients with azathioprine. Gut. 2003;52:140–2.

# 4
# Genotypes and phenotypes in clinical practice

## S. VERMEIRE

## INTRODUCTION

The past decade has been characterized by tremendous progress in the field of inflammatory bowel disease (IBD) genetics. The first gene identified was the NOD2/CARD15 gene on 16q in Crohn's disease (CD) in 2001[1,2]. CARD15 encodes a protein involved in the host's first-line defence against microbial invasion, and leads to apoptosis by the mediation of NFκB. Three major variants have been identified of which the frameshift mutation Leu1007insC leads to a truncated CARD15 protein with impaired ability to recognize microbial components. Very recently the gene underlying the IBD5 susceptibility locus on chromosome 5q has been identified as OCTN, a cation transporter[3], and at the same time the team of Schreiber and co-workers identified the gene underlying the observed linkage on chromosome 10 as DLG5, a gene encoding a scaffolding protein important in epithelial cell integrity[4]. Together with this rapid progress in IBD genetics the question is raised as to how all this information is translated for the clinician; the following questions are relevant in this respect:

1. Is genotyping clinically useful in a patient in whom the diagnosis of CD has been established in terms of: (a) predicting the localization of the disease; (b) predicting the behaviour of the disease; (c) defining the onset of the disease; (d) predicting the need for surgery; (e) predicting the occurrence of extraintestinal manifestations?

2. Is genotyping clinically useful in a patient with a diagnosis of IBD, that cannot further be classified into CD or UC (= the so-called indeterminate colitis patients)?

Each of these questions will now be discussed in detail.

# IS GENOTYPING USEFUL IN A PATIENT IN WHOM THE DIAGNOSIS OF CD HAS BEEN ESTABLISHED?

## Is genotyping useful in defining the localization of the disease?

Ahmad et al. studied 244 patients with CD recruited from a single centre in the UK, and showed that the clinical pattern of CD can be defined by specific genotypes[5]. In his series CARD15/NOD2 mutations were significantly associated with ileal disease, whereas certain HLA haplotypes were associated with colonic disease.

For CARD15 the relative risk increased from 4 with possession of more than one variant allele to >30 in the case of two mutations. The results from a French cohort of 453 CD patients showed similarly that the patients with two CARD15 mutations less frequently had colonic involvement than patients with no mutations (43% vs 62%, $p = 0.003$, OR 0.44)[6]. In a large European study on 444 CD patients the frequency of NOD2 mutations was the highest in patients with ileal involvement (26.9%), lower or intermediate in patients with ileocolonic involvement (19.7%) and the lowest in pure colitis (12.7%)[7]. Also in a North-American population from Quebec ($n$ = 231 CD patients) CARD15/NOD2 variants were inversely correlated with colonic disease (OR 2.86, 95% CI 1.41–5.78)[8]. At least another seven studies have confirmed these findings[9–15].

In conclusion, CARD15/NOD2 mutations are clearly associated with a disease localization of the ileum than of the colon. The relative risk in the case of one mutation is approximately 2–4[5,7–8,14]. An ulcerative colitis (UC)-like involvement is very unlikely in this group of patients.

As shown by Ahmad et al., colonic disease is more associated with HLA haplotypes. The same group previously showed that certain HLA alleles furthermore influence the extension of the colitis: presence of HLA-DRB1*0103 is associated with pancolitis whereas a decrease of HLA-DRB1*04 is associated with distal colitis[16–17].

## Is genotyping useful in defining the behaviour of the disease?

In contrast to the published results on disease localization, the associations with CARD15 mutations and disease behaviour are less consistent. The difficulty in interpreting the results on behaviour is that the definitions differ in the various studies. Some authors have taken stricturing and fistulizing disease together in their analyses, as opposed to inflammatory disease. Another difficulty in the interpretation is the changing nature of behaviour. The longer the disease exists, the more chance that the disease will evolve from inflammatory to stricturing or fistulizing, as shown by Louis et al.[18]. That group studied 163 patients with a firm diagnosis of CD who had non-penetrating non-stricturing disease at diagnosis. Five years after diagnosis there were still 67.5% of patients with non-stricturing non-penetrating disease, but already 11% had evolved towards stricturing disease and 21.5% towards penetrating disease. The NOD2/CARD15 genotype did not influence the progression of the disease. Several groups have, however, reported positive associations between stenosing or stricturing behaviour and CARD15 variants[6,9-12,19-20]. The patients with double-dose mutations in the French study were characterized by a more frequent stricturing phenotype (53% vs 28%; $p$ = 0.00003, OR 2.92)[6]. Abreu and colleagues studied two cohorts of consecutively identified patients referred to an IBD centre ($n$ = 142 collected between 1993 and 1996; $n$ = 59 collected between 1999 and 2001). CD-associated NOD2 variants were significantly associated with fibrostenosing disease in each cohort ($p$ = 0.049 and $p$ = 0.002, respectively)[19]. When both cohorts were analysed together the association between NOD2 variants and fibrostenosing disease was even more significant: 46% of CD patients with fibrostenotic disease carried CARD15 mutations compared to only 23% in the group without stenosis ($p$ = 0.001), and this difference was even more pronounced when carriers of two mutations were compared to non-carriers (85% vs 43%, $p$ = 0.004). In the study by Brant et al. the presence of double CARD15 mutations was associated with a 7.44-fold increased risk of stricturing disease, and increased to 17, in the case of concomitant ileal involvement[11]. A German study suggested that patients with CARD15/NOD2 mutations would be indicative for more aggressive disease, since an association was seen between CARD15/NOD2 mutations and the presence of fistula, the fibrostenotic phenotype and previous ileocaecal resection[9].

On the other hand, in the UK study the presence of a CARD15/NOD2 variant, and particularly the Leu1007fsinsC mutation, was protective against fistulizing disease. Stenotic disease was positively associated with the presence of a CARD15/NOD2 mutation, but this was not independent of the link with ileal disease[5].

The cytokine gene cluster on 5q31 (IBD5) has been associated with perianal fistulizing behaviour[21]. With the recent identification of the OCTN gene in this region it remains to be defined if patients with perianal fistulas have an increased prevalence of OCTN mutations[4].

## Is genotyping useful in defining the onset of the disease?

Before the identification of the CARD15/NOD2 gene it was shown that the linkage on 16q was higher in patients with a young age at onset, <22 years[22]. With the identification of the gene the association of CARD15/NOD2 mutations with an early age at onset has so far been confirmed by at least four groups[5-6,11,13]. In the study by Lesage et al. the age at onset of patients with two mutations was 16.9 years compared to 19.8 years in the patients who did not have mutations ($p$ = 0.00003)[6]. The Oxford group also found the same results: age at onset in the patients carrying double mutations was 23.5 years compared to 29 years in the patients without mutations ($p$ = 0.04)[5].

Mirza et al. furthermore showed that IBD5 and CARD15 have a synergistic effect on age at onset (CARD15$^+$/IBD5$^+$ median age at onset 24.1 years as compared to 30.1 years for CARD15$^-$/IBD5$^-$)[23].

## Is genotyping useful in predicting the need for surgery?

The only consistent genetic association for predicting surgery has been HLA-DRB1*0103. Presence of this allele has been associated with extensive colitis and the need for colectomy[16-17]. For NOD2/CARD15 there are insufficient data to answer this question. Ahmad et al. looked at predictors of need for surgical resection, but failed to show an association[5]. A German study, on the other hand, did find an association between mutations and previous ileocaecal resection[9]. Subsequent studies that have investigated the need for surgery show conflicting results, and it is not always clear if a particular association is independent from other clinical variables, for instance disease behaviour.

## Is genotyping useful in predicting the occurrence of extraintestinal manifestations (EIM)?

Joint manifestations occur in 10–20% of patients with IBD. Both axial involvement and peripheral arthropathies can be seen. In contrast to the idiopathic ankylosing spondylitis (AS), which is in almost all cases associated with HLA-B*27, the IBD-associated AS is characterized by a lower prevalence of HLA-B*27 (50–70% of patients). The isolated sacroiliitis which can also be observed has no association with HLA-B*27.

Orchard et al. proposed a classification system for IBD-related peripheral arthropathy without axial involvement[24]. Type 1 consists of a pauciarticular peripheral arthropathy, involving less than five joints and characterized by acute, self-limiting attacks. This type is associated with relapses of IBD and with other EIM of IBD. Type 2, on the other hand, is a polyarticular peripheral arthropathy involving more than five joints, and symptoms often persist for years. These would run a course independent of IBD and are associated with ocular manifestations, but not with other manifestations of disease. The authors clearly showed that type 1 is associated with HLA-B*27, HLA-B*35 and HLA-DRB1*0103 and type 2 with HLA-B*44[25].

For CARD15 a weak inverse relationship with the presence of EIM has been shown: in multiple logistic regression analysis, 27.7% of patients with CARD15 mutations had EIM, compared to 40.2% of patients without CARD15 mutations ($p$ = 0.031; OR 0.61; 95% CI 0.39–0.95)[14]. Other studies that included EIM in their analyses failed to show an association[5-6,8-10,15]; however, in most of these studies all EIM were taken together and analysed as such; therefore, more detailed studies may be needed before definitive conclusions can be made.

Concerning the other extraintestinal symptoms of eyes and skin, associations with HLA-B*15 and HLA-B*57 with erythema nodosum have been found[26]. Ocular manifestations such as uveitis are characterized by HLA-B1*0103, HLA-B*27 and HLA-B*58[26].

37

## IS CARD15/NOD2 GENOTYPING USEFUL IN A PATIENT WITH INDETERMINATE COLITIS?

Despite several classification systems and good clinical, endoscopic, radiological and histopathological tools, differentiation between CD and UC is still sometimes difficult. In approximately 10% of cases of colitis no differentiation can be made. Disease in these patients is classified as indeterminate colitis (IC). Excellent Scandinavian epidemiological studies have shown that 5–23% of all initial diagnoses of IBD are IC, and that the incidence of IC is 2.4 per 100 000. Despite the fact that IC is a rather temporary diagnosis, and that up to 80% of patients with this designation will be diagnosed with CD or UC over time, an early approximation of how the disease will progress could have important clinical usefulness. For example, patients with IC are more likely to undergo colectomy, and pouch failure rates are much higher than in patients with definite UC. Moreover, the prognosis in patients with IC is worse than in patients with UC. There is a higher frequency of relapse and an increased risk of colon cancer. So far no prognostic factors exist which could help to further classify IC patients into a definitive diagnosis. We recently showed in a prospective study that ASCA and atypical pANCA are helpful in patients with IC towards further precision of the disease phenotype[27]. We were able to obtain DNA from 68/97 patients from this initial cohort (collaborative effort between the Universities of Vienna, Austria and Leuven, Belgium) and genotyped these patients for the mutations Arg702Trp, Leu3020insC and Gly908Arg in NOD2/CARD15. Interestingly, a high percentage (14/68 or 20.6%) of these colitis patients carried at least one CARD15 mutation. Fifteen patients have reached a definitive diagnosis of CD ($n = 10$) or UC ($n = 5$) so far. Of the 10 patients who had their diagnosis changed into CD, only one carried NOD2 mutations, compared to 0/5 patients with a final diagnosis of UC and 13/53 of the remaining IC patients. This yielded a specificity and positive predictive value (PPV) of NOD2 variants towards CD of 77.6% and 7.1% respectively[28]. These preliminary data show that CARD15/NOD2 genotyping does not seem to offer an additive tool towards further categorizing IC patients. Interesting, however, and in contrast to the previously reported low prevalence of ASCA and atypical pANCA in patients with IC, was the high prevalence of NOD2 variants in the patients who remain indeterminate. Longer follow-up will tell us whether these patients will eventually prove to have CD.

## CONCLUSION

Although, currently, genetic testing has no additive value in the clinical practice of IBD patients, more and more molecular data consistently link certain genotypes to particular phenotypes such as location and behaviour of the disease and presence of EIM. Concerning disease location, NOD2/CARD15 is associated with ileal involvement whereas HLA genotypes are associated with colonic disease. A stricturing disease behaviour is also related to CARD15. On the other hand, distinct HLA alleles contribute to the occurrence of extraintestinal manifestations.

# References

1.  Hugot JP, Chamaillard M, Zouali H et al. Association of NOD2 leucine-rich repeat variants with susceptibility to Crohn's disease. Nature. 2001;411:599–603.
2.  Ogura Y, Bonen DK, Inohara N et al. A frameshift mutation in NOD2 associated with susceptibility to Crohn's disease. Nature. 2001;411:603–6.
3.  Peltekova VD, Wintle RF, Rubin LA et al. Functional variants of OCTN cation transporter genes are associated with Crohn disease. Nat Genet. 2004;36:471–5.
4.  Stoll M, Corneliussen B, Costello CM et al. Genetic variation in DLG5 is associated with inflammatory bowel disease. Nat Genet. 2004;36:476–80.
5.  Ahmad T, Armuzzi A, Bunce M et al. The molecular classification of the clinical manifestations of Crohn's disease. Gastroenterology. 2002;122:854–66.
6.  Lesage S, Zouali H, Cezard JP et al. CARD15/NOD2 mutational analysis and genotype–phenotype correlation in 612 patients with inflammatory bowel disease. Am J Hum Genet. 2002;70:845–57.
7.  Cuthbert AP, Fisher SA, Mirza MM et al. The contribution of NOD2 gene mutations to the risk and site of disease in inflammatory bowel disease. Gastroenterology. 2002;122:867–74.
8.  Vermeire S, Wild G, Kocher K et al. CARD15 genetic variation in a Quebec population: prevalence, genotype–phenotype relationship, and haplotype structure. Am J Hum Genet. 2002;71:74–83.
9.  Hampe J, Cuthbert A, Croucher PJ et al. Association between insertion mutation in NOD2 gene and Crohn's disease in German and British populations. Lancet. 2001;357:1925–8.
10. Mendoza JL, Murillo LS, Fernandez L et al. Prevalence of mutations of the NOD2/CARD15 gene and relation to phenotype in Spanish patients with Crohn disease. Scand J Gastroenterol. 2003;38:1235–40.
11. Brant SR, Picco MF, Achkar JP et al. Defining complex contributions of NOD2/CARD15 gene mutations, age at onset, and tobacco use on Crohn's disease phenotypes. Inflamm Bowel Dis. 2003;9:281–9.
12. Helio T, Halme L, Lappalainen M et al. CARD15/NOD2 gene variants are associated with familially occurring and complicated forms of Crohn's disease. Gut. 2003;52:558–62.
13. Bairead E, Harmon DL, Curtis AM et al. Association of NOD2 with Crohn's disease in a homogeneous Irish population. Eur J Hum Genet. 2003;11:237–44.
14. Esters N, Pierik M, van Steen K et al. Transmission of CARD15 (NOD2) variants within families of patients with inflammatory bowel disease. Am J Gastroenterol. 2004;99:299–305.
15. Tomer G, Ceballos C, Concepcion E, Benkov KJ. NOD2/CARD15 variants are associated with lower weight at diagnosis in children with Crohn's disease. Am J Gastroenterol. 2003;98:2479–84.
16. Satsangi J, Welsh KI, Bunce M et al. Contribution of genes of the major histocompatibility complex to susceptibility and disease phenotype in inflammatory bowel disease. Lancet. 1996;347:1212–17.
17. Roussomoustakaki M, Satsangi J, Welsh K et al. Genetic markers may predict disease behavior in patients with ulcerative colitis. Gastroenterology. 1997;112:1845–53.
18. Louis E, Michel V, Hugot JP et al. Early development of stricturing or penetrating pattern in Crohn's disease is influenced by disease location, number of flares, and smoking but not by NOD2/CARD15 genotype. Gut. 2003;52:552–7.
19. Abreu MT, Taylor KD, Lin YC et al. Mutations in NOD2 are associated with fibrostenosing disease in patients with Crohn's disease. Gastroenterology. 2002;123:679–88.
20. Heresbach D, Gicquel-Douabin V, Birebent B et al. NOD2/CARD15 gene polymorphisms in Crohn's disease: a genotype–phenotype analysis. Eur J Gastroenterol Hepatol. 2004;16:55–62.
21. Armuzzi A, Ahmad T, Ling KL et al. Genotype–phenotype analysis of the Crohn's disease susceptibility haplotype on chromosome 5q31. Gut. 2003;52:1133–9.
22. Brant SR, Panhuysen CI, Bailey-Wilson JE et al. Linkage heterogeneity for the IBD1 locus in Crohn's disease pedigrees by disease onset and severity. Gastroenterology. 2000;119:1483–90.
23. Mirza MM, Fisher SA, King K et al. Genetic evidence for interaction of the 5q31 cytokine locus and the CARD15 gene in Crohn disease. Am J Hum Genet. 2003;72:1018–22.

24. Orchard TR, Wordsworth BP, Jewell DP. Peripheral arthropathies in inflammatory bowel disease: their articular distribution and natural history. Gut. 1998;42:387–91.
25. Orchard TR, Thiyagaraja S, Welsh KI, Wordsworth BP, Hill Gaston JS, Jewell DP. Clinical phenotype is related to HLA genotype in the peripheral arthropathies of inflammatory bowel disease. Gastroenterology. 2000;118:274–8.
26. Orchard TR, Chua CN, Ahmad T, Cheng H, Welsh KI, Jewell DP. Uveitis and erythema nodosum in inflammatory bowel disease: clinical features and the role of HLA genes. Gastroenterology. 2002;123:714–18.
27. Joossens S, Reinisch W, Vermeire S et al. The value of serologic markers in indeterminate colitis: a prospective follow-up study. Gastroenterology. 2002;122:1242–7.
28. Vermeire S, Reinisch W, Joossens S et al. NOD2/CARD15 genotyping in patients with indeterminate colitis (IC): helpful towards definitive diagnosis? Gastroenterology. Suppl. 2002;M1416.

# Section II
# Microbiology

**Chair: J.-F. COLOMBEL and S. KOLACEK**

# 5
# Chronic inflammatory disorders, the gut and the 'Old Friends' hypothesis

G. A. W. ROOK and L. ROSA BRUNET

## INTRODUCTION

Several types of chronic inflammatory disorder are becoming much more common in the rich developed countries. These include inflammatory bowel diseases (ulcerative colitis and Crohn's disease), certain autoimmune diseases (for example type 1 diabetes and multiple sclerosis) and allergies. It will be argued below that the rapidly increasing prevalences of these disorders can be attributed to changing patterns of microbial exposure due to the modern lifestyle, leading to a failure of immunoregulatory mechanisms in individuals with certain genetic polymorphisms of the innate immune system. This view is referred to as the 'Old Friends' hypothesis, and is an updating of the previous 'Hygiene' hypothesis. A particular aim of this chapter is to point out that versions of the 'Hygiene' hypothesis that seek to implicate decreased exposure to disease-causing infections (i.e. true pathogens) are taking into account neither the epidemiology that led to the hypothesis, nor the available experimental evidence. The term 'Old Friends' is therefore used to highlight the fact that the bulk of the evidence implicates decreased exposure to harmless organisms that are part of mammalian evolutionary history, and can be shown to have an immunoregulatory role. We will show that from an immunological point of view this makes sense, and fits within well-established immunological mechanisms.

## THE 'HYGIENE' HYPOTHESIS AS PROPOSED IN THE LATE 1990s

The 'Hygiene' hypothesis as it exists in 2004 is very different from the hypothesis that emerged in the late 1990s. The view that changing patterns of infectious disease (and overall reduction in exposure to pathogens) might lead to an increase in the incidence or presentation of various immunological disorders has appeared intermittently in the medical literature for decades. In 1996 Strachan and colleagues gave new impetus to the idea when they observed

43

that allergies were less common in children from large families with older siblings, especially older brothers, who would be expected to expose a younger child to more infections in the home[1]. A flood of further epidemiological data, particularly studies of allergies in farming communities[2,3] and military personnel[4], provided circumstantial support for the view that diminished exposure to certain microorganisms might lead to an enhanced risk of allergy. It was suggested by many authors at that time that Th1 responses induced by pathogens might be needed to down-regulate Th2 responses, and that the modern hygienic way of life was providing too little stimulus for Th1 cells, with a consequent non-specific increase in Th2 activity. With hindsight it is clear that this view did not fit the facts in two clear ways. The untenable aspects were, first, the notion that the effect of diminished exposure to microorganisms was mediated by an imbalance between Th1 and Th2 (the Th1/Th2 see-saw), and secondly, the view that the most important organisms were disease-causing pathogens. These issues are discussed below.

## The Th1/Th2 see-saw; a discredited idea

The concept of a 'Th1/Th2 see-saw', put forward in early versions of the 'Hygiene' hypothesis, was greatly weakened by the realization that Th2- and Th1-mediated chronic inflammatory disorders were increasing simultaneously and in the same countries[5]. Thus there is a striking correlation between type 1 diabetes (Th1-mediated destruction of $\beta$ cells in the pancreas) and allergies (Th2-mediated), both within and outside Europe[6]. Similarly, inflammatory bowel disease (IBD) is increasing in the same regions (Th1-mediated, or mixed)[7,8]. These parallels have been extensively reviewed[9]. Clearly the explanation for the increased prevalence of allergic disorders cannot be a lack of stimuli for Th1 activity if Th1-mediated diseases are also increasing. It is just as clear that increasing Th1-mediated disorders (Crohn's disease (CD), multiple sclerosis (MS), type 1 diabetes) cannot be attributable to decreased Th2 activity, when there is a simultaneous rapid rise in the prevalence of allergies. The obvious 'umbrella' hypothesis is that something is causing a simultaneous increase in immune responses against targets that should be tolerated, irrespective of whether the attack is mediated by Th1 or by Th2 cells. These inappropriate targets include trivial levels of allergens, self antigens, and the contents of the gut (resulting in allergies, autoimmunity and IBD respectively). The current hypothesis is that what matters is the balance of regulatory T cells ($T_{reg}$) to effector T cells (whether Th1 or Th2) rather than the balance of Th1 to Th2.

Moreover Th1 (effector) responses may *not* be physiological regulators of Th2 activity *in vivo*. The earlier literature suggesting that Th1 cells can down-regulate Th2 must be reinterpreted in the light of the recent evidence proving the existence of several subsets of $T_{reg}$[10,11]. When the presence of $T_{reg}$ is rigorously excluded, and cloned polarized Th1 cells are used in cell transfer experiments, Th1 cells may even exacerbate ongoing Th2-mediated inflammation[12]. This is not contradicted by the recent observation that mice that lack T-bet, and so cannot develop effector Th1 cells, have dominantly Th2 responses and can develop some asthma-like pathology[13]. This is not a realistic parallel of

44

the physiological situation. The increased Th2 in T-bet knockouts is probably an inevitable default mechanism restoring T cell numbers with the only subsets available. Individuals with plenty of Th1 cells, but an almost complete lack of functional Th1 effector mechanisms due to congenitally defective IFN-$\gamma$ receptors, do not have any sign of increased Th2 activity[14]. Finally, it is often forgotten that IFN-$\gamma$ is in fact a dominant cytokine in human asthma[15].

In the context of IBD and Th1-mediated autoimmunity it has been suggested that, rather than a failure of regulation of Th2 by Th1 (the original 'Hygiene Hypothesis'), we are seeing the exact opposite: a failure of regulation of Th1 by Th2. However, clearly it makes no sense for allergologists and gastroenterologists to support opposite and mutually exclusive hypotheses. Still more important is the mounting evidence that the presence of a small Th2 component in a dominantly Th1 response does not necessarily counteract Th1-mediated tissue damage. In fact a superimposed Th2 response can dramatically *increase* immunopathology caused by Th1[16–18]. Again the problem lies in the failure to distinguish between Th2 and $T_{reg}$.

## Does Th1/Th2 balance play any role?

Although Th1-mediated and Th2-mediated inflammatory disorders are increasing in parallel within populations[6,9], there does seem to be some segregation of Th1 and Th2 at the level of the *individual* in some studies. Thus in some communities individuals suffering from the Th1-mediated autoimmune diseases, type 1 diabetes or multiple sclerosis, are less likely to be allergic[19,20]. However, the Th1/Th2 balance of the individual might reflect genetic background and past history of infections and vaccinations. This in turn might determine the type of immunoregulatory disorder that will occur in that individual when regulatory T cell activity is compromised.

## Imbalance between effector and regulatory T cells ($T_{reg}$); lessons from Foxp3

The ability of decreased activity of $T_{reg}$ to explain the simultaneous increases in all three major groups of chronic inflammatory disorder has now been repeatedly reviewed, and it is reasonable to start using the term chronic 'immunoregulatory' disorder[21–23]. Support for this view has come from recent analyses of genetic disorders in humans and in mice attributable to mutations in the transcription factor Foxp3[24,25]. Foxp3 is fundamental to the development of $T_{reg}$[10,26]. In both species the resulting pathology encompasses precisely the spectrum of diseases (allergies, autoimmunity and IBD) that is increasing in the developed world, irrespective of the Th1 or Th2 bias of the dominant effector mechanisms.

## THE 'OLD FRIENDS' HYPOTHESIS (FIGURE 1)

So, having turned away from the 'Th1/Th2 see-saw' in favour of a $T_{reg}/T_{effector}$ imbalance, what about the types of organism that are important for immunoregulation, and why are they important? Why do they drive $T_{reg}$? The epidemiology has highlighted exposure to cowsheds[2], and the presence of endotoxin in house dust[3] as clear correlates of protection from allergic disorders, and shown that childhood virus infections are not protective[4]. So what suggested the notion that the relevant organisms were pathogens? A single study, based only on skin-test responses, suggested that tuberculosis might be protective[27], but this has not been reproducible. It seems that the involvement of pathogens was an assumption, not based on the epidemiological data, which points to harmless organisms not associated with disease. We have no reason to fear cowsheds!

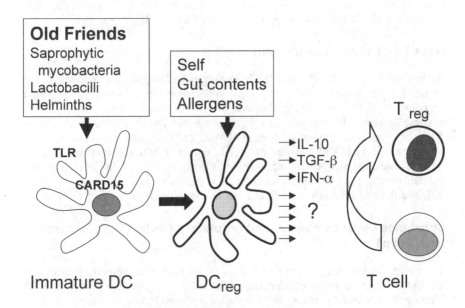

**Figure 1** The 'Old Friends' hypothesis as determined from work on *M. vaccae*. The multiple receptors of the innate immune system (only TLR2 and CARD15 are illustrated) recognize the 'Old Friend' as harmless, as a consequence of its presence throughout mammalian evolutionary history. This causes the dendritic cell (DC) to mature, while maintaining the ability to drive regulation ($DC_{reg}$). (Note that immature DC may drive $T_{reg}$ activation, but this property is usually lost with maturation of the DC.) These $DC_{reg}$ process antigens from the 'Old Friend' in addition to any other antigens present, including self, allergens and gut contents. This type of antigen presentation drives $T_{reg}$, some specific for the 'Old Friend', others specific for the potential targets of chronic inflammatory disorders. The overall result is two types of immunoregulation: (1) bystander regulation due to the continuous exposure to the 'Old Friend'; (2) regulation specific for self, allergens and gut contents. This scheme can also be deduced from the studies on helminths and lactobacilli

Similarly, when one considers why diminished exposure to microorganisms might result in inadequate priming of $T_{reg}$, one is again drawn for clear theoretical reasons to harmless organisms. In fact $T_{reg}$ are probably part of the normal immune response to all organisms, whether pathogenic or not. Where pathogens are concerned, after the effector response has reduced pathogen numbers to low levels, $T_{reg}$ may help to terminate the response[28]; but clearly, excessive early activation of $T_{reg}$ by pathogens would lead to progressive disease. Therefore the organisms most likely to have a highly developed ability to drive $T_{reg}$ rather than $T_{eff}$, and perhaps to bypass the phase of $T_{eff}$ induction altogether[29], are harmless 'Old Friends' and commensals that have been present throughout our evolutionary history, and that have become part of mammalian physiology. In addition to the unidentified (but clearly non-pathogenic) factor present in German cowsheds[2], four groups of organism have been highlighted in this context: lactobacilli[30], saprophytic environmental mycobacteria[31,32], the helminths[22], and certain viruses that are ubiquitous in developing countries[33]. These are not pathogens in the true sense. Each of these groups of organism has been implicated in down-regulation of allergic responses in humans, either by direct clinical studies or (less convincingly) by epidemiological correlation[4,30,34–36]. Moreover helminth components and *Mycobacterium vaccae* have been shown to induce $T_{reg}$ *in vitro*[37] and *in vivo* respectively[32].

## Lactobacilli

Lactobacilli were an integral part of the hunter–gatherer diet, and were common in poorly preserved vegetables and in beverages prepared by malolactic fermentation[38]. They are present in greater quantities in the guts of children from areas with a low incidence of allergies[39]. They are clearly not pathogens or 'infections' but they are part of our evolutionary history.

## Saprophytic environmental mycobacteria

Similarly the saprophytic mycobacteria are ubiquitous in mud and untreated water, and used to be consumed in such quantities that detectable levels of tuberculostearic acid accumulated in lymphoid tissue[40]. The presence of mycobacteria in the mammalian evolutionary past is indicated by the presence of CD1-restricted T cell subsets that appear to recognize only mycobacterial lipids and glycolipids[41]. Most of the population in developing countries are skin-test-positive to multiple mycobacterial species[42], whereas this is rare in the developed countries. This is because saprophytic mycobacteria are not part of the commensal flora, so exposure to them depends entirely upon how and where one lives. Several clinical studies with *M. vaccae* have yielded promising results, and further trial results are awaited[35,36].

## Helminths

Helminths are also present throughout life in most individuals in developing countries, and in a diminishing percentage of the populations of developed countries. Approximately one-third of the population of the world harbours organisms of this group, usually with little or no ill effect[43]. This may be achieved by the induction of $T_{reg}$. Indeed the ability of helminths to drive $T_{reg}$ activity is an integral component of the host–parasite relationship. Some helminths, such as schistosomes, drive granulomatous responses to ova; but others, such as the microfilariae, suppress the response to a remarkable extent. When this mechanism fails, an immune response to the parasite ensues, with massive lymphadenopathy and pathology[44,45].

## Viruses; hepatitis A

Virus infections characteristic of childhood (measles, rubella, mumps, influenza, colds) in the developed countries appear not to protect from allergies[4], which fits with the observation that children in inner cities are heavily exposed to these infections and still have a high and rising risk of allergies. However, it was noted that antibody to the infections that are transmitted by the faeco-oral route (for example human hepatitis A virus (HAV)) correlated with a lower incidence of allergic symptoms in Italian army recruits[4]. Subsequently it has emerged that a T cell membrane protein, TIM-1, that controls the development of airway hyperreactivity and production of Th2 cytokines in mice, is a homologue of the human receptor for HAV. This may explain the inverse relationship between HAV infection and the development of atopy[33].

## The induction of allergen-specific $T_{reg}$ by 'Old Friends'

Most of the work on the 'Old Friends' has been performed using models of allergy rather than models of IBD, so these will be considered first. $T_{reg}$ can be induced by the saprophytic environmental mycobacterium, *M. vaccae* (using SRP299, a GMP preparation also in use in human trials)[32]. The specificity of the $T_{reg}$ induced was tested in a cell transfer system. It was demonstrated that the $T_{reg}$ induced in this system have specificity for the allergen to which the animals were exposed. However, after the regulatory properties of the $T_{reg}$ have been triggered by specific allergen, they can exert bystander inhibition of the response to other allergens[32].

## Two roles for the 'Old Friends'

This bystander phenomenon may be of fundamental importance to these disorders of immunoregulation, particularly in the gut, where the crucial role of $T_{reg}$ has been extensively documented[46]. When SCID mice (which lack T lymphocytes) were reconstituted with effector T cells with a normal range of specificities, they developed severe colitis. The colitis persisted if they were given cloned transgenic $T_{reg}$ specific for a single epitope of ovalbumin. However, if ovalbumin was added to the diet, the bystander suppression exerted by the ovalbumin-specific $T_{reg}$ was able to alleviate the colitis, despite the fact that the colitis will have been triggered by multiple antigens within the food and gut flora[47]. These observations emphasize the point that the 'Old Friends' discussed earlier may have two distinct roles. Experiments *in vitro* and *in vivo* suggest that they can act as '$T_{reg}$ adjuvants', maturing antigen-presenting cells to a phenotype that tends to direct T cells towards specific regulatory function[32,37,48]. However the 'Old Friends' may also provide a crucial background of regulation, because their constant presence leads to a steady background activation of $T_{reg}$ that recognize the 'Old Friends' themselves, and therefore promote background levels of IL-10. Again this argument points towards organisms that will have been taken in orally throughout mammalian evolution (such as saprophytic mycobacteria and lactobacilli) and organisms that were permanently present in the tissues (such as some helminths and some strains of lactobacilli).

## $T_{reg}$ do not suppress when the innate immune system detects 'danger'

Why does bystander suppression not constantly incapacitate the immune system? Conceptually the answer must be that $T_{reg}$ down-regulate immune responses that should not be there, such as responses to self, gut contents, and trivial levels of harmless aero-allergens. Interestingly when 'danger signals' are present the $T_{reg}$ stop regulating. In an *in-vitro* system $T_{reg}$ failed to function when danger signals were provided by addition of CpG motifs or endotoxin[49]. Similarly, $T_{reg}$ can block graft-versus-host disease while permitting anti-tumour responses[50]. Finally, it has been noted that $T_{reg}$ express an unusual pattern of Toll-like receptors, and these may also play a role in the modulation of $T_{reg}$ function[51].

## Innate immunity and induction of $T_{reg}$

How then are the 'Old Friends' recognized as harmless, and how does that recognition lead to the activation of $T_{reg}$? Similarly how do danger signals sometimes stop $T_{reg}$ from developing, as outlined in the previous paragraph? The answer probably lies in the innate immune system. The pattern recognition receptors (PRR) of the innate immune system perform rapid 'taxonomy' on the organism before the specific immune response can be generated and then direct the adaptive immune response towards appropriate effector mechanisms (i.e. via Th1 or Th2) or towards immunoregulation via $T_{reg}$. In general, *immature* dendritic cells (DC) will drive $T_{reg}$, but this property is lost when the DC mature. The 'Old Friends' must therefore be able to induce maturation of DC, with preservation of the ability to drive $T_{reg}$. It is therefore of particular interest that polymorphisms of genes involved in the innate immune system are proving to be relevant to the incidence of disorders of immunoregulation. The parallels between the components of the innate immune system that are involved in recognition of bacteria and the components that have been shown to be relevant to some of the disorders of immunoregulation have been extensively reviewed elsewhere[52]. A few examples are given below.

## CARD15 (previously known as NOD2)

Bacterial cell wall muramyl peptides can signal via recently discovered intracellular components of the innate immune system, NOD1 (CARD4) and CARD15 (NOD2). Polymorphisms of CARD15 are associated with susceptibility to Crohn's disease, and predominantly Th1-mediated condition[53]; however, the polymorphism also increases susceptibility to allergies[54]. These are therefore examples of genetically impaired recognition of bacterial components leading to a variety of Th1-mediated or Th2-mediated immunoregulatory disorders in different individuals.

## Toll-like receptors

The Toll-like receptor (TLR) family acts as pattern-recognition receptors for pathogen-associated molecular patterns (PAMP). The use of the word 'pathogen' here is misleading since these receptors are also involved in recognition of commensals and even of self components such as heat-shock proteins. Ten TLR types are known and these may act as homo- or as heterodimers with other TLR. They bind conserved microbial derivatives such as lipopolysaccharide, lipoproteins, peptigoglycans, glycosylphosphatidylinositols, double-stranded RNA, CpG motifs in DNA and flagellin. All microorganisms contain ligands for many TLR. Nevertheless when whole live organisms are added to TLR-expressing cells, a more restricted pattern can emerge. Gram-negative bacteria tend to trigger TLR4, whereas mycobacteria tend to trigger TLR2[55]. Recent data suggest the possibility of links between polymorphisms of TLR2 and susceptibility to asthma. Exposure to the farming environment during early life significantly protects from allergic disorders[2]. Blood cells from the children from these farms express significantly higher amounts of mRNA encoding CD14 and TLR2 (but not TLR4) than do those from non-farmers' children[56]. Finally, the beneficial effect of exposure during early life to the cowsheds on these farms is dependent upon a polymorphism of TLR2 (von Mutius E, personal communication). Additional data supporting a role for TLR in the induction of immunoregulation has come from work on lactobacilli. Significantly, it is claimed that the protective effect of the lactobacilli in models of IBD discussed later turns out to depend upon TLR9, which binds CpG motifs.[57]. So are TLR sometimes involved in driving $T_{reg}$?

## TLR and $T_{reg}$?

Activation of human DC via TLR4 and TLR2 may result in different immunological outcomes[58]. Unlike TLR4, TLR2 does not lead to IL-12 release, but rather to inhibitory IL-12 p40 dimers[59], and also to IL-23, which is formed when p40 dimerizes with p19. IL-23 has a different spectrum of activities from IL-12 and, unlike IL-12, causes proliferation of murine CD45RB[lo] T cells, that might include a $T_{reg}$ component[60]. Human DC that had been preincubated with a schistosome extract caused T cells to secrete IL-10[61]. If TLR2 was blocked with an antibody the T cells secreted Th2 cytokines, but no IL-10[61]. It is not yet clear why the same TLR can sometimes drive potent inflammatory responses, and yet sometimes be essential for initiating regulation. As stated above, the anti-inflammatory effects of lactobacilli require TLR9, but there is nothing special about the DNA of the lactobacilli. It seems that the lactobacilli present their DNA in a 'package', or in the context of other signals to the innate immune system, that permit recognition of the lactobacilli as 'Old Friends', and consequently the activation of regulation[57]. Such findings imply that mimicking the effects of the 'Old Friends' with single molecules may be difficult. The 'code' for $T_{reg}$ induction is likely to be complex so that pathogens do not too easily acquire it.

## THE 'OLD FRIENDS' IN EXPERIMENTAL MODELS AND IN CLINICAL STUDIES: THE ROLE OF THE GUT

Animal experiments (and clinical studies) provide further support for the validity of the 'Old Friends' hypothesis, and for the view that we should consider the allergies, autoimmune disorders and IBD as representing different manifestations of overlapping immunoregulatory problems. These 'Old Friends' appear to be active in several models of chronic inflammatory disease, irrespective of the nature of the disorder, or whether it is mediated by Th1 or Th2. Moreover it seems that the oral route may be the optimal route of administration. This probably reflects a central role of the bowel flora and mucosal immune system in the priming of $T_{reg}$[62,63].

The involvement of helminths as regulators of allergic disorders was deduced from field studies of allergic sensitization in parts of Africa where helminth infections are endemic[22,64]. Experimental models support this concept[65]. However, experimental infection with the helminth *Schistosoma mansoni* also prevents inflammation during colitis[66] and reduces the severity of an auto-immune disease that mimics type 1 diabetes[67]. More recently it has proved possible to apply these experimental findings to the clinical setting, and clinical trials are in progress to determine whether exposure to helminths such as *Trichuris suis* can provide therapeutic alternatives or adjunct therapy for patients with IBD[43,68].

The lactobacilli are equally versatile. Interleukin 10-deficient (IL-10$^{-/-}$) mice remain healthy under germ-free conditions, but some IL-10$^{-/-}$ strains develop colitis when given a normal specific pathogen-free (SPF) bowel flora. This is due to a poorly controlled immune response to the bowel flora. (Interestingly, the colitis seen in SPF IL-2$^{-/-}$ mice also depends upon the presence of intestinal bacterial flora[69].) However, colonization of germ-free IL-10$^{-/-}$ mice with *Lactobacillus plantarum* does not induce colitis, and can protect from subsequent introduction of SPF flora. Moreover the lactobacillus could also alleviate the established colitis in SPF IL-10$^{-/-}$ mice[29]. It remains to be seen whether this protective effect is mediated via $T_{reg}$, which in this model would need to be able to function without IL-10.

As discussed above, these protective effects of the lactobacilli turn out to depend upon TLR9, which binds CpG motifs[57]. Similarly, antigen preparations from anaerobic bowel commensals given by gavage to Balb/c mice will alleviate colitis induced by dextran sulphate which, like that seen in IL-10$^{-/-}$ and IL-2$^{-/-}$ mice, is also attributable to an immune response to gut contents[70].

The immunoregulatory effects of lactobacilli, like those of helminths, are not confined to one type of chronic inflammatory disorder. Lactobacilli are also implicated in the control of allergic disorders[38], and recent clinical studies of large oral doses in atopic eczema in infants have shown promise[30]. Similarly lactobacilli are said to be effective in several models of autoimmune disease, including diabetes in the NOD mouse[71], and collagen arthritis[72].

The full range of activity of the saprophytic mycobacteria has not yet been defined. *M. vaccae* will inhibit a model of autoimmunity[73], but has not yet been tried in a model of IBD. Most work has concentrated on the models of allergy using subcutaneous administration. Used in this way *M. vaccae* can induce allergen-specific $T_{reg}$ (CD4$^+$, CD45RB$^{lo}$, IL-10$^+$) that will inhibit allergic symptoms, and that are active in a passive transfer model. Their function in the allergic recipient animals was blocked by a combination of anti-IL-10 and anti-TGF-$\beta$[32,74]. Recently it has emerged that the same material is equally active via the oral route[75]. It has not yet been formally shown that the material still acts via induction of $T_{reg}$ when used in this way, but there was no increase in Th1 activity, and a marked increase in secretion of IL-10, so regulatory pathways are strongly implicated[75].

The beneficial effect of oral treatment with *M. vaccae* has been observed in another disease model associated with a Th2 bias. Periodontal disease is characterized by a change in the periodontal flora and a switch in the local immune response towards Th2[76]. Experimentally, it can be induced in Wistar rats by placing a ligature in the gingival sulcus. In this model, as in the model of pulmonary allergy reported above, oral administration of *M. vaccae* was as effective as subcutaneous injection[77]. There was significantly reduced loss of bone and attachment fibres following oral treatment with *M. vaccae*.

Interestingly the doses of *M. vaccae* required for efficacy by the oral route are identical to those that are optimal by the subcutaneous route, despite the inevitable loss of most of the organisms in the faeces[75,77,78]. This surprising efficacy of oral delivery is most likely due both to the importance of the gut as a site for $T_{reg}$ induction, and to the size, charge and hydrophobicity of *Mycobacterium* spp. (whether living or dead) which lead to their rapid uptake into Peyer's patches[79,80]. Throughout mammalian evolution there will have been a constant heavy oral input of saprophytic mycobacteria of which there can be as many as $10^9$/litre in untreated water.

## GUT COMMENSALS AND MODULATION OF CHRONIC INFLAMMATORY DISORDERS

In addition to the direct experimental work and clinical studies with identified 'Old Friends' outlined above, the 'Old Friends' hypothesis is also supported by studies on the effects on chronic inflammatory disorders of varying the composition of the bowel flora. These experiments all emphasize the immunoregulatory role of harmless organisms that have a long evolutionary history of coexistence with the mammalian immune system. It may be very significant that the diversity of bowel flora in patients with CD and UC is reduced to 50% and 30% respectively of that seen in controls[81]. It is of course not yet clear whether this is a consequence of the disease, or a factor that contributes to it, but a role for bowel flora in the modulation of immune responses has been suggested by a number of studies. For instance it is known that IgE production cannot be switched off by oral induction of tolerance in germ-free mice[82], and neonatal antibiotic treatment can deviate the immune system towards Th2. These effects are prevented if commensal intestinal flora are reintroduced[82,83].

Many models of autoimmune disease are also influenced by the nature of the commensal microbial flora. For instance non-obese diabetic (NOD) mice reared in gnotobiotic environments have worsened diabetes[84]. Similarly female PVG/c strain rats maintained under specific pathogen-free conditions until weaning were found to be significantly less susceptible to the induction of autoimmune thyroiditis by thymectomy and irradiation than conventionally reared rats of the same strain[85].

However, the most striking effects are seen in various models of autoimmune arthritis. Although the mechanisms were not understood, it was known in the 1980s that germ-free rats were moderately susceptible to adjuvant arthritis, and that this susceptibility could be enhanced or decreased simply by reconstituting the bowel flora with different bacterial species[86]. Susceptibility to pristane-induced arthritis, which involves immune responses to heat-shock protein hsp60, requires exposure to other arthritic animals, presumably to something present within their microbial flora[87]. Similarly the germ-free state protects against ankylosing enthesopathy in B10.BR mice[88], and against inflammatory peripheral joint disease in HLA-B27 transgenic rats[89].

In general, however, bacterial flora appear to protect from autoimmune arthritis. For example collagen-induced arthritis is increased in germ-free DA rats compared to conventionally raised rats[90]. Similarly germ-free Wistar rats are much more susceptible to adjuvant arthritis than are conventionally raised Wistar rats[91]. Moreover, transfer of spleen cells from the 'dirty' rats to the clean rats is sufficient to reduce the susceptibility of the latter[92]. Thus in various rat strains background exposure to harmless microorganisms can influence Th1-mediated autoimmunity, and the last experiment quoted suggests a role for $T_{reg}$. The mechanism has been studied further in streptococcal cell wall arthritis. This is a chronic, erosive polyarthritis that can be induced in susceptible Lewis rats by a single intraperitoneal injection of an aqueous, sterile suspension of streptococcal cell wall. Interestingly, F344 rats housed under conventional conditions are resistant to this treatment because commensal bacterial bowel flora induce neonatally a state of tolerance to arthritogenic epitopes in the streptococcal cell wall material. This state of tolerance is maintained through-out life. However, germ-free F344 rats are as susceptible to streptococcal cell wall-induced arthritis as are Lewis rats. This susceptibility of germ-free F344 rats rapidly disappears if the bowel flora are reconstituted. In contrast, Lewis rats are arthritis-prone even when they have a normal bowel flora, because this tolerance is deficient and/or easily broken[93].

## CONCLUSIONS

In conclusion, the 'Old Friends' hypothesis provides a partial explanation for the increasing prevalence of chronic inflammatory disorders, including IBD. The hypothesis suggests that the underlying problem is a failure of immunoregulation secondary to decreased exposure to harmless microorganisms that have been present throughout mammalian evolutionary history. It is argued that these organisms are effectively part of our physiology, and are recognized as harmless by the innate immune system. The crucial point might be that 'Old Friends' can cause maturation of DC without the loss of ability to drive $T_{reg}$ that accompanies DC maturation driven by other stimuli. They consequently increase the background numbers of antigen presenting cells (APC) primed for the induction of $T_{reg}$. By activating these $APC_{reg}$ the 'Old Friends' act as $T_{reg}$ adjuvants for other antigens such as self, allergens and gut contents, and so drive $T_{reg}$ specific for these inappropriate targets, and limit the three classes of chronic immunoregulatory disorder. Moreover, in addition to this adjuvant effect, $T_{reg}$ that recognize the 'Old Friends' themselves will also be generated. Then continuous exposure to 'Old Friends' will activate bystander suppression and so lower the threshold for inflammation. These mechanisms can be revealed by altering the microbial exposure, or by studying the presence of polymorphisms of the innate immune system, since both factors can be shown to influence susceptibility. Together they provide a series of classical gene–environment interactions. In the modern world much of our exposure to microorganisms is via vaccination, but these vaccines are designed to evoke potent effector responses, not regulation. However, recent clinical studies suggest that it will be possible to exploit the 'Old Friends' as vaccines, probably administered orally, to restore the appropriate level of immunoregulation.

## References

1.  Strachan DP, Taylor EM, Carpenter RG. Family structure, neonatal infection, and hay fever in adolescence. Arch Dis Child. 1996;74:422–6.
2.  Riedler J, Braun-Fahrlander C, Eder W et al. Exposure to farming in early life and development of asthma and allergy: a cross-sectional survey. Lancet. 2001;358:1129–33.
3.  Braun-Fahrlander C, Riedler J, Herz U et al. Environmental exposure to endotoxin and its relation to asthma in school-age children. N Engl J Med. 2002;347:869–77.
4.  Matricardi PM, Rosmini F, Riondino S et al. Exposure to foodborne and orofecal microbes versus airborne viruses in relation to atopy and allergic asthma; epidemiological study. Br Med J. 2000;320:412–17.
5.  Rook GAW. Clean living increases more than atopic disease. Immunol Today. 2000;21:249.
6.  Stene LC, Nafstad P. Relation between occurrence of type 1 diabetes and asthma. Lancet. 2001;357:607.
7.  Lindberg E, Lindquist B, Holmquist L, Hildebrand H. Inflammatory bowel disease in children and adolescents in Sweden, 1984–1995. J Pediatr Gastroenterol Nutr. 2000;30: 259–64.
8.  Sawczenko A, Sandhu BK, Logan RF et al. Prospective survey of childhood inflammatory bowel disease in the British Isles. Lancet. 2001;357:1093–4.
9.  Bach JF. The effect of infections on susceptibility to autoimmune and allergic diseases. N Engl J Med. 2002;347:911–20.
10. Hori S, Nomura T, Sakaguchi S. Control of regulatory T cell development by the transcription factor Foxp3. Science. 2003;299:1057–61.

11. Roncarolo M, Levings MK. The role of different subsets of T regulatory cells in controlling autoimmunity. Curr Opin Immunol. 2000;12:676–83.
12. Hansen G, Berry G, DeKruyff RH, Umetsu DT. Allergen-specific Th1 cells fail to counterbalance Th2 cell-induced airway hyperreactivity but cause severe airway inflammation. J Clin Invest. 1999;103:175–83.
13. Finotto S, Neurath MF, Glickman JN et al. Development of spontaneous airway changes consistent with human asthma in mice lacking T-bet. Science. 2002;295:336–8.
14. Lammas DA, Casanova JL, Kumararatne DS. Clinical consequences of defects in the IL-12-dependent interferon-gamma (IFN-gamma) pathway. Clin Exp Immunol. 2000;121: 417–25.
15. Krug N, Madden J, Redington AE et al. T-cell cytokine profile evaluated at the single cell level in BAL and blood in allergic asthma. Am J Respir Cell Mol Biol. 1996;14:319–26.
16. Hernandez-Pando R, Aguilar D, Garcia Hernandez ML, Orozco H, Rook GAW. Pulmonary tuberculosis in Balb/c mice with non-functional IL-4 genes; changes in the inflammatory effects of TNF-α in the regulation of fibrosis. Eur J Immunol. 2004;34:174–83.
17. Lawrence CE, Paterson JC, Higgins LM, MacDonald TT, Kennedy MW, Garside P. IL-4-regulated enteropathy in an intestinal nematode infection. Eur J Immunol. 1998;28:2672–84.
18. Wynn TA, Cheever AW, Jankovic D et al. An IL-12-based vaccination method for preventing fibrosis induced by schistosome infection. Nature. 1995;376:594–6.
19. Douek IF, Leech NJ, Gillmor HA, Bingley PJ, Gale EA. Children with type-1 diabetes and their unaffected siblings have fewer symptoms of asthma. Lancet. 1999;353:1850.
20. Tremlett HL, Evans J, Wiles CM, Luscombe DK. Asthma and multiple sclerosis: an inverse association in a case-control general practice population. Q J Med. 2002;95:753–6.
21. Umetsu DT, McIntire JJ, Akbari O, Macaubas C, DeKruyff RH. Asthma: an epidemic of dysregulated immunity. Nat Immunol. 2002;3:715–20.
22. Yazdanbakhsh M, Kremsner PG, van Ree R. Allergy, parasites, and the hygiene hypothesis. Science. 2002;296:490–4.
23. Rook GA, Adams V, Hunt J, Palmer R, Martinelli R, Brunet LR. Mycobacteria and other environmental organisms as immunomodulators for immunoregulatory disorders. Springer Semin Immunopathol. 2004;25:237–55.
24. Brunkow ME, Jeffery EW, Hjerrild KA et al. Disruption of a new forkhead/winged-helix protein, scurfin, results in the fatal lymphoproliferative disorder of the scurfy mouse. Nat Genet. 2001;27:68–73.
25. Wildin RS, Ramsdell F, Peake J et al. X-linked neonatal diabetes mellitus, enteropathy and endocrinopathy syndrome is the human equivalent of mouse scurfy. Nat Genet. 2001;27: 18–20.
26. Fontenot JD, Gavin MA, Rudensky AY. Foxp3 programs the development and function of CD4(+)CD25(+) regulatory T cells. Nat Immunol. 2003;4:330–6.
27. Shirakawa T, Enomoto T, Shimazu S, Hopkin JM. The inverse association between tuberculin responses and atopic disorder. Science. 1996;275:77–9.
28. Belkaid Y, Piccirillo CA, Mendez S, Shevach EM, Sacks DL. CD4+CD25+ regulatory T cells control *Leishmania major* persistence and immunity. Nature. 2002;420:502–7.
29. Schultz M, Veltkamp C, Dieleman LA et al. *Lactobacillus plantarum* 299V in the treatment and prevention of spontaneous colitis in interleukin-10-deficient mice. Inflamm Bowel Dis. 2002;8:71–80.
30. Kalliomaki M, Salminen S, Arvilommi H, Kero P, Koskinen P, Isolauri E. Probiotics in primary prevention of atopic disease: a randomised placebo-controlled trial. Lancet. 2001; 357:1076–9.
31. Rook GAW, Stanford JL. Give us this day our daily germs. Immunol Today. 1998;19:113–16.
32. Zuany-Amorim C, Sawicka E, Manlius C et al. Suppression of airway eosinophilia by killed *Mycobacterium vaccae*-induced allergen-specific regulatory T-cells. Nat Med. 2002;8:625–9.
33. McIntire JJ, Umetsu SE, Akbari O et al. Identification of Tapr (an airway hyperreactivity regulatory locus) and the linked Tim gene family. Nat Immunol. 2001;2:1109–16.
34. van den Biggelaar AH, van Ree R, Rodrigues LC et al. Decreased atopy in children infected with *Schistosoma haematobium*: a role for parasite-induced interleukin-10. Lancet. 2000;356:1723–7.

35. Arkwright PD, David TJ. Intradermal administration of a killed *Mycobacterium vaccae* suspension (SRL 172) is associated with improvement in atopic dermatitis in children with moderate-to-severe disease. J Allergy Clin Immunol. 2001;107:531–4.
36. Camporota L, Corkhill A, Long H et al. The effects of *Mycobacterium vaccae* on allergen-induced airway responses in atopic asthma. Eur Respir J. 2003;21:287–93.
37. van der Kleij D, Latz E, Brouwers JF et al. A novel host–parasite lipid cross-talk. Schistosomal lyso-phosphatidylserine activates Toll-like receptor 2 and affects immune polarization. J Biol Chem. 2002;277:48122–9.
38. Kalliomaki M, Isolauri E. Role of intestinal flora in the development of allergy. Curr Opin Allergy Clin Immunol. 2003;3:15–20.
39. Sepp E, Julge K, Vasar M, Naaber P, Bjorksten B, Mikelsaar M. Intestinal microflora of Estonian and Swedish infants. Acta Paediatr. 1997;86:956–61.
40. Hanngren A, Odham G, Eklund A, Hoffner S, Stjernberg N, Westerdahl G. Tuberculostearic acid in lymph nodes from patients with sarcoidosis. Sarcoidosis. 1987;4:101–4.
41. Dutronc Y, Porcelli SA. The CD1 family and T cell recognition of lipid antigens. Tissue Antigens. 2002;60:337–53.
42. Stanford JL, Nye PM, Rook GA, Samuel N, Fairbank A. A preliminary investigation of the responsiveness or otherwise of patients and staff of a leprosy hospital to groups of shared or species antigens of mycobacteria. Lepr Rev. 1981;52:321–7.
43. Weinstock JV, Summers R, Elliott DE. Helminths and harmony. Gut. 2004;53:7–9.
44. Satoguina J, Mempel M, Larbi J et al. Antigen-specific T regulatory-1 cells are associated with immunosuppression in a chronic helminth infection (onchocerciasis). Microbes Infect. 2002;4:1291–300.
45. Steel C, Nutman TB. CTLA-4 in filarial infections: implications for a role in diminished T cell reactivity. J Immunol. 2003;170:1930–8.
46. Mottet C, Uhlig HH, Powrie F. Cutting edge: cure of colitis by CD4+CD25+ regulatory T cells. J Immunol. 2003;170:3939–43.
47. Groux H, O'Garra A, Bigler M et al. A CD4+ subset inhibits antigen-specific T cell responses and prevents colitis. Nature. 1997;389:737–42.
48. Adams VC, Hunt J, Martinelli R, Palmer R, Rook GAW, Rosa Brunet L. *Mycobacterium vaccae* induces a population of pulmonary antigen presenting cells that have regulatory potential in allergic mice. Eur J Immunol. 2004;34:631–8.
49. Pasare C, Medzhitov R. Toll pathway-dependent blockade of CD4+CD25+ T cell-mediated suppression by dendritic cells. Science. 2003;299:1033–6.
50. Edinger M, Hoffmann P, Ermann J et al. CD4+CD25+ regulatory T cells preserve graft-versus-tumour activity while inhibiting graft-versus-host disease after bone-marrow transplantation. Nat Med. 2003;9:1144–50.
51. Caramalho I, Lopes-Carvalho T, Ostler D, Zelenay S, Haury M, Demengeot J. Regulatory T cells selectively express toll-like receptors and are activated by lipopolysaccharide. J Exp Med. 2003;197:403–11.
52. Rook GA, Martinelli R, Brunet LR. Innate immune responses to mycobacteria and the downregulation of atopic responses. Curr Opin Allergy Clin Immunol. 2004;3:337–42.
53. Ogura Y, Bonen DK, Inohara N et al. A frameshift mutation in NOD2 associated with susceptibility to Crohn's disease. Nature. 2001;411:603–6.
54. Kabesch M, Peters W, Carr D, Leupold W, Weiland SK, von Mutius E. Association between polymorphisms in caspase recruitment domain containing protein 15 and allergy in two German populations. J Allergy Clin Immunol. 2003;111:813–17.
55. Heldwein KA, Fenton MJ. The role of Toll-like receptors in immunity against mycobacterial infection. Microbes Infect. 2002;4:937–44.
56. Lauener RP, Birchler T, Adamski J et al. Expression of CD14 and Toll-like receptor 2 in farmers' and non-farmers' children. Lancet. 2002;360:465–6.
57. Rachmilewitz D, Karmeli F, Takabayashi K et al. Immunostimulatory DNA ameliorates experimental and spontaneous murine colitis. Gastroenterology. 2002;122:1428–41.
58. Re F, Strominger JL. Toll-like receptor 2 (TLR2) and TLR4 differentially activate human dendritic cells. J Biol Chem. 2001;276:37692–9.
59. O'Neill LA. Toll-like receptor signal transduction and the tailoring of innate immunity: a role for Mal? Trends Immunol. 2002;23:296–300.

60. Oppmann B, Lesley R, Blom B et al. Novel p19 protein engages IL-12p40 to form a cytokine, IL-23, with biological activities similar as well as distinct from IL-12. Immunity. 2000;13:715–25.
61. Van der Kleij D, Van Remoortere A, Schuitemaker JH et al. Triggering of innate immune responses by schistosome egg glycolipids and their carbohydrate epitope GalNAc beta 1-4(Fuc alpha 1-2Fuc alpha 1-3)GlcNAc. J Infect Dis. 2002;185:531–9.
62. Kullberg MC, Jankovic D, Gorelick PL et al. Bacteria-triggered CD4(+) T regulatory cells suppress *Helicobacter hepaticus*-induced colitis. J Exp Med. 2002;196:505–15.
63. Maloy KJ, Powrie F. Regulatory T cells in the control of immune pathology. Nature Immunol. 2001;2:816–22.
64. Maizels RM, Yazdanbakhsh M. Immune regulation by helminth parasites: cellular and molecular mechanisms. Nat Rev Immunol. 2003;3:733–44.
65. Lima C, Perini A, Garcia ML, Martins MA, Teixeira MM, Macedo MS. Eosinophilic inflammation and airway hyper-responsiveness are profoundly inhibited by a helminth (*Ascaris suum*) extract in a murine model of asthma. Clin Exp Allergy. 2002;32:1659–66.
66. Moreels TG, Nieuwendijk RJ, De Man JG et al. Concurrent infection with *Schistosoma mansoni* attenuates inflammation induced changes in colonic morphology, cytokine levels, and smooth muscle contractility of trinitrobenzene sulphonic acid induced colitis in rats. Gut. 2004;53:99–107.
67. Cooke A, Tonks P, Jones FM, O'Shea H, Hutchings P, Fulford AJ et al. Infection with *Schistosoma mansoni* prevents insulin dependent diabetes mellitus in non-obese diabetic mice. Parasite Immunol. 1999;21:169–76.
68. Summers RW, Elliott DE, Qadir K, Urban JF Jr, Thompson R, Weinstock JV. *Trichuris suis* seems to be safe and possibly effective in the treatment of inflammatory bowel disease. Am J Gastroenterol. 2003;98:2034–41.
69. Contractor NV, Bassiri H, Reya T et al. Lymphoid hyperplasia, autoimmunity, and compromised intestinal intraepithelial lymphocyte development in colitis-free gnotobiotic IL-2-deficient mice. J Immunol. 1998;160:385–94.
70. Verdu EF, Bercik P, Cukrowska B et al. Oral administration of antigens from intestinal flora anaerobic bacteria reduces the severity of experimental acute colitis in BALB/c mice. Clin Exp Immunol. 2000;120:46–50.
71. Matsuzaki T, Nagata Y, Kado S et al. Prevention of onset in an insulin-dependent diabetes mellitus model, NOD mice, by oral feeding of *Lactobacillus casei*. APMIS. 1997;105:643–9.
72. Kato I, Endo-Tanaka K, Yokokura T. Suppressive effects of the oral administration of *Lactobacillus casei* on type II collagen-induced arthritis in DBA/1 mice. Life Sci. 1998;63:635–44.
73. Thompson SJ, Butcher PD, Patel VKR et al. Modulation of pristane-induced arthritis by mycobacterial antigens. Autoimmunity. 1991;11:35–43.
74. Zuany-Amorim C, Manlius C, Trifilieff A et al. Long-term protective and antigen-specific effect of heat-killed *Mycobacterium vaccae* in a murine model of allergic pulmonary inflammation. J Immunol. 2002;169:1492–9.
75. Hunt JR, Martinelli R, Adams V, Rook GAW, Rosa Brunet L. Oral *Mycobacterium vaccae* inhibits severe pulmonary allergic inflammation in a mouse model. 2004 (Submitted).
76. Tokoro Y, Matsuki Y, Yamamoto T, Suzuki T, Hara K. Relevance of local Th2-type cytokine mRNA expression in immunocompetent infiltrates in inflamed gingival tissue to periodontal diseases. Clin Exp Immunol. 1997;107:166–74.
77. Breivik T, Rook GA. Oral treatment with SRP299 (killed *Mycobacterium vaccae*) inhibits experimental periodontal disease in Wistar rats. J Clin Periodontol. 2003;30:931–6.
78. Breivik T, Rook GA. Treatment with SRL172 (heat-killed *Mycobacterium vaccae*) inhibits progression of established experimental periodontal disease in Wistar rats. J Periodontal Res. 2002;37:210–14.
79. Fujimura Y. Functional morphology of microfold cells (M cells) in Peyer's patches – phagocytosis and transport of BCG by M cells into rabbit Peyer's patches. Gastroenterol Jpn. 1986;21:325–35.
80. Momotani E, Whipple DL, Thiermann AB, Cheville NF. The role of M cells and macrophages in the entrance of *Mycobacterium paratuberculosis* into domes of ileal patches in calves. Vet Pathol. 1988;25:131–7.

81. Ott SJ, Musfeldt M, Wenderoth DF et al. Reduction in diversity of the colonic mucosa associated bacterial microflora in patients with active inflammatory bowel disease. Gut. 2004;53:685–93.
82. Sudo N, Sawamura S, Tanaka K, Aiba Y, Kubo C, Koga Y. The requirement of intestinal bacterial flora for the development of an IgE production system fully susceptible to oral tolerance induction. J Immunol. 1997;159:1739–54.
83. Sudo N, Yu XN, Aiba Y et al. An oral introduction of intestinal bacteria prevents the development of a long-term Th2-skewed immunological memory induced by neonatal antibiotic treatment in mice. Clin Exp Allergy. 2002;32:1112–16.
84. Kukreja A, Maclaren NK. NKT cells and type-1 diabetes and the 'hygiene hypothesis' to explain the rising incidence rates. Diabetes Technol Ther. 2002;4:323–33.
85. Penhale WJ, Young PR. The influence of the normal microbial flora on the susceptibility of rats to experimental autoimmune thyroiditis. Clin Exp Immunol. 1988;72:288–92.
86. Kohashi O, Kohashi Y, Takahashi T, Ozawa A, Shigematsu N. Reverse effect of gram-positive bacteria vs. gram-negative bacteria on adjuvant-induced arthritis in germfree rats. Microbiol Immunol. 1985;29:487–97.
87. Thompson SJ, Elson CJ. Susceptibility to pristane-induced arthritis is altered with changes in bowel flora. Immunol Lett. 1993;36:227–31.
88. Rehakova Z, Capkova J, Stepankova R et al. Germ-free mice do not develop ankylosing enthesopathy, a spontaneous joint disease. Hum Immunol. 2000;61:555–8.
89. Taurog JD, Richardson JA, Croft JT et al. The germfree state prevents development of gut and joint inflammatory disease in HLA-B27 transgenic rats. J Exp Med. 1994;180:2359–64.
90. Breban MA, Moreau MC, Fournier C, Ducluzeau R, Kahn MF. Influence of the bacterial flora on collagen-induced arthritis in susceptible and resistant strains of rats. Clin Exp Rheumatol. 1993;11:61–4.
91. van de Langerijt AG, van Lent PL, Hermus AR, Sweep CG, Cools AR, van den Berg WB. Susceptibility to adjuvant arthritis: relative importance of adrenal activity and bacterial flora. Clin Exp Immunol. 1994;97:33–8.
92. Moudgil KD, Kim E, Yun OJ, Chi HH, Brahn E, Sercarz EE. Environmental modulation of autoimmune arthritis involves the spontaneous microbial induction of T cell responses to regulatory determinants within heat shock protein 65. J Immunol. 2001;166:4237–43.
93. van den Broek MF, van Bruggen MC, Koopman JP, Hazenberg MP, van den Berg WB. Gut flora induces and maintains resistance against streptococcal cell wall-induced arthritis in F344 rats. Clin Exp Immunol. 1992;88:313–17.

# 6
# Innate immune receptors and inflammatory bowel disease

E. CARIO and D. K. PODOLSKY

## INTRODUCTION

The intestinal mucosa must rapidly recognize detrimental pathogenic threats of the lumen to initiate controlled immune responses, but maintain hyporesponsiveness to omnipresent harmless commensals. Janeway first suggested so-called pattern recognition receptors (PRR) may play an essential role in allowing innate immune cells to discriminate between self and microbial non-self based on the recognition of broadly conserved molecular patterns[1]. Toll-like receptors (TLR), which comprise a class of transmembrane PRR, play a key role in microbial recognition, induction of antimicrobial genes and the control of adaptive immune responses. NOD are structurally distinct family of intracellular receptors which presumably in the context of microbial invasion subserve similar functions. TLR and NOD are widely expressed on various cell types of the gastrointestinal mucosa participating in host defence against microbial pathogens in at least four ways: (1) recognition of molecular patterns present on pathogens; (2) expression at the interface with the 'environment' of the gastrointestinal lumen; (3) induction of secretion of pro/anti-inflammatory cytokines and chemokines that link to the adaptive immune system; (4) induction of antimicrobial effector pathways.

## TLR DISTRIBUTION AND EXPRESSION

Consistent with their roles in immune surveillance, TLR are mostly expressed at higher levels in tissues exposed to the external environment such as lung and gastrointestinal tracts, as well as in immunologically important settings such as peripheral blood leucocytes and spleen. TLR are expressed by many different cell types throughout the whole gastrointestinal tract *in vitro* and *in vivo*, including macrophages and epithelial cells of the small intestine and colon[2,3].

TLR2 and TLR4 are present at the apical pole of differentiated intestinal epithelial cells *in vitro* and *in vivo*, and thus are well positioned to monitor the lumenal milieu of bacterial products[4]. In contrast, TLR5 appears to be preferentially expressed at the basolateral pole[5]. TLR4 appears in 'vesicular' structures in the apical cytoplasm of intestinal epithelial cells (IEC) after exposure to lipopolysaccharide (LPS)[4] which was recently identified as being part of the Golgi apparatus, possibly functioning to sample and detoxify its cargo lipopolysaccharide (LPS). In response to LPS, TLR4 traffics from its apical location to intracytoplasmic compartments near the basolateral membrane in IEC[6].

## TLR LIGANDS AND MODULATORS: LINK TO DISEASE

Specific TLR may detect individual features that are common to different classes of microorganisms. In addition, several non-bacterial ligands have been identified selectively activating different TLR diseases. Many of the diverse factors recognized by TLR have been implicated in pathophysiological mechanisms of various gastrointestinal disorders, including infectious diseases, allergy/food intolerance and inflammatory bowel disease (IBD).

TLR2 has a broad specificity for the recognition of various patterns from Gram-positive bacteria and mycobacterial species, including bacterial lipopeptides and lipoteichoic acid. TLR2 may cooperate with TLR6 and TLR1, suggesting a mechanism for diversifying the repertoire of TLR-mediated responses[7]. Direct binding of TLR2 to peptidoglycan, a major cell wall component of Gram-positive bacteria, which may be increased by interaction with soluble CD14, has now been demonstrated[8]. TLR2 expression appears to be absent in normal gastroenterological mucosa but is present in monocytes and macrophages in active IBD[9]. Although several TLR2 agonists have been found to activate diverse immune responses in intestinal epithelial cells, no report has yet implicated a causal pathophysiological role for TLR2 in gastrointestinal or hepatic disorders.

Genetic and biochemical evidence strongly implicate TLR4 as the major receptor for LPS activation[10]. TLR4 signalling does not require additional TLR, but other accessory proteins, such as MD-2 or LBP, cooperatively increase LPS recognition. LPS from different enteric pathogens (e.g. *Salmonella*) induce cellular activation via TLR4[11]. However, TLR4 does not mediate responses to all LPS serotypes[12]; furthermore, although certain TLR4 mutations can be linked to a change in the LPS response phenotype, additional – yet to be identified – genes appear to be involved in determining responsiveness to LPS.

*Salmonella typhimurium* may translocate flagellin, the monomeric subunit of flagella, independent of bacterial invasion across intestinal epithelia which activates potent proinflammatory gene expression via TLR5 at the basolateral pole[5]. Flagellin induces intestinal epithelial chemokine secretion which will, in turn, initiate dendritic cell migration and recruitment to the mucosal site of inflammation[13], presumably via TLR5.

CpGDNA modulates TLR9[14]. Rachmilewitz et al.[15] have recently shown that administration of CpGDNA ameliorates the outcome of both chemically induced and spontaneously occurring colitis in mice. Immunostimulatory DNA sequences inhibited induction of various proinflammatory cytokines and chemokines and matrix metalloproteineases in both DSS- and TNBS-induced colitis. Presumably CpGDNA-induced anti-inflammatory immune responses are mediated via TLR9 in the intestinal mucosa. It remains to be determined whether suppression of colitis is mainly mediated by dendritic cells which have recently been identified as primary target cells for CpGDNA. TLR11 has recently been found to be neccesary to confer resistence to carcinogenic bacteria though the specific PAMP recognized remains to be determined, while TLR7 and TLR8 bind single-strand RNA from disparate viruses.

## PRO- AND ANTI-INFLAMMATORY MEDIATORS OF TLR RESPONSES

Binding of ligands to TLR leads to cell activation and release of cytokines and chemokines which orchestrate both adaptive and innate immune responses. Cytokines are produced in response to bacterial ligands via activation of distinct TLR, including IL-12, IFN-$\alpha$ and TNF. It follows that excessive activation of TLR could lead to tissue injury through cytokine production and apoptosis, although TLR activation may also lead to tissue repair. Li et al.[16] recently provided first evidence that necrotic cells may activate NF$\kappa$B via TLR2, leading to anti-inflammatory and repair gene expression, suggesting that, in addition to exogenous microbial ligands, TLR also recognize endogenous inflammatory stimuli released from damaged tissue[17].

TLR4 expression is significantly increased in IEC throughout the lower gastrointestinal tract in association with IBD[9]. Variants of TLR4 (defined by single nucleotide polymorphisms) have been associated with IBD in some studies but not others. The factors and mechanisms up-regulating TLR4 expression in IEC in IBD are not yet clear. TLR4 up-regulation could also result from ligands other than LPS. T-cell-derived cytokines, such as IFN-$\gamma$ and IL-4 which play significant pathophysiological roles in triggering IBD, have been found to regulate intestinal epithelial TLR4 expression[18,19]. Furthermore LPS significantly induces production of TGF-$\beta_1$[20], which itself also up-regulates TLR4 mRNA expression in IEC[18], thus enhancing expression of pro- and anti-inflammatory genes in a positive regulatory loop. The T-cell-derived cytokine macrophage migration inhibitory cytokine (MIF) which is released from IEC after bacterial infection, and has recently been identified as an essential mediator of chronic inflammation in IBD[21], activates NF$\kappa$B through TLR4, leading to TNF production in macrophages[22] of the underlying lamina propria. Chemokines downstream of pathogen-induced TLR activation include IL-8, MIP, RANTES and IP-10, which recruit further innate immune effector cells into the site of mucosal inflammation, to amplify the local immune response. In *Drosophila*, activation of Toll leads to production of antimicrobial peptides[23]. Recent studies have suggested that TLR may similarly mediate defensin release in intestinal epithelial cells in IBD.

## TLR SIGNALLING

TLR share conserved homologies in their cytoplasmic domains to Toll/IL-1R/ plant R gene homology (TIR) domains. It had been assumed that intracellular signalling mechanisms mediated by TIRs result in activation of one 'classical' pathway involving recruitment of the adaptor molecule MyD88, activation of the serine/threonine kinases of the IRAK family, subsequently leading to degradation of Iκ-B and translocation of NFκB to the nucleus. However, recent reports have demonstrated additional signalling molecules to interact with TIR signalling domains and other proteins, suggesting complex signal propagation downstream of distinct TLR: 'Toll-interleukin 1 receptor (TIR) domain-containing adapter protein' (TIRAP[24] – also known as Mal) has been shown to control activation of MyD88-*independent* signalling pathways downstream of TLR4. Of note, the double-stranded RNA-binding protein kinase PKR has been proposed to be a central downstream component of both the TIRAP/Mal- and MyD88-dependent signalling pathways[24], and could mediate crosstalk between these signalling brands. In contrast, Toll-interacting protein (Tollip) plays an inhibitory role in TLR2/4-mediated cell activation[25] by suppressing the activity of IRAK[26]. Notably, optimal downstream signal transduction in response to various TLR ligands appears to involve both IRAK-4 and IRAK-1. However, IRAK-2 also seems to be indispensable for activating NFκB through TIRAP/Mal.

Rip2[27] is a direct downstream signal transducer of both TLR and NOD, (which are intracellular PRR that bind fragments of peptidoglycan), thus possibly allowing regulatory crosstalk between these two distinct pathways in response to LPS and other ligands[28]. Dominant-negative Rip2 inhibited NFκB activation mediated by TLR4 and NOD1. Rip2 has been identified as the first common checkpoint downstream of TLR and NOD combining multiple signalling pathways of both the innate and adaptive immune systems. However, the regulatory role of Rip2 in coordinating TLRx/NOD-interdependent signalling, and possibly connecting to the TNFR pathways, remains to be further clarified in detail.

Genetic variation in NOD2 has been associated with increased susceptibility to some types of Crohn's disease (CD)[29,30]. However, the function of NOD2 and its mechanistic role in CD remain to be delineated. The CARD domain of NOD2 has been implicated to play an essential role in mediating apoptosis which may switch off exaggerated inflammatory responses, thus maintaining mucosal homeostasis. Loss of function could exacerbate intestinal inflammation by uncontrolled clonal expansion and activation of mucosal lymphocytes in response to unharmful resident antigens in CD. Importantly, NOD2 has been found to exert antibacterial activity in intestinal epithelial cells, limiting survival of enteric bacteria after invasion[31]. The CD-associated mutant exhibits defective antibacterial activity in intestinal epithelial cells. It will be essential to further determine how TLR/NOD may crosstalk with each other and, yet unknown, kinases in regulatory feedback loops, and how imbalance of signalling events may lead to impaired apoptosis in the mucosal immune system.

Activation of different TLR induces expression of different sets of cytokines and chemokines in different cell types. Recently, the IFN-STAT pathway has been implicated to mediate differential patterns of gene expression activated by distinct TLR. Toshchakov et al.[32] demonstrated that the TLR4 agonist, LPS, induces TIRAP-dependent IFN-β mRNA expression leading to phosphorylation of STAT1. Conversely, Dalpke et al.[33] showed that TLR9 agonist CpGDNA triggers synthesis of SOCS proteins that act as negative regulators of the JAK-STAT pathway.

The importance of TLR to control adaptive immune responses is reflected in the study by Medzhitov et al., which demonstrates profound defects in the activation of antigen-specific Th1 but not Th2 immune responses present in MyD88 deficient mice[34]. These results suggest that the innate immune system controls activation of the two effector arms of adaptive immunity.

## INTESTINAL MUCOSAL TLR 'TOLERANCE'

As the front line of the mucosal immune system the intestinal epithelium is constantly exposed to large amounts of lumenal LPS; however, it remains unclear how the intestinal epithelium is able to discriminate between microbial pathogens and indigenous commensals to maintain hyporesponsiveness to lumenal bacteria. The phenomenon of intestinal LPS tolerance is thought to be the result of regulation of TLR expression and downstream signalling events. TLR2 and TLR4 are normally present in small amounts on IEC *in vivo*, thus minimizing lumenal bacterial recognition in the healthy intestine[9.] Decreased expression of TLR4 surface protein correlates with inhibition of downstream cytokine production in intestinal epithelial LPS tolerance and overexpression of both TLR4 and MD-2 may confer LPS responsiveness to IEC[35]. Hornef et al.[6] previously demonstrated that continuous exposure to LPS may lead to a tolerant phenotype in IEC. Desensitization in response to LPS, LTA or Fimbrillin exposure may also inhibit responses to the other stimulus ('cross-tolerance')[36]. TLR expression is not altered in intestinal epithelial cross-tolerance, suggesting that functions of TLR and downstream elements of signalling pathways shared by TLR2 and TLR4 might be impaired[36]. Recent findings suggest that up-regulation of TOLLIP expression may mediate this effect and play a pivotal role in acquired IEC hyporesponsiveness in the context of extended PAMP exposure.

## CONCLUSIONS

Recent studies have started to elucidate the mechanisms by which TLR and NOD may regulate intestinal innate immunity controlling downstream effects. Adaptive immunity is emerging as an important feedback regulator of innate efferent responses. Cooperative as well as competitive interactions may occur between different bacterial and non-bacterial ligands via TLR and NOD and other components of the innate immune system, which may lead to differential pro- or anti-inflammatory immune responses resulting in predominance of either tissue injury or repair.

Future research must focus on the direct pathogenetic relevance and immune consequences of TLR and NOD (dys-)regulation in active IBD and other aberrant inflammatory processes in the gastrointestinal tract.

## References

1. Medzhitov R, Janeway CA Jr. Decoding the patterns of self and nonself by the innate immune system. Science. 2002;296:298–300.
2. Cario E, Rosenberg IM, Brandwein SL, Beck PL, Reinecker HC, Podolsky DK. Lipopolysaccharide activates distinct signaling pathways in intestinal epithelial cell lines expressing Toll-like receptors. J Immunol. 2000;164:966–72.
3. Smith PD, Smythies LE, Mosteller-Barnum M et al. Intestinal macrophages lack CD14 and CD89 and consequently are down-regulated for LPS- and IgA-mediated activities. J Immunol. 2001;167:2651–6.
4. Cario E, Brown D, McKee M, Lynch-Devaney K, Gerken G, Podolsky DK. Commensal-associated molecular patterns induce selective toll-like receptor-trafficking from apical membrane to cytoplasmic compartments in polarized intestinal epithelium. Am J Pathol. 2002;160:165–73.
5. Gewirtz AT, Navas TA, Lyons S, Godowski PJ, Madara JL. Cutting edge: bacterial flagellin activates basolaterally expressed TLR5 to induce epithelial proinflammatory gene expression. J Immunol. 2001;167:1882–5.
6. Hornef MW, Frisan T, Vandewalle A, Normark S, Richter-Dahlfors A. Toll-like receptor 4 resides in the Golgi apparatus and colocalizes with internalized lipopolysaccharide in intestinal epithelial cells. J Exp Med. 2002;195:559–70.
7. Hajjar AM, O'Mahony DS, Ozinsky A et al. Cutting edge: functional interactions between toll-like receptor (TLR) 2 and TLR1 or TLR6 in response to phenol-soluble modulin. J Immunol. 2001;166:15–19.
8. Iwaki D, Mitsuzawa H, Murakami S et al. The extracellular toll-like receptor 2 domain directly binds peptidoglycan derived from *Staphylococcus aureus*. J Biol Chem. 2002;277: 24315–20.
9. Cario E, Podolsky DK. Differential alteration in intestinal epithelial cell expression of toll-like receptor 3 (TLR3) and TLR4 in inflammatory bowel disease. Infect Immun. 2000;68:7010–17.
10. da Silva Correia J, Soldau K, Christen U, Tobias PS, Ulevitch RJ. Lipopolysaccharide is in close proximity to each of the proteins in its membrane receptor complex. transfer from CD14 to TLR4 and MD-2. J Biol Chem. 2001;276:21129–35.
11. Tapping RI, Akashi S, Miyake K, Godowski PJ, Tobias PS. Toll-like receptor 4, but not toll-like receptor 2, is a signaling receptor for *Escherichia* and *Salmonella* lipopolysaccharides. J Immunol. 2000;165:5780–7.
12. Pulendran B, Kumar P, Cutler CW, Mohamadzadeh M, Van Dyke T, Banchereau J. Lipopolysaccharides from distinct pathogens induce different classes of immune responses *in vivo*. J Immunol. 2001;167:5067–76.
13. Sierro F, Dubois B, Coste A, Kaiserlian D, Kraehenbuhl JP, Sirard JC. Flagellin stimulation of intestinal epithelial cells triggers CCL20-mediated migration of dendritic cells. Proc Natl Acad Sci USA. 2001;98:13722–7.
14. Hemmi H, Takeuchi O, Kawai T et al. A Toll-like receptor recognizes bacterial DNA. Nature. 2000;408:740–5.
15. Rachmilewitz D, Karmeli F, Takabayashi K et al. Immunostimulatory DNA ameliorates experimental and spontaneous murine colitis. Gastroenterology. 2002;122:1428–41.
16. Li M, Carpio DF, Zheng Y et al. An essential role of the NF-kappa B/Toll-like receptor pathway in induction of inflammatory and tissue-repair gene expression by necrotic cells. J Immunol. 2001;166:7128–35.
17. Vabulas RM, Ahmad-Nejad P, da Costa C et al. Endocytosed HSP60s use toll-like receptor 2 (TLR2) and TLR4 to activate the toll/interleukin-1 receptor signaling pathway in innate immune cells. J Biol Chem. 2001;276:31332–9.
18. Suzuki M, Hisamatsu T, Podolsky DK. Gamma interferon augments the intracellular pathway for lipopolysaccharide (LPS) recognition in human intestinal epithelial cells

through coordinated up-regulation of LPS uptake and expression of the intracellular Toll-like receptor 4-MD-2 complex. Infect Immun. 2003;71:3503–11.

19. Abreu MT, Arnold ET, Thomas LS et al. TLR4 and MD-2 expression is regulated by immune-mediated signals in human intestinal epithelial cells. J Biol Chem. 2002;277: 20431–7.

20. Yoshioka T, Morimoto Y, Iwagaki H et al. Bacterial lipopolysaccharide induces transforming growth factor beta and hepatocyte growth factor through toll-like receptor 2 in cultured human colon cancer cells. J Int Med Res. 2001;29:409–20.

21. de Jong YP, Abadia-Molina AC, Satoskar AR et al. Development of chronic colitis is dependent on the cytokine MIF. Nat Immunol. 2001;2:1061–6.

22. Roger T, David J, Glauser MP, Calandra T. MIF regulates innate immune responses through modulation of Toll-like receptor 4. Nature. 2001;414:920–4.

23. Birchler T, Seibl R, Buchner K et al. Human Toll-like receptor 2 mediates induction of the antimicrobial peptide human beta-defensin 2 in response to bacterial lipoprotein. Eur J Immunol. 2001;31:3131–7.

24. Horng T, Barton GM, Medzhitov R. TIRAP: an adapter molecule in the Toll signaling pathway. Nat Immunol. 2001;2:835–41.

25. Bulut Y, Faure E, Thomas L, Equils O, Arditi M. Cooperation of Toll-like receptor 2 and 6 for cellular activation by soluble tuberculosis factor and *Borrelia burgdorferi* outer surface protein A lipoprotein: role of Toll-interacting protein and IL-1 receptor signaling molecules in Toll-like receptor 2 signaling. J Immunol. 2001;167:987–94.

26. Zhang G, Ghosh S. Negative regulation of toll-like receptor-mediated signaling by Tollip. J Biol Chem. 2002;277:7059–65.

27. McCarthy JV, Ni J, Dixit VM. RIP2 is a novel NF-kappaB-activating and cell death-inducing kinase. J Biol Chem. 1998;273:16968–75.

28. Kobayashi K, Inohara N, Hernandez LD et al. RICK/Rip2/CARDIAK mediates signalling for receptors of the innate and adaptive immune systems. Nature. 2002; 416:194–9.

29. Ogura Y, Bonen DK, Inohara N et al. A frameshift mutation in NOD2 associated with susceptibility to Crohn's disease. Nature. 2001;411:603–6.

30. Hugot JP, Chamaillard M, Zouali H et al. Association of NOD2 leucine-rich repeat variants with susceptibility to Crohn's disease. Nature. 2001;411:599–603.

31. Hisamatsu T, Suzuki M, Reinecker HC, Nadeau WJ, McCormick BA, Podolsky DK. CARD15/NOD2 functions as an antibacterial factor in human intestinal epithelial cells. Gastroenterology. 2003;124:993–1000.

32. Toshchakov V, Jones BW, Perera PY et al. TLR4, but not TLR2, mediates IFN-beta-induced STAT1alpha/beta-dependent gene expression in macrophages. Nat Immunol. 2002;3:392–8.

33. Dalpke AH, Opper S, Zimmermann S, Heeg K. Suppressors of cytokine signaling (SOCS)-1 and SOCS-3 are induced by CpG-DNA and modulate cytokine responses in APCs. J Immunol. 2001;166:7082–9.

34. Schnare M, Barton GM, Holt AC, Takeda K, Akira S, Medzhitov R. Toll-like receptors control activation of adaptive immune responses. Nat Immunol. 2001;2:947–50.

35. Abreu MT, Vora P, Faure E, Thomas LS, Arnold ET, Arditi M. Decreased expression of Toll-like receptor-4 and MD-2 correlates with intestinal epithelial cell protection against dysregulated proinflammatory gene expression in response to bacterial lipopolysaccharide. J Immunol. 2001;167:1609–16.

36. Otte J-M, Cario E, Podolsky DK. Mechanisms of cross hyporesponsiveness to Toll-like receptor bacterial ligands in intestinal epithelial cells. Gastroenterology. 2004;126:1054–1070.

# 7
# Manipulation of enteric flora in clinical practice

## M. A. KAMM

## INTRODUCTION

Inflammatory bowel diseases (IBD) result from a complex interaction between the intestinal contents (gut flora and food), and the mucosal barrier (epithelial and immune cells). This interaction is influenced by intrinsic factors such as genetic predisposition, triggering factors such as enteric infection, and external factors such as smoking. Ultimately, however, it is the gut flora that provide the main antigenic drive to the inflammatory process. Altering the gut flora in an effort to diminish this antigenic drive is an attractive therapeutic possibility. This can be achieved in several ways: by the use of antibiotics, prebiotics and changes in food or bacterial substrates, probiotics, introduction of other organisms such as helminths, and alterations in other flora-modifying factors such as stress.

## ANTIBIOTICS AND PROBIOTICS

### Pouchitis

In patients who have had a proctocolectomy and ileoanal restorative procedure ('pouch'), inflammation ('pouchitis') is one of the major complications. This occurs in about 50% of patients, is a major cause of morbidity, and necessitates pouch removal in up to 5–10% of patients in the long term.

Although there has been speculation regarding the cause of this condition, including ischaemia and stasis of enteric contents, it is now clear that the condition relates at least in large part to a mucosal or immune response to the enteric bacteria. The condition is clearly responsive to treatment with antibiotics. Many patients respond to treatment with metronidazole; this drug has been shown to be effective in a double-blind placebo-controlled study of patients with active pouchitis treated with metronidazole 400 mg three times daily or placebo[1]. A small proportion of patients have frequently recurrent or chronic active pouchitis which is more troublesome. For these patients we have recently demonstrated that 4 weeks of treatment with combined ciprofloxacin 500 mg twice daily and metronidazole 400 mg or 500 mg twice daily will result in 82% of patients entering remission, and the remainder improving substantially[2].

Patients with previous pouchitis can be effectively maintained in remission using probiotics. A double-blind placebo-controlled study compared placebo with VSL#3, a mixture of eight bacterial probiotic strains: four strains of lactobacilli, three strains of bifidobacteria, and one strain of *Streptococcus salivarius*[3]. Patients received medication once per day for 12 months. After 12 months the relapse rate was 15% in those receiving probiotic and 100% in those on placebo.

This same probiotic mixture has also been shown to prevent the development of pouchitis after pouch formation, in a double-blind study[4].

## Ulcerative colitis (UC)

There is no convincing evidence that antibiotics are effective in the treatment of active UC.

There is preliminary evidence that probiotics may be effective in acute colitis. In an open study of acute mild to moderate acute colitis, 6 g (two sachets) of VSL#3 was administered twice per day[5]). Eighty three per cent of patients had left-sided disease and 17% had total colitis. Assessment was performed clinically and endoscopically. Remission was achieved in 63%, and a response in a further 23%, of patients.

In an open study of the maintenance of remission with VSL#3, 15 of 20 treated patients remained in remission after 1 year[6].

The most impressive trial to date with respect to maintaining remission of UC is that from Kruis et al.[7]. Three hundred and twenty-seven patients in remission were randomized in a double-blind study to receive mesalazine or the probiotic *Escherichia coli* Nissle for 1 year. After 1 year 34% on mesalazine and 36% on *E. coli* Nissle had relapsed, demonstrating equal efficacy.

## Anal Crohn's disease (CD)

The effective treatment of anal CD depends on adequate surgical drainage of sepsis, antibiotics for both acute infection and to diminish the bacterial contribution to inflammation, and immunosuppression.

Metronidazole achieved fistula closure in 83% of 21 patients treated in an open study[8]. Four subsequent uncontrolled studies have shown complete closure in 34–50% of patients. Improvement usually occurs within 6–8 weeks of therapy. The problem remains that metronidazole cannot be used in the long term due to side-effects such as peripheral neuropathy. Discontinuation of antibiotics usually results in fistula recurrence.

Although ciprofloxacin is widely used to treat anal CD, and the clinical impression is that it makes a therapeutic contribution, there is little hard evidence to support this use. Ciprofloxacin has been evaluated in only one open study of 10 patients[9]. Greater than 3 months treatment with 1.0–1.5 g per day resulted in healing of fistulas in 2 patients, intermittent recurrence requiring further antibiotics in five patients, two patients had severe persistent rectal disease, and one patient had to discontinue treatment.

## Lumenal CD

Six randomized trials assessing the value of metronidazole in lumenal CD have been mostly negative. In a study by Sutherland et al.[10] of 105 patients only 56 completed. Metronidazole 10 mg/kg per day and 20 mg/kg per day were compared to placebo. Metronidazole resulted in a significant reduction in CDAI and CRP, but the remission rate achieved was not significant. Benefit appeared to be seen predominantly in patients with ileocolonic or colonic disease.

Studies of the value of ciprofloxacin are contradictory. Arnold et al.[11] studied 47 patients, and demonstrated a remission rate of 76% vs 25% with placebo.

In an open study of 72 patients with active CD of the ileum ($n = 27$), ileocolon ($n = 22$) or colon ($n = 23$), patients were treated with ciprofloxacin 500 mg twice daily and metronidazole 250 mg three times daily for a mean of 10 weeks[12]. Clinical remission was observed in 49 patients (68%), and 55 patients (76%) showed a clinical response. A clinical response was noted in 29 of 43 patients (67%) who were not taking concurrent prednisolone treatment and in 26 of 29 patients (90%) receiving prednisolone (mean dose of 15 mg/day). A clinical response also occurred in a greater proportion of patients with colonic disease, with or without ileal involvement (84%), compared with patients with ileal disease alone (64%), and in patients without resection (86%) compared with those with previous resection (61%). After a mean follow-up of 9 months, clinical remission was maintained in 26 patients off treatment and in 12 patients who continued antibiotic therapy. Ciprofloxacin in combination with metronidazole may play a beneficial role in achieving clinical remission for patients with active CD, particularly when there is involvement of the colon.

Antibiotics do not appear to provide additional benefit to treatment with steroids. In a double-blind study of patients with active CD of the ileum, right colon, or both, patients were randomized to receive oral ciprofloxacin and metronidazole, both 500 mg twice daily, or placebo for 8 weeks[13]. All patients also received oral budesonide 9 mg once daily. In patients with active CD of the ileum the addition of ciprofloxacin and metronidazole to budesonide was an ineffective intervention, but the antibiotics may have improved the outcome in those with colonic inflammation.

Colombel et al.[14], in a 6-week study in 40 patients, found that ciprofloxacin 1 g/day was equal in efficacy to mesalazine 4 g/day for patients with mild to moderate active CD, with about 55% remission in both groups.

Prantera et al.[15] found that ciprofloxacin with metronidazole gave an inferior remission rate of 46% compared to 63% for methylprednisolone in a small study of 41 patients.

In summary, in treating lumenal CD there may be a modest benefit from antibiotics. The best evidence is for metronidazole; however, there is a lack of convincing large studies.

In relation to probiotics there are no convincing data for benefit in CD.

## MECHANISM OF ACTION OF PROBIOTICS

Probiotic bacteria modulate a diverse range of functions, all of which may contribute to their anti-inflammatory effect. Evidence exists for an effect on the composition of the bacterial flora, changes in epithelial cell function, and effects on immune cell function.

Borruel et al.[16] have shown that various *E. coli* and *Lactobacilli* probiotic strains produce a reduction in TNF-alpha production in cultured mucosal biopsies from Crohn's patients and controls.

Madsen et al.[17] have shown that the probiotic mixture VSL#3 normalizes barrier function, decreases production of TNF-α and IFN-γ, and improves histology in IL-10-deficient mice. The effect appeared to be produced by a soluble factor produced by the probiotic bacteria.

Joint work from the groups in Kiel (Germany), Bologna (Italy) and ourselves (London, England) has shown that in maintenance of remission of pouchitis probiotic bacteria modify NFkB activity, increase bacterial diversity, and increase mucosal levels of IL-10[18].

To further define which probiotic bacteria may be most therapeutically helpful *in-vitro* mechanistic studies comparing different single bacteria are required. Hooper et al.[19] used gene array analysis to study the effect on epithelial gene expression of *Bacteroides thetaiotaomicron*, a normal commensal in mice and humans. This bacterium demonstrated effects on food absorption, barrier function, drug handling, and postnatal gut development.

Our own group has studied the effect of probiotic bacteria on dendritic cell function. The dendritic cell plays a key role in sampling enteric bacteria and orchestrating the subsequent T cell inflammatory immune response. We have shown that bifidobacteria stimulate dendritic cell IL-10 production and diminish IL-12 production, that combined bacterial preparations have cumulative effects, and that the effects are seen with both cell wall and bacterial DNA preparations[20].

Bacterial DNA, in the absence of live bacteria, may be sufficient to have some immune effects. Rachmilewitz et al. have shown that bacterial CPG motifs modify immune function, and our own group has shown that bacterial DNA activates dendritic cells which produce IL-10[21]. These effects are mediated via toll-like receptor 9 (TLR-9).

Animal studies have demonstrated that immunoregulatory effects occur with non-enteric intravenous or peritoneal bacterial exposure. Some effects of probiotic bacteria therefore occur through non-enteric immune mechanisms.

Probiotic bacteria can be genetically modified to produce anti-inflammatory cytokines[22]. Such organisms are currently undergoing therapeutic evaluation.

## OTHER APPROACHES TO GUT FLORA MANIUPULATION

### Gut flora transplantation

Borody et al.[23] used transplanted gut flora to treat resistant UC. Six patients with severe recurrent colitis had bowel cleansing followed by daily faecal suspension enemas for 5 days. Patients had complete relief by 4 months and have had no recurrence 1–13 years after treatment. This approach remains to be reproduced.

### Helminths

The group in Iowa have used helminth infection to treat patients with UC and CD. This was based on the low incidence of IBD in countries where helminth infection is common, the positive therapeutic effect of helminth infection in animal models of colitis, and the reciprocal down-regulation of Th1 inflammation by the Th2 immune response elicited by helminth infection.

In their first published report treatment with eggs of *Trichuris suis* resulted in six of seven patients with CD and UC going into remission[24]. In an open study of 29 patients with active CD, by 24 weeks 79% had responded and 72% were in remission with a fall in the mean Crohn's Disease Activity Index (CDAI) from 296 to 90[25]. In 54 patients with active UC, 48% had a response compared to 15% on placebo, although many patients did not go into remission[26].

## FOOD AS AN ANTIGENIC DRIVER OF INFLAMMATION

Food undoubtedly plays some role in driving the inflammatory response. Liquid elemental and polymeric diets have both been shown to diminish inflammation in CD. Faecal diversion diminishes inflammation in lumenal CD, while re-establishing the enteric stream causes recurrent inflammation.

*In-vitro* lymphocyte proliferation in response to six different groups of food yeast and bacterial antigens was tested in 31 Crohn's patients and 22 healthy controls[27]. An abnormal response to multiple food antigens was much more common in CD patients.

The *in-vivo* response to food antigens was tested in 10 CD patients[28]. There was an abnormal response to all food antigens in Crohn's patients, and this was highly significant for yeast and citrus groups.

## THE EFFECT OF STRESS ON GUT FLORA AND INFLAMMATION

Stress is associated with modified bacterial mucosal adherence (which is associated with increased permeability), decreased lumenal lactobacilli, and decreased secretory IgA[29]. Stress reduction may therefore have a role to play in diminishing IBD recurrence, and part of this effect may be mediated by changes in enteric flora.

## CONCLUSIONS

The modification of the gut bacterial environment has already been proven to play an important role in the prevention and treatment of IBD. Further work in this area is certain to bear fruit in the management of these conditions.

## References

1. Madden MV, McIntyre AS, Nicholls RJ. Double-blind crossover trial of metronidazole versus placebo in chronic unremitting pouchitis. Dig Dis Sci. 1994;39:1193–6.
2. Mimura T, Rizzello F, Helwig U et al. Four-week open-label trial of metronidazole and ciprofloxacin for the treatment of recurrent or refractory pouchitis. Aliment Pharmacol Ther. 2002;16:909–17.
3. Mimura T, Rizzello F, Helwig U et al. Once daily, high dose probiotic therapy (VSL#3) for maintaining remission in recurrent or refractory pouchitis. Gut. 2004;53:108–14
4. Gionchetti P, Rizzello F, Helwig U et al. Prophylaxis of pouchitis onset with probiotic therapy: a double-blind, placebo-controlled trial. Gastroenterology. 2003;124:1202–9.
5. Fedorak RN, Gionchetti P, Campieri M et al. VSL3 probiotic mixture induces remission in patients with active ulcerative colitis. Gastreonterology. 2003;124(Suppl. 1):A377.
6. Venturi A, Gionchetti P, Rizzello F et al. Impact on the composition of the faecal flora by a new probiotic preparation: preliminary data on maintenance treatment of patients with ulcerative colitis. Aliment Pharmacol Ther. 1999;13:1103–8.
7. Kruis W, Frik P, Pokrotnieks J et al. Maintaining remission of ulcerative colitis with the probiotic E coli Nissle 1917 is as effective as with standard mesalamine. Gut. 2004 (In press).
8. Bernstein LH, Frank MS, Brandt LJ, Boley SJ. Healing of perineal Crohn's disease with metronidazole. Gastroenterology. 1980;79:357–65.
9. Turunen U, Farkkila M, Valtonen V, Seppala K. Long term outcome of ciprofloxacin treatment in severe perianal or fistulous Crohn's disease. Gastroenterology. 1993;104:A793.
10 Sutherland L, Singleton J, Sessions J et al. Double blind, placebo controlled trial of metronidazole in Crohn's disease. Gut. 1991;32:1071–5.
11. Arnold GL, Beaves MR, Pryjdun VO, Mook WJ. Preliminary study of ciprofloxacin in active Crohn's disease. Inflamm Bowel Dis. 2002;8:10–15.
12. Greenbloom SL, Steinhart AH, Greenberg GR. Combination ciprofloxacin and metronidazole for active Crohn's disease. Can J Gastroenterol. 1998;12:53–6.
13. Steinhart AH, Feagan BG, Wong CJ et al. Combined budesonide and antibiotic therapy for active Crohn's disease: a randomized controlled trial. Gastroenterology. 2002;123:33–40.
14. Colombel JF, Lemann M, Cassagnou M et al. A controlled trial comparing ciprofloxacin with mesalazine for the treatment of active Crohn's disease. Groupe d'Etudes Therapeutiques des Affections Inflammatoires Digestives (GETAID). Am J Gastroenterol. 1999;94: 674–8.
15. Prantera C, Zannoni F, Scribano ML et al. An antibiotic regimen for the treatment of active Crohn's disease: a randomized, controlled clinical trial of metronidazole plus ciprofloxacin. Am J Gastroenterol. 1996;91:328–32.
16. Borruel N, Carol M, Casellas F et al. Increased mucosal tumour necrosis factor alpha production in Crohn's disease can be downregulated ex vivo by probiotic bacteria. Gut. 2002;51:659–64.
17. Madsen K, Cornish A, Soper P et al. Probiotic bacteria enhance murine and human intestinal epithelial barrier function. Gastroenterology. 2001;121:580–91.
18. Kuehbacher T, Ott S, Mimura T et al. Bacterial diversity and normal anaerobic bacterial flora is restored in pouchitis through probiotic therapy. Gastroenterology. 2004;126:4(Suppl. 2):A577.
19. Hooper LV, Wong MH, Thelin A, Hansson L, Falk PG, Gordon JI. Molecular analysis of commensal host–microbial relationships in the intestine. Science. 2001;291:881–4.
20. Hart AL, Lammers K, Brigidi P et al. Modulation of human dendritic cell phenotype and function by probiotic bacteria. Gut. 2004 (In press).

21. Rachmilewitz D, Katakura K, Karmeli F et al. Toll-like receptor 9 signaling mediates the anti-inflammatory effects of probiotics in murine experimental colitis. Gastroenterology. 2004;126:520–8.
22. Steidler L, Neirynck S, Huyghebaert N et al. Biological containment of genetically modified Lactococcus lactis for intestinal delivery of human interleukin 10. Nat Biotechnol. 2003;21: 785–9.
23. Borody TJ, Warren EF, Leis S, Surace R, Ashman O. Treatment of ulcerative colitis using fecal bacteriotherapy. J Clin Gastroenterol. 2003;37:42–7.
24. Summers RW, Elliott DE, Qadir K, Urban JF Jr, Thompson R, Weinstock JV. *Trichuris suis* seems to be safe and possibly effective in the treatment of inflammatory bowel disease. Am J Gastroenterol. 2003;98:2034–41.
25. Summers RW, Elliott DE, Thompson R, Urban JF, Weinstock JV. Trial of helminth ova in active Crohn's disease. Gastroenterology. 2004;126:4(Suppl. 2): A75.
26. Summers RW, Elliott DE, Thompson R, Urban JF, Weinstock JV. Double-blind, placebo-controlled trial of helminth ova therapy in active ulcerative colitis (UC). Gastroenterology. 2004;126:4(Suppl. 2):A83.
27. van den Bogaerde J, Kamm MA, Knight SC. Immune sensitisation to food, yeast and bacteria in Crohn's disease. Aliment Pharmacol Ther. 2001;15:1647–53.
28. van den Bogaerde J, Kamm MA, Cahill J et al. Gut mucosal response to food in Crohn's disease. Aliment Pharmacol Ther. 2002;16:1903–15.
29. Hart AH, Kamm MA. Mechanisms of initiation and perpetuation of gut inflammation by stress. Aliment Pharmacol Ther. 2002;16:2017–28.

# Section III
# Immunoregulation: novel therapeutic agents

Chair: R. LÖFBERG and M. LUKÁŠ

# 8
# Lymphocyte integrins and the pathogenesis of experimental colitis

K. P. PAVLICK and M. B. GRISHAM

## INTRODUCTION

The inflammatory bowel diseases (IBD) (Crohn's disease (CD) and ulcerative colitis (UC)) are a collection of chronic, idiopathic inflammatory disorders of the intestine and/or colon that are characterized by rectal bleeding, severe diarrhoea, abdominal pain, fever and weight loss. Biopsies obtained from patients with active disease reveal infiltration of large numbers of leucocytes such as polymorphonuclear leucocytes (PMN; neutrophils), monocytes, and lymphocytes in the intestinal and/or colonic interstitium. Coincident with this inflammatory infiltrate is extensive mucosal and/or transmural injury including oedema, loss of goblet cells, decreased mucus production, crypt cell hyperplasia, epithelial erosions and ulcerations resulting in loss of barrier function. Although the aetiology of IBD remains to be defined, recent experimental and clinical studies suggest that the initiation and pathogenesis of these diseases are multi-factorial, involving interactions among genetic, environmental and immune factors[1,2]. A recent conceptual advance has arisen from experimental studies suggesting that chronic gut inflammation results from a dysregulated immune response to components of the normal gut flora[2-5].

The intestinal mucosal interstitium is continuously exposed to large amounts of exogenous (i.e. dietary) and endogenous (e.g. bacteria) antigens. As one would predict, the mucosal immune system has developed efficient mechanisms to distinguish between potentially pathogenic bacterial, parasitic, viral and dietary antigens from non-pathological antigens and resident gut flora. Specific subsets of $CD4^+$ T lymphocytes play a major role in mediating and regulating these effector functions *in vivo* and are called T-helper (Th) cells. A number of different studies have shown that Th1 cells are primarily involved in cell-mediated immunity (CMI) which is mounted against certain infectious agents such as intracellular bacteria, fungi and protozoa. This protective immune response involves not only activation of Th1 cells and the subsequent release of their cytokines but also Th1 cytokine-mediated activation of intestinal macrophages (M$\phi$) and other phagocytic leucocytes to release additional proinflammatory cytokines. The net result of Th1 cell activation and CMI is the

75

recruitment of phagocytic leucocytes (e.g. PMN, Mϕ, etc.) into the gut interstitium to engulf and destroy invading microorganisms. Together, Th1- and macrophage-derived cytokines and mediators activate the microvasculature endothelium within the gut to enhance expression of adhesion molecules, thereby promoting the recruitment of potentially injurious phagocytic leucocytes such as neutrophils, monocytes and macrophages[6,7]. Thus, activation and unregulated recruitment of $CD4^+$ T cells into the gut is important for the perpetuation of chronic inflammation and tissue injury. It is currently thought that lymphocyte accumulation observed in IBD occurs by increased trafficking, decreased cell death of activated cells and enhanced proliferation in the tissue. Several different T-cell-associated adhesion molecules have been identified as important molecular determinants for T-cell activation and recruitment to target tissues. As a result, the cell–cell interactions involved in the activation and migration of lymphocytes into the gut mucosa have been the focus area for the development of many therapeutic strategies[8,9]. This review will focus on lymphocyte-associated integrins and their counter-receptors involved in the recruitment and activation of lymphocytes which are thought to play a role in the initiation and perpetuation of chronic gut inflammation.

## LYMPHOCYTE HOMING AND IMMUNE ACTIVATION

The initiation of any cell-mediated immune response starts with the activation of naive T cells that arise from the bone marrow and mature in primary lymphoid tissue such as the thymus. These naive cells continuously disseminate the body in search of different antigens by recirculating between the vasculature and secondary lymphoid tissues. The ability to traverse between these tissues is mediated by specific adhesive interactions between molecules on the lymphocyte surface that interact with cognate molecules on the endothelium lining localized on either the vasculature or lymphatic vessels[10,11]. Naive lymphocytes express L-selectin and utilize this molecule to interact with sialyl-Lewis[x]-like glycoproteins located on high endothelial venules (HEV) to initiate lymphocyte tethering to the vessel wall in the lymph node. The interactions are transient such that new bonds are formed downstream as upstream bonds dissociate, resulting in a rolling motion that keeps the leucocyte near the endothelium. A rolling cell must encounter additional adhesion molecules to be able to firmly adhere to the endothelium. Firm adherence is mediated by the integrin family of adhesion molecules, specifically the alpha4 integrins ($\alpha4\beta1$ and $\alpha4\beta7$) and lymphocyte function-associated-1 (LFA-1; $\alpha L\beta2$). Once firmly adhered, lymphocytes rearrange their cytoplasmic material to facilitate migration between the epithelial junctions in order to enter into tissue.

Once in the lymph tissue, naive T cells can become activated and proliferate into effector cells if antigen, which is bound and displayed by the major histocompatibility complex (MHC) class II molecule on the surface of antigen-presenting cells, is recognized by the T-cell antigen receptor (TcR). This interaction is stabilized by additional adhesion molecules that help form the immunological synapse[12]. Activated cells then shed L-selectin, while the expression of other adhesion and activation molecules is modulated, enter into the circulation and migrate to the original source of the antigen. Effector lymphocytes leave the circulation and enter into the tissue containing the antigen at post-capillary venules. Thus, T cells require strict regulation in their adhesive interactions at every step in immune activation in order to elicit the correct and proper immune response.

## $\beta_2$ INTEGRINS AND THEIR ROLE IN GUT INFLAMMATION

Activated/effector lymphocytes are able to leave the circulation and enter into inflamed tissue through adhesive interactions involving intracellular adhesion molecules (ICAM-1 and -2)[13,14]. These proteins are members of the immuno-globulin superfamily characterized by having multiple extracellular Ig-like domains. ICAM-1 is expressed at low levels on endothelium under normal conditions; however, upon activation of the endothelium by Th1- and/or macrophage-derived cytokines ICAM-1 expression is increased dramatically[15]. ICAM-2 is actually a truncated form of ICAM-1 that is constitutively expressed but at a 10-fold higher level compared to ICAM-1 under basal conditions[16]. In addition, ICAM-2 expression does not change upon stimula-tion of the endothelium. It is known that the expression of ICAM-1 is increased in the intestinal mucosa in active episodes of both UC and CD patients[17,18]. Furthermore, ICAM-1 expression in the mucosa is higher in a T-cell-dependent model of chronic colitis and also in IL-10$^{-/-}$ mice with enterocolitis compared to control tissue[19–21]. This has led to the development of blocking strategies against ICAM-1 which have been employed in animal models of IBD. For example, anti-ICAM-1 antibodies attenuate disease in the acetic acid-induced model of colitis and in the SAMP/Yit adoptive transfer model of ileitis[22,23]. In the latter, attenuation of disease was in the acute phase and combination treatment with anti-VCAM-1 displayed a better outcome, suggesting redun-dancy in the adhesive interactions between leucocytes and the endothelium. Inhibiting the expression of ICAM-1 by the use of antisense phosphorothioate oligonucleotides reduced the clinical signs of colitis in the dextran sulphate sodium (DSS)-induced injury in mice[24]. However, the use of ICAM-1 antisense inhibitors for the treatment of human IBD has been unsuccessful in placebo-controlled multicentre clinical trials[8,9]. The ineffectiveness of ICAM-1 anti-sense technology could be several-fold; however, one simple explanation may be due to the redundancy in the adhesion pathway provided by not only VCAM-1 but also ICAM-2[25]. Whereas a significant amount of work has focused on ICAM-1, very little information is available pertaining to the role of ICAM-1 cognate receptor, LFA-1, in the pathogenesis of IBD.

LFA-1 is a heterodimeric protein consisting of common beta chain (CD18, $\beta_2$) subunit and an alphaL (CD11a, $\alpha_L$) subunit, thus LFA-1 is a member of the $\beta_2$-integrin family of cell adhesion molecules which also include the additional family members $\alpha M\beta_2$, $\alpha x\beta_2$ and $\alpha d\beta_2$ (see ref. 26). LFA-1 is expressed on T and B cells, granulocytes, monocytes and macrophages[26,27]. LFA-1 interacts with the ICAM-1 or ICAM-2 expressed on antigen-presenting cells and the endothelium[10,28,29]. Interaction of LFA-1 with ICAM-1 or ICAM-2 has been shown to promote migration of leucocytes out of the vasculature and into the tissue interstitium as well as T-cell homing and activation/polarization. Mabs to LFA-1 almost completely block the migration of lymphocytes into peripheral lymph nodes and severely reduce recirculation to Peyer's patches[30]. However, mAbs blocking LFA-1 interactions or lymphocytes deficient in LFA-1 by genetic mutation only slightly reduce the homing to mesenteric lymph nodes in the mouse, and migration to the gut is unaffected[31–33]. It has also been observed that CD patients have increased LFA-1 expression on CD4+ lamina propria cells, compared to normal tissue, providing anecdotal evidence for the role of integrins in intestinal inflammation[34–36]. In addition, recent data from our laboratory demonstrate that LFA-1 expression on CD4+ T cells is necessary to induce chronic colitis in a T-cell-dependent model of murine colitis. We found that transfer of LFA-1-deficient naive T cells into immunodeficient RAG-1 recipient mice failed to induce colitis in this model. Lack of T-cell-associated LFA-1 resulted in a reduction in the numbers of Th1 type CD4+ T cells in the colonic interstitium. However, normal peripheral numbers of T cells were found in the spleens of colitic and non-colitic LFA-1$^{-/-}$-reconstituted mice, suggesting that both wild-type and LFA-1$^{-/-}$ T cells become activated and proliferate to the same extent *in vivo*. Taken together, these data suggest that the integrin LFA-1 is critical for the recruitment of effector T cells into the colonic interstitium where these cells initiate chronic gut inflammation. Furthermore, it has been shown that LFA-1-deficient mice have attenuated disease in the DSS model of colitis (personal communication). Specific blocking of LFA-1/ICAM interactions has not been studied in humans to date. However, reports by d'Agata et al. and Uzel et al. suggest that these molecules may not be involved[37,38]. These groups present a case study of a leucocyte adhesion deficiency (LAD; loss of the $\beta_2$ integrins) patient who has ileocolitis, and in whom bone marrow transplantation completely resolved the intestinal inflammation. Therefore, they suggest that marrow-derived leucocyte dysfunction may be a contributing factor in the disease. To date, direct targeting of LFA-1 in human IBD has not been done; however, a humanized murine anti-CD11a antibody, efalizumab, has shown promise in attenuating the T-cell-mediated disease of psoriasis[39–41].

## α4 INTEGRINS AND THEIR ROLE IN GUT INFLAMMATION

The α4 integrins are also involved in the adhesion, migration, and activation of immune cells, suggesting that these integrins play a significant role in inflammation. The α4 integrins are heterodimeric proteins consisting of a common α4 subunit that interacts with either of two beta subunits, $\beta_1$ and $\beta_7$, referred to as α4β1 (very late antigen-4; VLA-4) and α4β7 (lymphocyte–Peyer's patch adhesion molecule-1; LPAM-1). The expression of the α4 integrins is predominantly found on lymphocytes, monocytes, eosinophils and basophils, and not neutrophils[42]. These integrins interact with cognate receptors expressed on post-capillary venule endothelial cells; vascular cell adhesion molecule-1 (VCAM-1) interacts with α4β1 and mucosal addressin cell adhesion molecule-1 (MAdCAM-1) interacts with α4β7. The α4β1 integrin has also been shown to mediate interaction with the extracellular matrix (ECM) proteins fibronectin and osteopontin[42,43].

Expression of $\alpha4\beta1$ is found on most leucocytes excluding neutrophils, and surface expression of $\alpha4\beta1$ increases on both effector and memory T cells that migrate to areas of inflammation. The ligand, VCAM-1, is absent on most non-activated endothelial cells but is induced upon stimulation with cytokines. Interaction of $\alpha4\beta1$ with VCAM-1 has multiple outcomes for lymphocytes, including activation, migration and proliferation. In addition to its role in adhesion, blocking $\alpha4\beta1$/VCAM-1 interaction increases programmed cell death (apoptosis) of T cells under certain conditions[44,45]. Interaction of $\alpha4\beta1$ with fibronectin can increase the expression of matrix metalloproteinases (MMP), which degrade and modify the ECM, which may aid in the migration of leucocytes within inflammatory tissues[46,47]. The integrin $\alpha4\beta7$ interacts with MAdCAM-1 to promote the homing of leucocytes to mucosal tissues[14]. Gut-homing effector and memory T cells have been shown to have increased surface expression of $\alpha4\beta7$. MAdCAM-1 is specifically expressed on the gut endothelium and is thought to mediate lymphocyte homing to the gut. Blocking of $\alpha4\beta7$/MAdCAM interactions with antibodies blocks the migration of lymphocytes to Peyer's patches, mesenteric lymph nodes and the gut, without influencing the homing to other lymphoid or non-mucosal tissue[48,49].

Interference between the $\alpha4$ integrins and their counter-receptors has been an interest in studies of therapeutic strategies for chronic inflammatory disorders for quite some time. Studies in mice have shown that MAdCAM-1 or $\alpha4\beta7$ antibodies block leucocyte recruitment and attenuate the severity of chronic colitis in mouse models of colitis[50,51]. Blocking of the $\alpha4$ integrins also attenuated spontaneous colitis in the cottontop tamarin[52,53]. Interestingly, loss of the $\beta7$ subunit by genetic mutation did not alter the development of spontaneous colitis in IL-2$^{-/-}$ mice, but loss of $\beta7$ on T cells did significantly delay the onset of colitis in an adoptive transfer model[54]. These studies have led to the development of humanized mouse antibodies to $\alpha4$ integrins or specifically for $\alpha4\beta7$ for use in treating IBD[8,9]. To date the targeting of $\alpha4$ integrins in clinical trials, inhibiting both $\alpha4\beta1$/VCAM-1 and $\alpha4\beta7$/MAdCAM-1 interactions, has shown the best promise for treating IBD.

## CONCLUDING REMARKS

A major conceptual advancement in the understanding of the pathogenesis of IBD has been the realization that a dysregulation of the normal cell-mediated immune response to commensal enteric antigens promotes chronic gut inflammation. Many of the new therapeutic strategies designed to treat patients with IBD have focused on modulating or inhibiting the chronic inflammatory response resulting from this break in tolerance to enteric bacteria. These strategies have been developed to either reduce the production of proinflammatory cytokines (e.g. TNF-$\alpha$), or inhibit the recruitment of pathogenic cells into the gut interstitium through employment of selective adhesion molecule (SAM) inhibitors. In particular, anti-adhesion therapy has focused on the $\alpha4$ integrins ($\alpha4\beta1$ and $\alpha4\beta7$) and the $\beta_2$/ICAM-1 interactions with varying outcomes. To date, blocking the interaction between the $\alpha4$ integrins and their cognate ligands has shown the most promise in clinical IBD trials compared to strategies involving ICAM-1. The reasons for these differing results may be simply due to the technology employed for ICAM-1 versus $\alpha4$ targeting (antisense versus monoclonal antibodies), or the utilization of additional pathways or ligands. This may be the case for targeting of ICAM-1 in view of the fact that ICAM-2 and $\alpha4$-adhesion pathways are still available. The existence of multiple adhesion molecules involved in lymphocyte trafficking may provide a mechanistic redundancy to ensure for proper function. Consequently, pathway redundancies may complicate a specific therapeutic adhesion strategy such that treatment with several SAM inhibitors may be required for more beneficial effects in the treatment of IBD. However, single or multiple SAM inhibitors may result in additional abnormalities involving lymphocyte or leucocyte function such that migration into additional tissue or abnormal cell activation may be affected, resulting in unwanted side-effects or infections. Anti-adhesion therapies may need to be developed that target specific cell types or specific pathways instead of targeting general leucocyte adhesion. For the future, further characterization of cell–cell adhesion events in the initiation of the immune response will be needed in order to fully understand the mechanisms related to the initiation and perpetuation of IBD, which would significantly aid in the development of improved anti-adhesion therapies.

## References

1.  Fiocchi C. Inflammatory bowel disease: etiology and pathogenesis. Gastroenterology. 1998;115:182–205.
2.  Powrie F. T cells in inflammatory bowel disease: protective and pathogenic roles. Immunity. 1995;3:171–4.
3.  Elson CO, Sartor RB, Tennyson GS, Riddell RH. Experimental models of inflammatory bowel disease. Gastroenterology. 1995;109:1344–67.
4.  Maloy KJ, Powrie F. Regulatory T cells in the control of immune pathology. Nature Immunol. 2001;2:816–22.
5.  Singh B, Read S, Asseman C et al. Control of intestinal inflammation by regulatory T cells. Immunol Rev. 2001;182:190–200.
6.  Laroux FS, Grisham MB. Immunological basis of inflammatory bowel disease: role of the microcirculation. Microcirculation. 2001;8:283–301.
7.  Pavlick KP, Laroux FS, Fuseler J et al. Role of reactive metabolites of oxygen and nitrogen in inflammatory bowel disease. Free Rad Biol Med. 2002;33:311–22.
8.  Rutgeerts P, Van Deventer S, Schreiber S. Review article: The expanding role of biological agents in the treatment of inflammatory bowel disease – focus on selective adhesion molecule inhibition. Aliment Pharmacol Ther. 2003;17:1435–50.
9.  Van Assche G, Rutgeerts P. Antiadhesion molecule therapy in inflammatory bowel disease. Inflamm Bowel Dis. 2002;8:291–300.
10. von Andrian UH, Mackay CR. Advances in immunology: T-cell function and migration: two sides of the same coin. N Engl J Med. 2000;343:1020–33.
11. von Andrian UH, Mempel TR. Homing and cellular traffic in lymph nodes. Nature Rev Immunol. 2003;3:867–78.
12. Huppa JB, Davis MM. T-cell-antigen recognition and the immunological synapse. Nature Rev Immunol. 2003;3:973–83.
13. Carlos TM, Harlan JM. Leukocyte–endothelial adhesion molecules. Blood. 1994;84:2068–101.
14. Springer TA. Traffic signals for lymphocyte recirculation and leukocyte emigration: the multistep paradigm. Cell. 1994;76:301–14.
15. Dustin ML, Rothlein R, Bhan AK, Dinarello CA, Springer TA. Induction by IL 1 and interferon-gamma: tissue distribution, biochemistry, and function of a natural adherence molecule (ICAM-1). J Immunol. 1986;137:245–54.

16. Staunton DE, Dustin ML, Springer TA. Functional cloning of ICAM-2, a cell adhesion ligand for LFA-1 homologous to ICAM-1. Nature. 1989;339:61–4.
17. Jones SC, Banks RE, Haidar A et al. Adhesion molecules in inflammatory bowel disease. Gut. 1995;36:724–30.
18. Nakamura S, Ohtani H, Watanabe Y et al. *In situ* expression of the cell adhesion molecules in inflammatory bowel disease. Evidence of immunologic activation of vascular endothelial cells. Lab Invest. 1993;69:77–85.
19. Kawachi S, Cockrell A, Laroux FS et al. Role of inducible nitric oxide synthase in the regulation of VCAM-1 expression in gut inflammation. Am J Physiol. 1999;277:G572–6.
20. Kawachi S, Morise Z, Jennings SR et al. Cytokine and adhesion molecule expression in SCID mice reconstituted with CD4+ T cells. Inflamm Bowel Dis. 2000;6:171–80.
21. Kawachi S, Jennings S, Panes J et al. Cytokine and endothelial cell adhesion molecule expression in interleukin-10-deficient mice. Am J Physiol Gastrointest Liver Physiol. 2000; 278:G734–43.
22. Burns RC, Rivera-Nieves J, Moskaluk CA, Matsumoto S, Cominelli F, Ley K. Antibody blockade of ICAM-1 and VCAM-1 ameliorates inflammation in the SAMP-1/Yit adoptive transfer model of Crohn's disease in mice. Gastroenterology. 2001;121:1428–36.
23. Wong PY, Yue G, Yin K et al. Antibodies to ICAM-1 ameliorate inflammation in acetic acid induced inflammatory bowel disease. Adv Prostaglandin Thromboxane Leukot Res. 1995;23:337–9.
24. Bennett CF, Kornbrust D, Henry S et al. An ICAM-1 antisense oligonucleotide prevents and reverses dextran sulfate sodium-induced colitis in mice. J Pharmacol Exp Ther. 1997;280:988–1000.
25. Lehmann JC, Jablonski-Westrich D, Haubold U, Gutierrez-Ramos JC, Springer T, Hamann A. Overlapping and selective roles of endothelial intercellular adhesion molecule-1 (ICAM-1) and ICAM-2 in lymphocyte trafficking. J Immunol. 2003;171:2588–93.
26. Harris ES, McIntyre TM, Prescott SM, Zimmerman GA. The leukocyte integrins. J Biol Chem. 2000;275:23409–12.
27. Springer TA, Dustin ML, Kishimoto TK, Marlin SD. The lymphocyte function-associated LFA-1, CD2, and LFA-3 molecules: cell adhesion receptors of the immune system. Annu Rev Immunol. 1987;5:223–52.
28. Dustin ML, Shaw AS. Costimulation: building an immunological synapse. Science. 1999; 283:649–50.
29. Monks CR, Freiberg BA, Kupfer H, Sciaky N, Kupfer A. Three-dimensional segregation of supramolecular activation clusters in T cells. Nature. 1998;395:82–6.
30. Hamann A, Jablonski-Westrich D, Duijvestijn A et al. Evidence for an accessory role of LFA-1 in lymphocyte-high endothelium interaction during homing. J Immunol. 1988;140: 693–9.
31. Berlin-Rufenach C, Otto F, Mathies M et al. Lymphocyte migration in lymphocyte function-associated antigen (LFA)-1-deficient mice. J Exp Med. 1999;189:1467–78.
32. Schmits R, Kundig TM, Baker DM, Shumaker G, Simard JJ, Duncan G et al. LFA-1-deficient mice show normal CTL responses to virus but fail to reject immunogenic tumor. J Exp Med. 1996;183:1415–26.
33. Shier P, Otulakowski G, Ngo K et al. Impaired immune responses toward alloantigens and tumor cells but normal thymic selection in mice deficient in the beta2 integrin leukocyte function-associated antigen-1. J Immunol. 1996;157:5375–86.
34. Bernstein CN, Sargent M, Rector E. Alteration in expression of beta 2 integrins on lamina propria lymphocytes in ulcerative colitis and Crohn's disease. Clin Immunol. 2002;104:67–72.
35. Kirman I, Nielsen OH. LFA-1 subunit expression in ulcerative colitis patients. Dig Dis Sci. 1996;41:670–6.
36. Vainer B, Nielsen OH, Horn T. Comparative studies of the colonic *in situ* expression of intercellular adhesion molecules (ICAM-1, -2, and -3), beta2 integrins (LFA-1, Mac-1, and p150,95), and PECAM-1 in ulcerative colitis and Crohn's disease. Am J Surg Pathol. 2000; 24:1115–24.
37. D'Agata ID, Paradis K, Chad Z, Bonny Y, Seidman E. Leucocyte adhesion deficiency presenting as a chronic ileocolitis. Gut. 1996;39:605–8.
38. Uzel G, Kleiner DE, Kuhns DB, Holland SM. Dysfunctional LAD-1 neutrophils and colitis. Gastroenterology. 2001;121:958–64.

39. Dedrick RL, Walicke P, Garovoy M. Anti-adhesion antibodies – efalizumab, a humanized anti-CD11a monoclonal antibody. Transplant Immunol. 2002;9:181–6.
40. Gottlieb AB, Krueger JG, Wittkowski K, Dedrick R, Walicke PA, Garovoy M. Psoriasis as a model for T-cell-mediated disease – immunobiologic and clinical effects of treatment with multiple doses of efalizumab, an anti-CD11a antibody. Arch Dermatol. 2002;138:591–600.
41. Lebwohl M, Tyring SK, Hamilton TK et al. A novel targeted T-cell modulator, efalizumab, for plaque psoriasis. N Engl J Med. 2003;349:2004–13.
42. Lobb RR, Hemler ME. The pathophysiologic role of alpha 4 integrins in vivo. J Clin Invest. 1994;94:1722–8.
43. Bayless KJ, Meininger GA, Scholtz JM, Davis GE. Osteopontin is a ligand for the alpha4beta1 integrin. J Cell Sci. 1998;111:1165–74.
44. Leussink VI, Zettl UK, Jander S, Pepinsky RB, Lobb RR, Stoll G et al. Blockade of signaling via the very late antigen (VLA-4) and its counterligand vascular cell adhesion molecule-1 (VCAM-1) causes increased T cell apoptosis in experimental autoimmune neuritis. Acta Neuropathol (Berl). 2002;103:131–6.
45. Tchilian EZ, Owen JJ, Jenkinson EJ. Anti-alpha 4 integrin antibody induces apoptosis in murine thymocytes and staphylococcal enterotoxin B-activated lymph node T cells. Immunology. 1997;92:321–7.
46. Madri JA, Graesser D, Haas T. The roles of adhesion molecules and proteinases in lymphocyte transendothelial migration. Biochem Cell Biol. 1996;74:749–57.
47. Yakubenko VP, Lobb RR, Plow EF, Ugarova TP. Differential induction of gelatinase B (MMP-9) and gelatinase A (MMP-2) in T lymphocytes upon alpha(4)beta(1)-mediated adhesion to VCAM-1 and the CS-1 peptide of fibronectin. Exp Cell Res. 2000;260:73–84.
48. Hamann A, Andrew DP, Jablonski-Westrich D, Holzmann B, Butcher EC. Role of alpha 4-integrins in lymphocyte homing to mucosal tissues in vivo. J Immunol. 1994;152:3282–93.
49. Issekutz TB. Inhibition of in vivo lymphocyte migration to inflammation and homing to lymphoid tissues by the TA-2 monoclonal antibody. A likely role for VLA-4 in vivo. J Immunol. 1991;147:4178–84.
50. Picarella D, Hurlbut P, Rottman J, Shi X, Butcher E, Ringler DJ. Monoclonal antibodies specific for beta 7 integrin and mucosal addressin cell adhesion molecule-1 (MAdCAM-1) reduce inflammation in the colon of scid mice reconstituted with CD45RBhigh CD4+ T cells. J Immunol. 1997;158:2099–106.
51. Shigematsu T, Specian RD, Wolf RE, Grisham MB, Granger DN. MAdCAM mediates lymphocyte-endothelial cell adhesion in a murine model of chronic colitis. Am J Physiol Gastrointest Liver Physiol. 2001;281:G1309–15.
52. Hesterberg PE, Winsor-Hines D, Briskin MJ et al. Rapid resolution of chronic colitis in the cotton-top tamarin with an antibody to a gut-homing integrin alpha 4 beta 7. Gastroenterology. 1996;111:1373–80.
53. Podolsky DK, Lobb R, King N et al. Attenuation of colitis in the cotton-top tamarin by anti-alpha 4 integrin monoclonal antibody. J Clin Invest. 1993;92:372–80.
54. Sydora BC, Wagner N, Lohler J et al. Beta7 Integrin expression is not required for the localization of T cells to the intestine and colitis pathogenesis. Clin Exp Immunol. 2002;129:35–42.

# 9
# Role of cytokines and their receptors in inflammatory bowel diseases

H. TILG

## INTRODUCTION

Inflammatory bowel diseases (IBD) such as Crohn's disease (CD) and ulcerative colitis (UC) are common diseases in the western world, of unknown origin[1,2]. Mediators of the immune system are fundamentally involved in the regulation of chronic inflammation. It is commonly believed that the cytokine milieu and the predominant T helper cell phenotype (Th) critically affect the phenotype of chronic inflammation in the gut leading to the assumption that CD is a more Th1- and UC a more Th2-linked disease[3,4]. IBD are thought to result from inappropriate and ongoing activation of the mucosal immune system driven by the presence of normal lumenal flora. This aberrant response is most likely facilitated by defects in both barrier function of the intestinal epithelium and the mucosal immune system. The recent identification of an association between mutations of the *NOD2* gene and a subset ($\sim 20\%$ in the western world) of CD further substantiates this view: NOD2 encodes a putative intracellular pattern recognition receptor primarily binding peptidoglycans, which activates the nuclear factor $\kappa$B (NF$\kappa$B) signalling pathway. Interestingly, the NOD2 variants identified in CD patients appear to result in reduced macrophage activation of NF$\kappa$B. Consequently, an inappropriate $CD4^+$ T cell response is mounted, resulting in tissue damage and the aforementioned clinical sequelae of IBD. This inappropriate T cell activation seems to be the critical step in the pathophysiology of CD as well as UC, and seems to be the obvious result of the breakdown of tolerogenic mechanisms in the gut. $CD4^+$ T cells in CD are characterized by the expression of interferon gamma (IFN-$\gamma$) and tumour necrosis factor alpha (TNF-$\alpha$), resembling a Th1 immune response, while an atypical Th2 pattern with increased expression of IL-5 and transforming growth factor beta (TGF-$\beta$), but not interleukin 4 (IL-4), has been ascribed to UC[5].

The mucosal surface is the largest surface of the body exposed to foreign antigens; correspondingly, the gut is the largest 'immune tissue'. Alimentary antigens as well as the commensal flora have to be discriminated from pathogenic organisms with highest precision. An estimated $10^{12}$ commensal bacteria, belonging to about 400 species, are present per gram of stool. Several types of suppressive and regulatory T cells are involved in the limitation of a T cell response, with antigen-specific as well as antigen-non-specific actions in place. IL-10, TGF-β and CTLA4 play major functional roles in several of these regulatory or suppressive pathways[5]. Notably, tolerance is not only exerted towards 'real' self-antigens, but might also be exerted toward 'self' commensal flora. However, in IBD there is an obvious breakdown of these mechanisms, as outlined above, which could prove a valuable therapeutic target.

## PROINFLAMMATORY CYTOKINES

There continues to be a rapid expansion in the knowledge of regulatory peptides and their effects in mucosal inflammation. At present more than 50 regulatory peptides have been found within IBD mucosa. In contrast to immunoregulatory cytokines such as IL-2, IL-7 or IL-15, other cytokines such IL-1 and TNF-α have primarily proinflammatory features enhancing almost all aspects of inflammation. IL-1 family cytokine members participate in essentially all responses of the host to the surrounding environment, especially to infectious organisms, and IL-1-type cytokines are prominently involved in several disease states. Besides IL-1, TNF-α is the most potent proinflammatory and also pleiotropic cytokine. TNF-α is a member of a growing family of peptide mediators now comprising more than 20 cytokines, which include TNF-β, Fas ligand, nerve growth factor, and CD40 ligand. As a general rule members of this superfamily are primarily involved in the regulation of cell proliferation and apoptosis, although several members of the family, including TNF-α, TNF-β, FasL, CD30 ligand and CD40 ligand, have proinflammatory properties. TNF is an extremely potent proinflammatory cytokine that orchestrates inflammatory responses by activating a wide range of cells. TNF is also produced by, and acts upon, an enormous variety of cell types, causing a vast array of responses, and is especially produced by macrophages, an important cell type for immune and inflammatory responses.

## TNF-α IN CD

The role of TNF-α has been investigated in various preclinical and clinical situations including infection, immune-mediated disorders, and malignant disorders, and this cytokine has been the target for immunointervention more than any other cytokine. Two independent lines of investigation, including experimental and clinical studies, have strongly implicated TNF-α in the pathogenesis of CD[6-8]. TNF-α, as well as IL-1, TGF-β, and IFN-γ, were detected in large quantities in the inflamed mucosa of CD. Furthermore TNF-α levels are increased in serum and stool of patients with CD. In the intestine TNF-α mRNA can be detected by Northern blotting, reverse transcription and the polymerase chain reaction (RT-PCR), and by in-situ hybridization. Immunohistochemical localization of TNF-α demonstrated predominant expression in macrophages. Similar studies using whole tissue cultures and/or studies with lamina propria mononuclear cells isolated from intestinal tissue demonstrated increased TNF-α expression, and biologically active TNF-α has been recovered from those cultures.

The second line of evidence suggesting that TNF-α expression is not only increased in CD, but that it directly contributes to the pathogenesis of the disease, comes from studies in animals with experimentally induced disease. In various animal models, including the TNBS colitis model and the CD45RB[hi] transfer disease model, TNF-α expression has been documented in the diseased intestine. Inhibition of TNF-α appears to prevent disease onset and can reduce its severity. Studies in TNF-α transgenic mice have confirmed these antibody and inhibitor studies. Furthermore, the final proof for its role in CD came from a large series of human studies using TNF-neutralizing antibodies[8].

## IL-12 AND RELATED CYTOKINES

The data that support the theory that CD is a more Th1-directed disease is substantial. Various immunohistological studies indicate that IL-12 is over-produced in CD, mainly by macrophages[9]. This seems not to be the case in UC. Furthermore, macrophages isolated from inflammatory areas of CD patients synthesize increased amounts of IL-12 in contrast to UC (decreased synthesis of IL-12). In addition, nuclear extracts of T cells from affected areas of CD patients contain increased amounts of activated STAT4 and the transcription factor T-bet, which reflects IL-12 signalling. This circumstantial evidence for a key role of IL-12 in this Th1-oriented disease has been further supported by the recent evidence that patients treated with a specific antibody for the p40 chain of IL-12 show marked improvement of inflammation. In accordance with this finding, Stallmach et al. have recently demonstrated that an IL-12 p40–IgG2b fusion protein abrogates T-cell-mediated inflammation in experimental colitis and induces apoptosis in lamina propria mononuclear cells from patients with CD *in vitro*[10].

IL-23 (p19/p40) is another IL-12-related cytokine and a potent regulator of adaptive immune responses. It has been recently demonstrated that high constitutive IL-12 p40 expression in transgenic animals is mainly observed in the terminal ileum; this expression was paralleled by high expression of IL-23 p19/p40 proteins. The cells constitutively expressing IL-12 p40 were identified as lamina propria dendritic cells (LPDC). Interestingly, little or no p40 protein expression was found in LPDC in the terminal ileum of germfree mice, indicating a key role of the intestinal flora for p40 expression. These exciting data suggest a predisposition of the terminal ileum to develop chronic inflammatory responses through p40/IL-23, and might explain why CD shows clinical manifestation predominantly in this part of the gut[11].

## IL-18 AND IL-18 BINDING PROTEIN (IL-18bp)

IL-18 was initially described as an IFN-γ-inducing factor. The cytokine plays an important role in the induction of Th1 responses. IL-18 is related to the IL-1 family in terms of its structure, receptor family and signal transduction pathways. Mainly produced by macrophages, monocytes, intestinal epithelial cells, dendritic cells, adrenal cortex cells, microglial cells and synovial fibroblasts, IL-18 acts in synergy with IL-12 for Th1 differentiation. It also exerts proinflammatory properties by inducing the production of IL-1β, TNF-α, IL-6, IFN-γ, chemokines, nitric oxide and prostaglandins. The pleiotropic activities of IL-18 suggest an important role of this cytokine in triggering and polarization of the immune response.

IL-18 has been shown to play an important role in IBD. In CD up-regulation of IL-18 in active bowel inflammation has been demonstrated. Also in this disease IL-18 is predominantly expressed by intestinal epithelial cells, macrophages and dendritic cells, whereas IL-18 was not overexpressed in UC. Elevated plasma levels of IL-18 in CD patients have been reported. Intestinal mucosal lymphocytes from CD patients express functional IL-18 receptor, and freshly isolated mucosal lymphocytes from these patients also show a significant proliferative response to recombinant IL-18 compared to lymphocytes from healthy controls. In a CD-like animal model, blockade of IL-18 activity resulted in reduced inflammatory activity.

IL-18 binding protein (IL-18bp) is a readily secreted antagonist of IL-18. IL-18bp was purified from human urine[12] and cloned from cDNA deposit in the human genome DNA bank[13]. The genomic sequence of IL-18bp does not contain any exon coding for a transmembrane domain, suggesting that this soluble protein is not derived from cleavage of a membrane-bound receptor. IL-18bp mRNA is constitutively expressed in peripheral blood lymphocytes (PBL), spleen, thymus, colon, small intestine and prostate. *In-vitro* experiments demonstrated that this protein impairs the Th1 response, abolishes the induction of IL-18-induced IFN-γ and IL-8 production and activation of NFκB[12]. These results suggest that IL-18bp acts as a soluble decoy receptor for IL-18 and inhibits its biological activity[12]. A modulating role of IL-18bp in human IBD has been shown by Corbaz et al.[14]. The importance of IL-18 in IBD was first established by Siegmund et al. in a model of DSS colitis[15] and later corroborated by studies performed by Ten Hove et al. In these latter studies TNBS-induced colitis in mice could be ameliorated by blockade of IL-18 by IL-18bp[16,17]. These results also point to a potential therapeutic role of IL-18bp in IBD. We recently observed elevated serum levels of IL-18 and IL-18bp in CD and UC; however, calculated free IL-18 was significantly elevated only in CD compared to healthy controls.

## IL-1 FAMILY OF CYTOKINES

TNF-α and IL-1 play a major role in the development of IBD. TNF-α and IL-1β induce synthesis of chemokines, including IL-8, a potent neutrophil chemoattractant. Activation of neutrophils recruited to the intestinal mucosa results in the synthesis of proinflammatory cytokines, and the release of neutrophil granule enzymes involved in oxidative burst and tissue damage, such as myeloperoxidase. Accordingly, tissue levels of IL-1 closely correlate with the degree of mucosal inflammation and necrosis. These observations suggest that IL-1 is one of the critical mediators of intestinal inflammation in IBD[18].

Activities of IL-1 are, in part, regulated by the naturally occurring inhibitor IL-1Ra. IL-1Ra specifically inhibits IL-1 activities by binding to IL-1 receptors, but does not display agonist activity. In the intestinal mucosa the epithelial cells and lamina propria mononuclear cells are the major sources of IL-1Ra. An imbalance between the production of IL-1 and IL-1Ra has been described in freshly isolated intestinal mucosal cells and in colonic mucosal biopsies obtained from inflamed intestinal tissue of IBD patients. Administration of recombinant IL-1Ra prevents mucosal inflammation and necrosis in a rabbit model of dextran-induced colitis. Conversely, neutralization of endogenous IL-1Ra increases the severity of intestinal inflammation, indicating that endogenous IL-1Ra plays an anti-inflammatory role. The importance of IL-1 and IL-1Ra in the pathogenesis of IBD has been corroborated by the association between carriage of IL-1RN allele 2, low production of IL-1Ra and severity of disease in UC patients of Jewish or Hispanic genetic background.

All activities of IL-1 are mediated via the IL-1 receptor type I (IL-1RI). Myeloid cells also express IL-1 receptor type II (IL-1RII), a decoy receptor that presents a high degree of homology with the extracellular domain of IL-1RI. However, IL-1RII has a short (29 residue) intracytoplasmic domain that does not bind MyD88 and IRAK, signalling units otherwise recruited to members of the Toll-like receptor/IL-1R family. IL-1RII is therefore a naturally occurring functional antagonist of IL-1. IL-1 receptors play an additional role in the control of IL-1 activities through the proteolytic cleavage of their extracellular domains. Shedding of IL-1RII yields the IL-1 soluble receptor type II (IL-1sRII) which binds IL-1$\beta$ avidly, but IL-1$\alpha$ and IL-1Ra with low affinity. Thus, IL-1sRII contributes to IL-1 antagonism through the preferential neutralization of IL-1$\beta$ activity. In contrast, IL-1 soluble receptor type I (IL-1sRI) retains the ability of membrane-bound IL-1RI to bind IL-1Ra and IL-1$\alpha$ with greater affinity than IL-1$\beta$, and is therefore regarded as a proinflammatory moiety. To date the release and functional importance of IL-1 soluble receptors in intestinal inflammation remain poorly understood. However, there has been no report on surface expression or shedding of IL-1RI and IL-1RII by intestinal epithelial cells or lamina propria mononuclear cells from patients with CD or UC.

We recently demonstrated elevated plasma IL-1Ra levels in CD and UC compared to healthy subjects. IL-1$\alpha$ and IL-1$\beta$ were not detected. IL-1sRII levels were marginally lower in CD and UC whereas IL-1sRI levels were elevated in CD only. These results indicate that: (a) the proinflammatory moiety IL-1sRI is a systemic marker of inflammation and activity in CD, and (b) local shedding of the functional antagonist IL-1sRII may dampen colonic inflammation in CD, but not in UC.

## SUMMARY AND CONCLUSIONS:

Enormous progress has been made recently in understanding the pathogenesis of IBD. Through the study of patients and mouse models it has emerged that CD is driven by the production of IL-12 and IFN-$\gamma$, whereas UC is probably driven by the production of IL-13. Recently it has been demonstrated that in an

UC-like animal model of mucosal inflammation (oxazolone colitis) colitis is primarily mediated by natural-killer T cells (NK T cells) that produce mainly IL-13, and that elimination of NK T cells or blockade of IL-13 prevents the development of colitis.

More importantly, the dichotomy between Th1- and Th2-like mucosal inflammation is seen in human IBD (CD, Th1-like; UC, Th2-like). IL-12 is overproduced in CD, but not in UC. The importance of IL-12 in CD has been recently supported by the finding that antibodies specific for the p40 chain of IL-12 led to marked improvement of inflammation in most treated patients. Another IL-12-related cytokine (IL-23: consists of IL-12p40 and IL-23p19 chain) seems to be an even more potent Th1-proposing cytokine than IL-12, at least in mouse studies. Whether IL-23 has a crucial role in sustaining the Th1-mediated inflammation in CD remains to be seen. Despite the fact that T cells isolated from lesions in UC do not produce increased amounts of IL-4, UC is considered a Th2-like disease. This assumption is mainly based on the over-production of IL-5 and IL-13. As mentioned, lamina propria cells from patients with UC produce increased amounts of IL-13, but not cells isolated from patients with CD. The final proof for the potential importance of IL-13 in UC will, however, need respective clinical trials neutralizing endogenous IL-13. In addition, UC is associated with synthesis of autoantibodies, a feature which might be more indicative of a Th2- than a Th1-driven disease.

Other cytokines that do not necessarily belong to the Th1 or Th2 group, such as IL-1, IL-15, IL-16, IL-18, are also produced at higher levels in CD and UC. The role of these cytokines in IBD is less understood. IL-1 is a key mediator in the pathogenesis of IBD. Recent data suggest that the proinflammatory moiety IL-1sRI is a systemic marker of inflammation and activity in CD, and local shedding of the functional antagonist IL-1sRII may dampen colonic inflammation in CD, but not in UC. IL-18 is a proinflammatory cytokine and a member of the IL-1 family. Animal models and investigations in humans point to an important role for this cytokine in IBD. IL-18bp is a naturally occurring antagonist of IL-18. Endogenous IL-18 overproduction might be counteracted by IL-18bp insufficiently, especially in CD. A better understanding of these pleiotropic cytokines and their receptors might lead to more specific therapies in the future.

## References

1. Farrel RJ, Peppercorn MA. Ulcerative colitis. Lancet. 2003;359:331–40.
2. Podolsky DK. Inflammatory bowel disease. N Engl J Med. 2002;347:417–29.
3. Fiocchi C. Inflammatory bowel disease: etiology and pathogenesis. Gastroenterology. 1998; 115:182–205.
4. Sadlack B, Merz H, Schorle H et al. Ulcerative colitis-like disease in mice with a disrupted interleukin-2 gene. Cell. 1993;75:253–61.
5. Bouma G, Strober W. The immunological and genetic basis of inflammatory bowel disease. Nat Immunol. 2003;3:521–33.
6. Shanahan F. Inflammatory bowel disease: immunodiagnostics, immunotherapeutics, and ecotherapeutics. Gastroenterology. 2001;120:622–35.
7. Neurath MF, Fuss I, Pasparakis M et al. Predominant pathogenic role of tumor necrosis factor in experimental colitis in mice. Eur J Immunol. 1997;27:1743–50.

8. Van Dullemen HM, Van Deventer SJ, Hommes DW et al. Treatment of Crohn's disease with anti-tumor necrosis factor chimeric monoclonal antibody (cA2). Gastroenterology. 1995;109:129–35.
9. Neurath MF, Fuss I, Kelsall BL, Stuber E, Strober W. Antibodies to interleukin-12 abrogate established experimental colitis in mice. J Exp Med. 1995;182:1281–90.
10. Stallmach A, Marth T, Weiss B et al. An interleukin 12 p40-IgG2b fusion protein abrogates T cell-mediated inflammation: anti-inflammatory activity in Crohn's disease and experimental colitis *in vivo*. Gut. 2004;53:339–45.
11. Becker C, Wirtz S, Blessing M et al. Constitutive p40 promoter activation and IL-23 production in the terminal ileum mediated by dendritic cells. J Clin Invest. 2003;112:693–706.
12. Novick D, Kim SH, Fantuzzi G et al. Interleukin-18 binding protein: a novel modulator of the Th1 cytokine response. Immunity. 1999;10:127–36.
13. Aizawa Y, Akita K, Taniai M, et al. Cloning and expression of interleukin-18 binding protein. FEBS Lett. 1999;445:338–42.
14. Corbaz A, Ten Hove T, Herren S, et al. IL-18-binding protein expression by endothelial cells and macrophages is up-regulated during active Crohn's disease. J Immunol. 2002;168:3608–16.
15. Siegmund B, Fantuzzi G, Rieder F, et al. Neutralization of interleukin-18 reduces severity in murine colitis and intestinal IFN-gamma and TNF-alpha production. Am J Physiol Regul Integr Comp Physiol. 2001;281:R1264–73.
16. Novick D, Schwartsburd B, Pinkus R et al. A novel IL-18bp ELISA shows elevated serum IL-18bp in sepsis and extensive decrease of free IL-18. Cytokine. 2001;14:334–42.
17. Corbaz A, Ten Hove T, Herren S et al. IL-18-binding protein expression by endothelial cells and macrophages is up-regulated during active Crohn's disease. J Immunol. 2002;168:3608–16.
18. Dinarello CA. The role of the interleukin-1-receptor antagonist in blocking inflammation mediated by interleukin-1. N Engl J Med. 2000;343:732–4.

# 10
# State-of-the-art: targeting immunoregulation – biologicals

**W. J. SANDBORN**

## INTRODUCTION

This chapter will review the use of novel biologic therapies (other than infliximab) for the treatment of inflammatory bowel disease (IBD).

## TUMOUR NECROSIS FACTOR (TNF) ANTAGONISTS

Infliximab is a chimeric murine/human monoclonal antibody to TNF. Infliximab is an important advance in the treatment of Crohn's disease, but in some patients treatment is limited by formation of human anti-chimeric antibodies (immunogenicity) with resulting loss of response or intolerance[1]. Humanized and fully human biologic TNF antagonists have been developed to provide an alternative to infliximab. These agents include CDP571, CDP870, adalimumab (D2E7), etanercept, and onercept.

### CDP571

CDP571 is a humanized monoclonal antibody to TNF. It has a half-life of 12–14 days and is administered intravenously. CDP571 is an IgG4 antibody which does not fix complement or mediate antibody-dependent cytotoxicity. It is unclear whether CDP571 mediates apoptosis or not; probably not.

A phase II dose-finding study of CDP571 in patients with active Crohn's disease demonstrated efficacy for induction of clinical response (decrease in CDAI score $\geqslant 70$ points from baseline) at week 2 for 10 mg/kg but not 20 mg/kg compared to placebo[2]. This benefit was not sustained at 6 months. A phase III study of CDP571 in patients with active Crohn's disease enrolled 396 patients[3]. The primary endpoint was clinical response (decrease in CDAI of $\geqslant 100$ points or CDAI $< 150$ points) at week 28. CDP571 was effective for inducing a clinical response at week 2 but not at week 28. A *post-hoc* exploratory analysis of a subgroup of 40% of patients with a baseline CRP $\geqslant 10$ mg/L demonstrated efficacy for CDP571 compared to placebo at week 2

(50% versus 15%) and at week 28 (29% versus 12%). Another study with CDP571 demonstrated no benefit in 270 patients with steroid-dependent Crohn's disease[4].

## CDP870

CDP870 is a pegylated humanized Fab antibody fragment to TNF. It is administered subcutaneously. CDP870 does not fix complement or mediate antibody-dependent cytotoxicity. It is unclear whether CDP870 mediates apoptosis or not; probably not.

A phase II dose-finding study of CDP870 in 292 patients with active Crohn's disease enrolled 292 patients[5]. The primary endpoint was clinical response (decrease in CDAI of $\geqslant 100$ points or CDAI $< 150$ points) at week 12. CDP870 100 mg, 200 mg, and 400 mg was effective for inducing a clinical response at week 2 but not at week 12. A *post-hoc* exploratory analysis of a subgroup of 40% of patients with a baseline CRP $\geqslant 10$ mg/L demonstrated efficacy for CDP870 400 mg compared to placebo at week 2 (48% versus 11%) and at week 12 (52% versus 18%). A phase II dose-finding study of intravenous CDP870 in patients with active Crohn's disease was negative[6]. Phase III studies are under way.

## Adalimumab (D2E7)

Adalimumab (formerly known as D2E7) is a fully human monoclonal antibody to TNF. It has a half-life of 12–14 days and is administered subcutaneously. Adalimumab is an IgG1 antibody which fixes complement, mediates antibody-dependent cytotoxicity, and induces T cell apoptosis.

A pilot study of adalimumab in patients with Crohn's disease with prior loss of response or intolerance to infliximab has been conducted[7]. Twenty-four patients were treated with subcutaneous adalimumab 80 mg at week 0 and then 40 mg every other week for 10 weeks. Nineteen patients underwent dose escalation to 40 mg weekly after week 4 (for incomplete response). All 24 patients were able to tolerate adalimumab (including 14 who previously experienced treatment-limiting acute hypersensitivity reactions and six who previously experienced delayed hypersensitivity reactions with infliximab). Clinical remission and response rates at week 4 were 12% and 41%. Fistula improvement and fistula closure rates at week 4 were 33% and 44%. At week 12, clinical remission and response rates were 29% and 59%, and fistula improvement and fistula closure rates were 33% and 56%. A phase II dose-finding study has been completed and the results are pending, and a phase III trial is under way.

## Etandercept

Etanercept is a fully human fusion protein composed of an IgG1 Fc antibody fragment and two soluble p75 receptors to TNF. A controlled trial of etanercept at a dose of 25 mg subcutaneously twice weekly (the dose effective for rheumatoid arthritis) failed to show efficacy in patients with active Crohn's disease[8].

## Onercept

Onercept is a recombinant human soluble p55 TNF receptor. A pilot study of onercept for active Crohn's disease showed numerically higher rates of remission and response in patients who received a higher dose compared to a lower dose[9]. However, a subsequent phase II placebo-controlled trial failed to demonstrate efficacy for onercept in patients with active Crohn's disease (unpublished data).

## SELECTIVE ADHESION MOLECULE (SAM) INHIBITORS

Selective adhesion molecule inhibitors include natalizumab and MLN-02.

## Natalizumab

Natalizumab is a humanized IgG4 monoclonal antibody against the adhesion molecule α4 integrin. Natalizumab is administered intravenously every 4 weeks. A phase II dose-finding trial of natalizumab in active Crohn's disease demonstrated efficacy for induction of remission (CDAI < 150) for natalizumab compared to placebo[10]. The optimal natalizumab dose appeared to be 3 mg/kg every 4 weeks. A subsequent phase III trial of natalizumab 300 mg in 906 patients with active Crohn's disease failed to demonstrate efficacy, primarily due to an unexpectedly high placebo response rate[11]. A *post-hoc* exploratory analysis of remission at week 10 in subgroups of patients with increased concentrations of C-reactive protein and patients using immunosuppressive medications at baseline did show efficacy. A phase III maintenance trial of natalizumab in 339 patients who achieved response or remission to natalizumab in the phase III induction trial was also conducted[12]. In this maintenance trial (natalizumab withdrawal trial), a significant reduction in the frequency of patients losing response to natalizumab was observed at 6 months in natalizumab-treated patients.

## MLN-02 (LDP-02)

MLN-02 is an IgG1 humanized monoclonal antibody against the adhesion molecule α4β7 integrin. The Fc receptor recognition and binding are deleted, thus eliminating complement fixation and cytokine release. MLN-02 is administered intravenously every 4 weeks. A phase II trial of MLN-02 in 181 patients with active ulcerative colitis demonstrated significantly higher rates of remission for MLN-02-treated patients (33% and 34%) compared to placebo-treated patients (15%)[13]. Significantly higher rates of endoscopic remission and decrease in clinical symptom scores were also noted. A phase II study of MLN-02 in 185 patients with active Crohn's disease failed to show a significant difference in response rates (decrease in CDAI score ⩾ 70 points) but did show a significant difference in remission (CDAI < 150) at 8 weeks with the highest MLN-02 dose[14].

## MISCELLANEOUS BIOTECHNOLOGY AGENTS

A variety of miscellaneous biotechnology agents are undergoing evaluation for the treatment of IBD, including anti-interleukin-12 antibody, fontolizumab (anti-gamma interferon antibody), RDP58, anti-interleukin-6 receptor antibody, somatropin, sargramostim, daclizumab, visilizumab, and epidermal growth factor enemas.

### Anti-interleukin-12 antibody (J695, ABT-874)

A phase II trial of anti-interleukin-12 antibody was undertaken in patients with active Crohn's disease[15]. Patients received seven weekly subcutaneous injections of J695 1 or 3 mg/kg or placebo. Remission rates (CDAI <150) at week 7 were 0% for placebo, 8% for 1 mg/kg, and 38% for 3 mg/kg. Response rates (decrease in CDAI score ≥ 100 points) at week 7 were 25% for placebo, 27% for 1 mg/kg, and 75% for 3 mg/kg.

### Fontolizumab (anti-interferon-γ antibody)

A phase II trial of fontolizumab (a humanized antibody to interferon-γ) was undertaken in 42 patients with active Crohn's disease[16]. This study failed to demonstrate efficacy, possibly due to an unexpectedly high placebo response rate. Two subsequent phase II trials also failed to meet the primary endpoint, again due to a high placebo response rate, but a subgroup of patients undergoing multiple administrations of higher doses did appear to benefi[17].

### RDP58

RDP58 is an anti-inflammatory peptide which blocks the p38 and JNK map kinase pathways and inhibits the synthesis of TNF-α, interferon-γ, and interleukin-12 in animal models. RDP58 is not systemically bioavailable. A phase II study of RDP58 in 127 patients with active ulcerative colitis demonstrated efficacy for the highest two doses of RDP58 compared to placebo[18]. A phase II study of RDP58 in 104 patients with active Crohn's disease failed to demonstrate statistically significant difference in rates of response or remission compared to placebo[19].

### Anti-interleukin-6 receptor antibody (MRA)

MRA is a humanized antibody to the interleukin-6 receptor. Interleukin-6 is an important proinflammatory cytokine that stimulates the production of C-reactive protein. Thirty-six patients with active Crohn's disease and an elevated C-reactive protein were treated with MRA or placebo. Response and remission rates were greater in patients treated with high-dose MRA compared to placebo[20].

## Somatropin (growth hormone)

The rationale for use of somatropin (growth hormone) in the treatment of IBD is to reverse the catabolic process associated with inflammation. A placebo-controlled trial of somatropin plus a high-protein diet in 37 patients with active Crohn's disease demonstrated a significantly greater decrease in the mean CDAI score from baseline for somatropin compared to placebo[21].

## Sargramostim (GMCSF)

Gut inflammation phenotypically similar to Crohn's disease occurs in patients with chronic granulomatous disease, glycogen storage disease, and Chediac–Higashi syndrome[22,23]. Sargramostim is beneficial for patients with these disorders. A pilot study of sargramostim in patients with active Crohn's disease suggested benefit[24]. Subsequently, a controlled trial was conducted in which 124 patients with active Crohn's disease (not receiving concomitant therapy with steroids, immunosuppressive agents, or infliximab) were randomized to treatment with sargramostim 6 μg/kg or placebo daily for 8 weeks[25]. Sargramostim resulted in greater rates of remission and response than placebo.

## Daclizummab (anti-interleukin-2 receptor antibody)

Daclizumab is a humanized anti-interleukin-2 receptor antibody. An open-label study in 10 patients with active ulcerative colitis was performed, and patients were treated with 1 mg/kg at weeks 0 and 4. Remission was observed at week 8 in 50% of patients, and response was observed in 80% of patients[26]. A phase II dose-finding trial is under way.

## Visilizumab (anti-CD3 antibody)

Visilizumab is a humanized anti-CD3 antibody which is being investigated in a phase II trial for hospitalized patients with steroid-refractory ulcerative colitis. To date the results in five patients have been reported, with all patients experiencing clinical and endoscopic remission[27]. A moderate cytokine-release syndrome was observed in four of the five patients.

## Epidermal growth factor enemas for active ulcerative colitis

A controlled trial of epidermal growth factor enemas was conducted in 24 patients with active left-sided ulcerative colitis[28]. Patients were treated for 12 days. Oral mesalamine was continued at a dose of 1.2 g/day or the dose was increased by 1.2 g/day. Eighty-three per cent of patients experienced remission with epidermal growth factor enemas compared to 8% of placebo-treated patients.

## BIOTECHNOLOGY AGENTS THAT HAVE FAILED IN IBD

Biotechnology agents that have failed in IBD include interleukin-10 for both ulcerative colitis[29] and Crohn's disease[30–32], interleukin-11 for Crohn's disease[33,34], intravenous alicaforsen (Isis 2302) for Crohn's disease[35,36], the p38 MAP kinase inhibitor BIRB796BS for Crohn's disease, the kerotinocyte growth factor-2 (repifermin) for ulcerative colitis[37], and platelet-activating factor antagonists for ulcerative colitis[38].

## ADDITIONAL BIOTECHNOLOGY THERAPIES IN DEVELOPMENT FOR IBD

Additional biotechnology therapies in development for IBD include alicaforsen administered at high doses intravenously for Crohn's disease[39] and administered as an enema for patients with ulcerative colitis[40] and pouchitis[41], oral interleukin-11 for Crohn's disease[42], and the intravenous MAP kinase inhibitor CNI-1493 for active Crohn's disease[43].

## CONCLUSIONS

Biotechnology agents targeted against tumour necrosis factor, leucocyte adhesion, and other miscellaneous therapies are being evaluated for the treatment of Crohn's disease and ulcerative colitis. Based on the demonstrated efficacy of infliximab and natalizumab for Crohn's disease, and MLN-02 for ulcerative colitis, it seems certain that biologic therapies will play an important role in the future treatment of IBD.

### References

1. Baert F, Noman M, Vermeire S et al. Influence of immunogenicity on the long-term efficacy of infliximab in Crohn's disease. N Engl J Med. 2003;348:601–8.
2. Sandborn WJ, Feagan BG, Hanauer SB et al. An engineered human antibody to TNF (CDP571) for active Crohn's disease: a randomized double-blind placebo-controlled trial. Gastroenterology. 2001;120:1330–8.
3. Sandborn W, Feagan B, Radford-Smith G, Kovacs A, Enns R, Patel J. A randomized, placebo-controlled trial of CDP571, a humanized monoclonal antibody to TNF-α, in patients with moderate to severe Crohn's disease. Gastroenterology. 2003;124:A-61.
4. Celltech announces results from CDP 571 Phase III studies in Crohn's disease. Internet Press Release 2002.
5. Schreiber S, Rutgeerts P, Fedorak R, Khaliq-Kareemi M, Kamm MA, Patel J and the CDP870 Crohn's Disease Study Group. CDP870, a humanized anti-TNF antibody fragment, induces clinical response with remission in patients with active Crohn's disease (CD). Gastroenterology. 2003;124:A-61.
6. Winter T, Wright J, Ghosh S, Jahnsen J, Patel J. Intravenous CDP870, a humanized anti-TNF antibody fragment, in patients with active Crohn's disease – an exploratory study. Gastroenterology. 2003;124:A-377.
7. Sandborn WJ, Hanauer S, Loftus EV et al. An open-label study of the human anti-TNF monoclonal antibody adalimumab in subjects with prior loss of response or intolerance to infliximab for Crohn's disease. Gastroenterology. 2004;126:A53–4.

8. Sandborn WJ, Hanauer SB, Katz S et al. Etanercept for active Crohn's disease: a randomized, double-blind, placebo-controlled trial. Gastroenterology. 2001;121:1088–94.

9. Rutgeerts P, Lemmens L, Van Assche G, Noman M, Borghini-Fuhrer I, Goedkoop R. Treatment of active Crohn's disease with onercept (recombinant human soluble p55 tumour necrosis factor receptor): results of a randomized, open-label, pilot study. Aliment Pharmacol Ther. 2003;17:185–92.

10. Ghosh S, Goldin E, Gordon FH et al. Natalizumab for active Crohn's disease. N Engl J Med. 2003;348:24–32.

11. Rutgeerts P, Colombel J, Enns R et al. Subanalyses from a phase 3 sudy on the evaluation of natalizumab in active Crohn's disease tharapy-1 (ENACT-1). Gut. 2003;52(Suppl. VI):A239 (abstract: OP-G-297).

12. Elan and Biogen Idec announce ANTEGREN – natalizumab – phase III maintenance trial in Crohn's disease met its primary endpoint. Internet Press Release 2004.

13. Feagan B, Greenberg G, Wild G et al. A randomized trial of a humanized a4ß7 antibody in ulcerative colitis. Gastroenterology. 2003;125:606–7.

14. Feagan BG, Greenberg G, Wild G et al. Efficacy and safety of a humanized a4ß7 antibody in active Crohn's disease (CD). Gastroenterology. 2003;124:A25–6.

15. Mannon P, Fuss I, Mayer L et al. Group ftI-iCsDS. Anti-interleukin-12 treats active Crohn's disease. Gastroenterology. 2004;126:A22–3.

16. Rutgeerts P, Reinisch W, Colombel JF et al. Preliminary results of a phase I/II study of Huzaf, an anti-INF-gamma monoclonal antibody, in patients with moderate to severe active Crohn's disease. Gastroenterology. 2002;122:A-61.

17. Protein Design Labs Reports Progress on Two Humanized Antibodies at International Organization of Inflammatory Bowel Disease. Internet Press Release 2004.

18. Travis SPL et al., RDP Investigators Study Group. RDP-58: novel and effective therapy for ulcerative colitis. Results of parallel, prospective, placebo-controlled trials. Am J Gastroenterol. 2003;98:S239 (abstract 721).

19. Preliminary results of Sangstat's phase 2 studies of RDP58 show peak response of 77% and a 71% remission rate in ulcerative colitis patients. Additional investigation needed to determine efficacy in Crohn's disease. Internet Press Release 2003.

20. Ito H, Takazoe M, Fukuda Y et al. A pilot randomized trial of a human anti-interleukin-6 monoclonal antibody in active Crohn's disease. Gastroenterology. 2004;126:989–96.

21. Slonim AE, Bulone L, Damore MB, Goldberg T, Wingertzahn MA, McKinley MJ. A preliminary study of growth hormone therapy for Crohn's disease. N Engl J Med. 2000;342: 1633–7.

22. Korzenik JR, Dieckgraefe BK. Is Crohn's disease an immunodeficiency? A hypothesis suggesting possible early events in the pathogenesis of Crohn's disease. Dig Dis Sci. 2000; 45:1121–9.

23. Dieckgraefe BK, Korzenik JR, Husain A, Dieruf L. Association of glycogen storage disease 1b and Crohn disease: results of a North American survey. Eur J Pediatr. 2002; 161(Suppl. 1):S88–92.

24. Dieckgraefe BK, Korzenik JR. Treatment of active Crohn's disease with recombinant human granulocyte–macrophage colony-stimulating factor. Lancet. 2002;360:1478–80.

25. Berlex Laboratories announces results from randomized, placebo-controlled study of Leukine(R) for Crohn's disease. Internet Press Release 2003.

26. Van Assche G, Dalle I, Noman M et al. A pilot study on the use of the humanized anti-interleukin-2 receptor antibody daclizumab in active ulcerative colitis. Am J Gastroenterol. 2003;98:369–76.

27. Plevy S, Salzberg B, Regueiro M et al. A humanized anti-CD3 monoclonal antibody, visilizumab, for treatment of severe steroid refractory ulcerative colitis: preliminary results of a phase I study. Gastroenterology. 2003;124 (abstract).

28. Sinha A, Nightingale J, West KP, Berlanga-Acosta J, Playford RJ. Epidermal growth factor enemas with oral mesalamine for mild-to-moderate left-sided ulcerative colitis or proctitis. N Engl J Med. 2003;349:350–7.

29. Schreiber S, Fedorak R, Wild G et al. Ulcerative Colitis IL-10 Cooperative Study Group. Safety and tolerence of rHuIL-10 treatment in patients with mild/moderate active ulcerative colitis. Gastroenterology. 1998;114:A1080–1.

30. Schreiber S, Fedorak RN, Nielsen OH et al. Safety and efficacy of recombinant human interleukin 10 in chronic active Crohn's disease. Crohn's Disease IL-10 Cooperative Study Group. Gastroenterology. 2000;119:1461–72.
31. Fedorak RN, Gangl A, Elson CO et al. Recombinant human interleukin 10 in the treatment of patients with mild to moderately active Crohn's disease. The Interleukin 10 Inflammatory Bowel Disease Cooperative Study Group. Gastroenterology. 2000;119:1473–82.
32. Fedorak R, Nielsen O, Williams N et al. Human recombinant interleukin-10 is safe and well tolerated but does not induce remission in steroid dependent Crohn's disease. Gastroenterology. 2001;120:A-127.
33. Sands BE, Bank S, Sninsky CA et al. Preliminary evaluation of safety and activity of recombinant human interleukin 11 in patients with active Crohn's disease. Gastroenterology. 1999;117:58–64.
34. Sands BE, Winston BD, Salzberg B et al. Randomized, controlled trial of recombinant human interleukin-11 in patients with active Crohn's disease. Aliment Pharmacol Ther. 2002;16:399–406.
35. Schreiber S, Nikolaus S, Malchow H et al. Absence of efficacy of subcutaneous antisense ICAM-1 treatment of chronic active Crohn's disease. Gastroenterology. 2001;120:1339–46.
36. Yacyshyn BR, Chey WY, Goff J et al. Double blind, placebo controlled trial of the remission inducing and steroid sparing properties of an ICAM-1 antisense oligodeoxynucleotide, alicaforsen (ISIS 2302), in active steroid dependent Crohn's disease. Gut. 2002;51: 30–6.
37. Sandborn WJ, Sands BE, Wolf DC et al. Repifermin (keratinocyte growth factor-2) for the treatment of active ulcerative colitis: a randomized, double-blind, placebo-controlled, dose-escalation trial. Aliment Pharmacol Ther. 2003;17:1355–64.
38. Stack WA, Jenkins D, Vivet P, Hawkey CJ. Lack of effectiveness of the platelet-activating factor antagonist SR27417A in patients with active ulcerative colitis: a randomized controlled trial. The Platelet Activating Factor Antagonist Study Group in Ulcerative Colitis. Gastroenterology. 1998;115:1340–5.
39. Yacyshyn BR, Barish C, Goff J et al. Dose ranging pharmacokinetic trial of high-dose alicaforsen (intercellular adhesion molecule-1 antisense oligodeoxynucleotide) (ISIS 2302) in active Crohn's disease. Aliment Pharmacol Ther. 2002;16:1761–70.
40. Phase II study of antisense drug ISIS 2302 demonstrates significant and long-lasting improvement of symptoms in patients with ulcerative colitis. Internet Press Release 2001.
41. Miner P, Wedel M, Bane B, Bradley J. An enema formulation of alicaforsen, an antisense inhibitor of intercellular adhesion molecule-1, in the treatment of chronic, unremitting pouchitis. Aliment Pharmacol Ther. 2004;19:281–6.
42. Cotreau MM, Stonis L, Schwertschlag US. A phase 1, randomized, double-blind, placebo-controlled, dose-escalating, safety, tolerability, pharmacokinetic, and pharmacodynamic study of oral recombinant human interleukin eleven (O-rhIL-11) in normal healthy subjects. Gastroenterology. 2003;124:A-377.
43. Hommes D, van den Blink B, van de Heisteeg B et al. A new MAP-kinase inhibitor, CNI-1493, is safe and effective in the treatment of moderate to severe Crohn's disease. Gastroenterology. 2001;120:A452.

# Section IV
# Manipulation of the environment

**Chair: M. GASSULL and E. BUTRUK**

# 11
# Environmental epidemiology

## A. TIMMER

---

## INTRODUCTION

In 1767 a study was reported by G. Baker, in Devonshire, looking into the causes of a disease which seemed endemic to the region. Physicians had noticed for several decades that a syndrome of cramps, diarrhoea, seizures, and sometimes death occurred in cider drinkers. Cider drinkers in other regions of the country were not affected. Baker examined the apparatus used for cider making in various distilleries, and found it to contain an excess of lead in Devon as compared to other regions. Chemical analyses confirmed the presence of lead in Devon cider samples only. Thus, the 'endemic colic of Devonshire' turned out to be due to lead poisoning and could subsequently be effectively prevented.

This historical example opens the section on environmental epidemiology in a standard textbook of epidemiology[1]. Unfortunately, epidemiological research in inflammatory bowel diseases (IBD) has not been this straightforward and successful. However, the steps followed have been similar:

1. Observe regional (and temporal) differences.
2. Try to identify a necessary cause.
3. Examine circumstances of exposure.
4. Try to confirm the presence of the suspected agent.

In this chapter, environmental factors will be defined as any external factors, or anything not genetically determined, which may impact on disease manifestation or prognosis.

## TEMPORAL AND REGIONAL PATTERNS

A steep rise in the incidence of Crohn's disease (CD) over several decades, beginning in the 1960s, was described in several centres in Scandinavia, the UK, and Iceland from the 1960s[2] (Figure 1). Incidence data from other centres are available only from the 1980s or later. From the US, as well as within

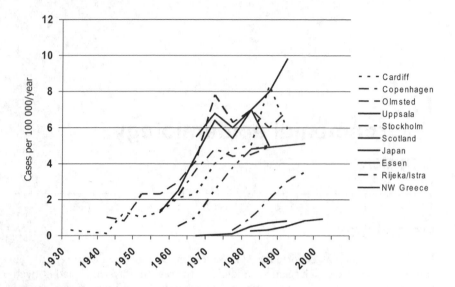

**Figure 1** Incidence of Crohn' disease (for references see text)

Europe, a north–south gradient was reported[3,4]. In countries outside Europe and the US the incidence seemed to be much lower[5]. However, even in these low-incidence countries the incidence seems to increase. Temporal trends in countries such as Japan or Croatia seem to follow those previously reported, but delayed for up to several decades[5,6] (more recent data for Croatia presented at Dubrovnik, Falk Meeting, 2004). In ulcerative colitis (UC) the incidence figures are similar, although somewhat less consistent (Figure 2). In most regions UC is about twice as common as CD.

The temporal trends and geographic differences may in part be explained methodologically. Advances in diagnosing IBD, increased awareness, improved access to the health-care system, more complete case ascertainment and, possibly, the application of statistical methods such as age standardization, may all explain some of the variability in the incidence figures. However, the extent and the consistency of the trends described are striking, and suggest the influence of environmental factors; specifically factors associated with a Western lifestyle.

## ENVIRONMENT VERSUS GENETICS

The advances in our knowledge of the molecular genetics of IBD over the past few years have been fascinating. However, it is obvious that genetic suscept-ibility is not sufficient to cause the disease. Environmental factors seem to be necessary to trigger the disease. There are several observations to support this

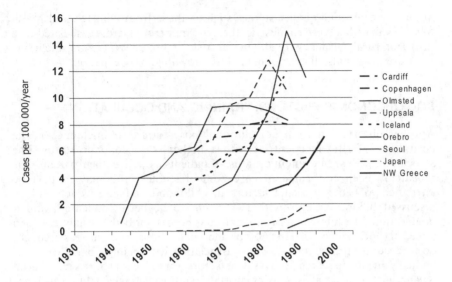

**Figure 2**   Incidence of ulcerative colitis (for references see text, as Figure 1)

view. First, genetic risk factors take much more than a lifetime to occur, and spontaneous mutations are rare. Therefore, temporal trends as described above cannot be explained genetically. Also, familial occurrence is rare in IBD – only about 10% of all patients have a first-degree affected relative.

Twin studies have been important in furthering our understanding of the role of genetic factors in IBD. On the other hand, concordance in monozygotic twins, although much higher than expected by chance, is far from complete – the pooled rates from several studies are 37% for CD and 10% for UC[7]. Lastly, to use the most prominent and successful example of the molecular genetics of IBD, the presence of CARD15 mutations (homozygosity, or compound heterozygosity) is neither a necessary nor a sufficient cause of CD. Due to the rarity of this allele in the population, CARD15 mutations account for only a small proportion of CD patients – possibly around 10%. In contrast, the attributable risk for smoking has been estimated to be about 30%[8].

## URBAN VERSUS RURAL RESIDENCY

Based on the geographical patterns described, urban residency has been associated with the occurrence of IBD. Originally some of this may have been due to methodological problems as described above. For example, a failure to use age standardization will almost invariably result in lower incidence figures in rural populations as compared to urban areas, as rural populations tend to be older. Health-care access and completeness of case ascertainment may also be a problem. On the other hand, an urban–rural gradient has also been shown in more recent, high-quality studies such as those from Olmsted and Manitoba[9,10]. As with a Western lifestyle, there are numerous factors associated with urban or rural residency, qualifying as risk or protective factors, including exposure to animals, diet, smoking habits, crowding, or occupation.

## SOCIOECONOMIC FACTORS, TRAINING AND OCCUPATION

Another observation from correlational observations of incidence is the occurrence of IBD in countries with a higher socioeconomic status. The association with socioeconomic status, of patients as well as their parents, has repeatedly been confirmed in case–control studies, and may in part be attributed to better hygienic factors in childhood, a factor which will be discussed below. Patients with IBD seem to be better educated, and a predilection for white-collar occupations has been described[11]. Analyses which linked the occurrence of IBD to specific occupations, such as less frequent occurrence in farmers, or a positive association with being a baker, seem of limited value in the search for causative factors. First, these analyses were based on administrative databases, using mortality or life insurance data – which are not likely to give a representative picture of IBD patients in general. Furthermore, since these analyses are retrospective, it is difficult to distinguish causes and consequences of the disease. In Figure 3, sex-adjusted age-specific incidence rates from two time periods are shown, based on a representative patient sample[12]. Patients with UC are, on average, slightly older at disease

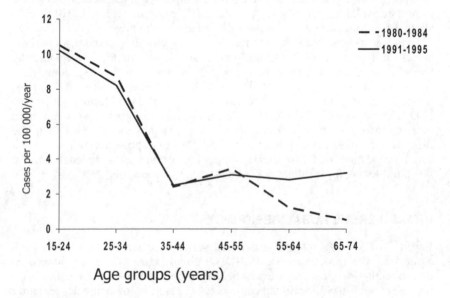

**Figure 3** Age-specific incidence, Crohn's disease, based on a regional patient group from Essen/Mulheim, Germany[12]

onset. Overall, the age distribution is strikingly similar across different time periods and regions. Considering the typical age of manifestation in IBD in early adulthood it is not likely that occupational factors play a major role in the aetiopathogenesis. Similarly, occupational diseases would be expected to be more common in males. In contrast, there is slight female preponderance in CD (RR 1.2)[3]. It seems more plausible to assume an influence of disease symptoms on career choices. Of note, while environmental epidemiology in general today has a strong focus on physical and chemical agents, no specific exposure possibly associated with working indoors, such as asbestos, or radio frequency radiation, has been examined in IBD.

## PSYCHOSOCIAL FACTORS

When patients with IBD were asked what they thought caused their disease, 'emotional factors' were named first[13]. There is a tradition to ascribe certain behavioural patterns or personality traits to patients with IBD, especially in UC. Although this assumption continues to be popular among patients and physicians, there has been no valid evidence to support the role of psychological factors in the aetiopathogenesis of IBD. Emotional factors have been described to interact with immunological factors; however, these interactions are complex and difficult to study. Specifically, in rare, chronic diseases such as IBD it is difficult, if not impossible, to differentiate between the causes and consequences of the disease with respect to coping strategies and psychosocial development.

This is different in the study of prognosis, as more frequently occurring endpoints, such as relapse, can be studied. Recently there have been some higher-quality prospective studies examining the effects of mood, psychosocial distress, stressful life events, and other concepts on outcomes in IBD, such as relapse. For example, Bitton et al. recently showed that stressful life events were associated with an increased risk for subsequent relapse[14]. Sewitch et al. found, besides the number of stressful life events, disease activity, and a shorter duration of the disease to be positively associated with psychological distress[15]. In this study satisfaction with social support was shown to be particularly important for coping in situations with high levels of perceived stress. In another study, from Vienna, depressive mood was shown to be associated with risk for relapse (Table 1)[16]. The results of these studies are not entirely consistent, mostly due to small study size and subsequent failure to replicate positive results from previous studies. However, there seems to be enough evidence now to stress the importance of considering psychosocial factors in the management of IBD patients.

## DIETARY FACTORS

Dietary changes may seem the most obvious candidate for a lifestyle-associated factor causing gastrointestinal disease. In the patient survey of perceived causes for IBD, 'diet' was placed third, following 'emotional factors' and 'genetics'.

**Table 1**  Dietary factors examined (for reference see review by Cashman and Shanahan[37])

| Risk factor? | Protective factor? |
|---|---|
| Sugar | Fibre |
| Coke, carrageenan, sweets | Fruits, citrus fruits |
| Polyunsaturated fatty acids | Fish oil |
| Margarine | Coffee |
| Cow's milk | Vegetables |
| Vitamin $B_6$ | |
| Baker's yeast | *No association* |
| Pasta, red meat, pork, rabbit, poultry | Toothpaste |
| | Alcohol |
| | Cereal |

However, the role of dietary factors in IBD is also difficult to study. Components of the diet, such as antigenic load or eicosanoids, may play a role in the aetiopathogenesis of the disease, as immunomodulating properties have been ascribed to these factors. On the other hand, abdominal symptoms are likely to influence dietary habits, possibly long before the diagnosis is established. Recall bias is another major problem in this context. Lastly, the role of nutrient deficiencies caused by the disease may be complex, interacting with disease manifestation and prognosis on the one hand, as well as dietary requirements and choices on the other hand.

An example for such a complex system of interactions is the role of sugar. An increased intake of sugar – or Coke, sweets, added sugar to drinks, chocolate, cake, etc., – by patients with CD as compared to patients with UC or healthy controls is one of the earliest and most consistent findings in the epidemiological literature on IBD[17,18]. However, whether sugar consumption is a cause or a consequence of the disease remains obscure. An attractive recent hypothesis suggests taste changes caused by zinc deficiency as a possible explanation for a higher sucrose intake in patients with CD[19]. Data on other dietary components have been less consistent (Table 1). However, while the role of dietary components is difficult to study in retrospective designs, this factor is amenable to interventional research, and a variety of dietary components, especially lipids, have been studied with respect to possible therapeutic effects. These are dealt with in another chapter of this volume.

## SMOKING AND ORAL CONTRACEPTIVES

Smoking is the most important external factor associated with the occurrence of IBD as well as the prognosis. Interestingly, the effect differs by type of IBD – CD occurs more often in smokers, UC is primarily a disease of non-smokers, or, rather, ex-smokers[20]. Similarly, smoking is associated with a poor prognosis in CD, but nicotine has been used as a therapeutic agent in UC[21]. These associations are firmly established by epidemiological data. In CD the consistency of the effect of smoking irrespective of study design and chosen endpoint, demonstration of a dose–response effect, and the reversibility of the effect after smoking cessation seem sufficient evidence to assume a causal relationship, even though the exact mechanism is not known. The chapter by J. Cosnes will deal with this subject in more detail.

Other than smoking, the effect of oral contraceptives is still a matter of debate. A meta-analysis showed a significant but small association of the use of oral contraceptives for both UC and CD[22]. Most studies on the role of oral contraceptives in the course of the disease were small, and results have been inconsistent. So far there is insufficient evidence to advise against the use of oral contraceptives in IBD.

## INFECTIOUS AGENTS

### Seasonality of flares

There has been an impression that IBD flares tend to peak at specific times during a year, hinting at a role of viral infections, such as gastroenteritis, for triggering disease activity. Peaks have been reported for several seasons – spring, autumn, winter. A very large, thoroughly done analysis on the UK General Practitioners Research Database (GPRD) has not shown any such association[23]. Although based on secondary data this study seems to be sufficient evidence against consistent seasonal peaks.

### Antibiotics

The role of the intestinal bacterial flora in the aetiopathogenesis of IBD has gained particular importance since the discovery of the NOD2/CARD15/NFκB pathways, and has consequently been covered by other chapters in this volume. As a sidenote in this context, in an interesting recent case–control study the use of antibiotics was shown to be associated with subsequent development of IBD[24].

### Pathogenic organisms

Several pathogenic infectious agents have been suspected to constitute external causative factors in IBD. In the epidemiological literature, hypotheses involving the role of measles have featured prominently. Neither the isolation of virus particles representing a persisting slow virus infection, nor the association with perinatal measles infection, nor a correlation with vaccination compaigns has been convincingly replicated. This interesting and long-lived topic is reviewed in detail in an excellent article by Robertson and Sandler[25]. In contrast, there are hardly any epidemiological studies of other suspected causative organisms. For example, most studies on *Mycobacterium* were concerned with the isolation of the organism from CD tissue[26].

## Perinatal and early childhood infections, appendectomy

There have been several reports on birth cohort phenomena in IBD and consequently the suspicion was raised that IBD could be due to some, possibly epidemic, infection taking place around birth[27]. Several investigators described a positive association with early childhood infections, both viral and bacterial, and subsequent development of CD, or, less pronounced, UC[28–30]. In this context Gilat et al. were the first to use appendectomy, tonsillectomy, and use of antibiotics as markers of bacterial infection[31]. They found appendectomy in childhood to be positively associated with CD. Unexpectedly, there was a strong negative association of appendectomy with UC. On this occasion this unexplained and unexpected finding did not receive much attention. Since then, however, the negative association between UC and appendectomy has been confirmed many times and is now firmly established. The results in CD have been less consistent. However, based on recent analyses from Sweden, among others, there is now convincing evidence for an increased prevalence of appendectomy in CD as opposed to a protective role in UC[32,33]. The age at which the appendix is removed seems to be an important determinant of the subsequent risk for disease. Similarly, Andersson et al. reported differences by underlying pathology – there was no association in the absence of inflammatory changes in the removed appendix. This was taken to indicate that CD or UC, and appendicitis represent different patterns of inflammatory responses, rather than being causally related. More experimental evidence is needed.

## OTHER CHILDHOOD FACTORS

### Domestic hygiene

At first sight, in contrast to the 'infections in childhood hypothesis', IBD have been shown to predominantly affect children growing up in affluent societies, specifically in improved housing conditions. As for atopic diseases, a sheltered environment, including advanced domestic hygiene, has been shown to be associated with subsequent CD[31,34,35]. The immunological concepts underlying the 'sheltered child hypothesis' are discussed in detail in the chapter by G. Rook.

### Early weaning

As early as 1961, early weaning was suspected to increase the risk of UC in later life[36]. Since then many studies have examined the effect of breastfeeding on IBD, as recently summarized by Cashman and Shanahan[37]. The results were not entirely consistent; however, overall, there seems to be a small protective effect by breastfeeding on the development of both UC and CD. This observation is compatible with several aetiological hypotheses, as it may lead to protection from gastrointestinal disease in infancy, as well as stimulation of mucosal development and maturation, and delayed exposure to aetiological agents, e.g. to those contained in cow's milk.

## THE COLD CHAIN HYPOTHESIS

A very intriguing hypothesis has recently been published in the *Lancet* by Hugot et al. – the cold chain hypothesis[38]. According to the authors there is a striking geographic and temporal coincidence of the introduction of refrigerators and the incidence of CD. As discussed above, CD seems to be linked to an environmental risk factor related to modern lifestyle and improved hygiene. Familial clusters have also been described. In addition, there is a propensity to ascribe the disease to dietary factors. From the bench perspective it has been shown that CARD15 mutations result in a more intense reaction to virulent bacteria due to loss of NFκB inhibition. The presence of a cold-resistant pathogen surviving in refrigerated food could represent the missing link between these observations. In this concept CD would represent a chronic infection by a cold-resistant bacterium, e.g. *Yersinia pseudotuberculosis*. Unfortunately, this pathogen has not yet been isolated from the gut of patients with CD, nor will the association of having a refrigerator with later development of CD ever be proven – nowadays there are simply not enough people in Westernized societies growing up without a refrigerator, for comparison; what a pity. Still, this paper makes very interesting reading, leading us back into the dark ages of the plague. Even more, it is a rare example of the combination of insights from bench research and epidemiological observations into a single hypothesis.

## OUTLOOK

As an introduction into the lines of thinking in environmental epidemiology I used a historical example from a leading textbook on epidemiology. Quoted in the same text are several challenges to environmental epidemiology formulated by Saracci in 1979[1]. These three challenges still hold, and have not always been sufficiently appreciated in the study of IBD. Closing this chapter, I have added a fourth.

1. Improve exposure assessment.

2. Separate the combined effects of multiple exposures: (a) consider induction periods, (b) examine dose–response effects, (c) assess the interactive effects of various exposures.

3. Integrate experimental and epidemiological evidence.

4. Consider variations in susceptibility.

## References

1. Rothman KJ, Greenland S. Modern Epidemiology, 2nd edn. Philadelphia: Lippincott-Raven, 1998.
2. Irvine EJ, Farrokhyar F, Swarbrick ET. A critical review of epidemiological studies in inflammatory bowel disease. Scand J Gastroenterol. 2001;36:2–15.

3.  Shivananda S, Lennard-Jones JE, Logan R et al. Incidence of inflammatory bowel disease across Europe: is there a difference between North and South? Results of the European collaborative study on inflammatory bowel disease (EC-IBD). Gut. 1996;39:690–7.
4.  Sonnenberg A, McCarty DJ, Jacobson SJ. Geographic variation of inflammatory bowel disease within the United States. Gastroenterology. 1991;100:143–9.
5.  Yang SK, Loftus EV Jr, Sandborn WJ. Epidemiology of inflammatory bowel disease in Asia. Inflamm Bowel Dis. 2001;7:260–70.
6.  Jovanovic Z. [Epidemiology of Crohn's disease in the Rijeka-Istra region]. Lijec Vjesn. 1999;121:8–13.
7.  Ahmad T, Satsangi J, McGovern D, Bunce M, Jewell DP. The genetics of inflammatory bowel disease. Aliment Pharmacol Ther. 2001;15:731–48.
8.  Corrao G, Tragnone A, Caprilli R et al. Risk of inflammatory bowel disease attributable to smoking, oral contraception and breastfeeding in Italy: a nationwide case–control study. Cooperative Investigators of the Italian Group for the Study of the Colon and the Rectum (GISC). Int J Epidemiol. 1998;27:397–404.
9.  Loftus EV Jr, Silverstein MD, Sandborn WJ, Tremaine WJ, Harmsen WS, Zinsmeister AR. Ulcerative colitis in Olmsted County, Minnesota, 1940–1993: incidence, prevalence, and survival. Gut. 2000;46:336–43.
10. Blanchard JF, Bernstein CN, Wajda A, Rawsthorne P. Small-area variations and socio-demographic correlates for the incidence of Crohn's disease and ulcerative colitis. Am J Epidemiol. 2001;154:328–35.
11. Sonnenberg A. Occupational distribution of inflammatory bowel disease among German employees. Gut. 1990;31:1037–40.
12. Timmer A, Breuer-Katschinski B, Goebell H. Time trends in the incidence and disease location of Crohn's disease 1980 to 1995. A prospective analysis in an urban population in Germany. Inflamm Bowel Dis. 1999;5:79–84.
13. Theis MK, Boyko EJ. Patient perceptions of causes of inflammatory bowel disease. Am J Gastroenterol. 1994;89:1920.
14. Bitton A, Sewitch MJ, Peppercorn MA et al. Psychosocial determinants of relapse in ulcerative colitis: a longitudinal study. Am J Gastroenterol. 2003;98:2203–8.
15. Sewitch MJ, Abrahamowicz M, Bitton A et al. Psychological distress, social support, and disease activity in patients with inflammatory bowel disease. Am J Gastroenterol. 2001;96:1470–9.
16. Mittermaier C, Dejaco C, Waldhoer T et al. Impact of depressive mood on relapse in patients with inflammatory bowel disease: a prospective 18-month follow-up study. Psychosom Med. 2004;66:79–84.
17. Martini GA, Brandes JW. Increased consumption of refined carbohydrates in patients with Crohn's disease. Klin Wochenschr. 1976;54:367–71.
18. Russel MG, Engels LG, Muris JW et al. 'Modern life' in the epidemiology of inflammatory bowel disease: a case–control study with special emphasis on nutritional factors. Eur J Gastroenterol Hepatol. 1998;10:243–9.
19. Schutz T, Drude C, Paulisch E, Lange KP, Lochs H. Sugar intake, taste changes and dental health in Crohn's disease. Dig Dis. 2003;21:252–7.
20. Calkins BM. A meta-analysis of the role of smoking in inflammatory bowel disease. Dig Dis Sci. 1989;34:1841–54.
21. Sandborn WJ. Nicotine therapy for ulcerative colitis: a review of rationale, mechanisms, pharmacology, and clinical results. Am J Gastroenterol. 1999;94:1161–71.
22. Godet PG, May GR, Sutherland LR. Meta-analysis of the role of oral contraceptive agents in inflammatory bowel disease. Gut. 1995;37:668–73.
23. Lewis JD, Aberra FN, Lichtenstein GR, Bilker WB, Brensinger C, Strom BL. Seasonal variation in flares of inflammatory bowel disease. Gastroenterology. 2004;126:665–73.
24. Card T, Logan RF, Rodrigues LC, Wheeler JG. Antibiotic use and the development of Crohn's disease. Gut. 2004;53:246–50.
25. Robertson DJ, Sandler RS. Measles virus and Crohn's disease: a critical appraisal of the current literature. Inflamm Bowel Dis. 2001;7:51–7.
26. Hubbard J, Surawicz CM. Etiological role of *Mycobacterium* in Crohn's disease: an assessment of the literature. Dig Dis. 1999;17:6–13.
27. Ekbom A, Zack M, Adami HO, Helmick C. Is there clustering of inflammatory bowel disease at birth? Am J Epidemiol. 1991;134:876–86.

28. Whorwell PJ, Holdstock G, Whorwell GM, Wright R. Bottle feeding, early gastroenteritis, and inflammatory bowel disease. Br Med J. 1979;1:382.
29. Wurzelmann JI, Lyles CM, Sandler RS. Childhood infections and the risk of inflammatory bowel disease. Dig Dis Sci. 1994;39:555–60.
30. Ekbom A, Adami HO, Helmick CG, Jonzon A, Zack MM. Perinatal risk factors for inflammatory bowel disease: a case–control study. Am J Epidemiol. 1990;132:1111–19.
31. Gilat T, Hacohen D, Lilos P, Langman MJ. Childhood factors in ulcerative colitis and Crohn's disease. An international cooperative study. Scand J Gastroenterol. 1987;22:1009–24.
32. Andersson RE, Olaison G, Tysk C, Ekbom A. Appendectomy and protection against ulcerative colitis. N Engl J Med. 2001;344:808–14.
33. Andersson RE, Olaison G, Tysk C, Ekbom A. Appendectomy is followed by increased risk of Crohn's disease. Gastroenterology. 2003;124:40–6.
34. Gent AE, Hellier MD, Grace RH, Swarbrick ET, Coggon D. Inflammatory bowel disease and domestic hygiene in infancy. Lancet. 1994;343:766–7.
35. Duggan AE, Usmani I, Neal KR, Logan RF. Appendicectomy, childhood hygiene, *Helicobacter pylori* status, and risk of inflammatory bowel disease: a case–control study. Gut. 1998;43:494–8.
36. Acheson ED, Truelove SC. Early weaning in the aetiology of ulcerative colitis. Br Med J. 1961;2:929–33.
37. Cashman KD, Shanahan F. Is nutrition an aetiological factor for inflammatory bowel disease? Eur J Gastroenterol Hepatol. 2003;15:607–13.
38. Hugot JP, Alberti C, Berrebi D, Bingen E, Cezard JP. Crohn's disease: the cold chain hypothesis. Lancet. 2003;362:2012–15.

# 12
# Understanding the opposite effects of smoking in inflammatory bowel disease

J. COSNES

## INTRODUCTION

It has been known for more than 20 years that smoking exerts opposite effects in inflammatory bowel disease (IBD), being beneficial in ulcerative colitis (UC) and harmful in Crohn's disease (CD). Conversely, smoking cessation aggravates UC and improves CD. The reasons why smoking has an opposite effect in two conditions which share many similarities remain obscure. The understanding of this duality is complicated by the variability of the effects according to the populations studied, the absence of an appropriate animal model, and finally the complexity of tobacco smoke, which contains hundreds of different substances, including nicotine, free radicals, and carbon monoxide.

## SMOKING AND UC

### Smoking protects against UC

Ulcerative colitis affects predominantly non-smokers and former smokers[1]. The percentage of current smokers in a group of patients with UC is about 10–15%[2–4]. The meta-analysis of Calkins yielded a pooled odds ratio of 0.41 (0.34–0.48) for current smokers compared with lifetime non-smokers[1]. The effect of smoking is only suspensive, as the relative risk of UC is not decreased in former smokers.

### Smoking improves the course of UC

Ulcerative colitis runs a more benign disease course in smokers when compared to non-smokers: flare-up and hospitalization rates[5], annual disease activity (Figure 1), the need for oral steroids[2] and, more importantly, colectomy rate, are lower in smokers than non-smokers. A meta-analysis of several large

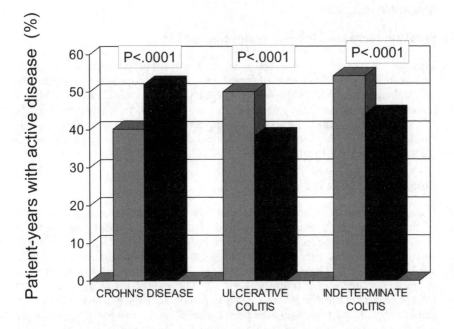

**Figure 1** Comparison of annual disease activity in smokers (black columns) vs non-smokers (grey columns) in Crohn's disease (2832 patients, 13 953 patient-years), ulcerative colitis (1264 patients, 5303 patient-years), and indeterminate colitis (154 patients, 660 patient-years). Result of the 1995–2003 prospective follow-up of 4250 patients at St-Antoine Hospital, Paris

series of UC totalling 1489 patients[2,5-9] gives an odds ratio of 0.57 (95% confidence intervals 0.38–0.85) for total colectomy in current smokers when compared to non-smokers.

## Smoking cessation increases the risk of UC and worsens its course

Stopping smoking increases the risk of developing UC when compared to never smokers. This increased risk, about 1.64 (1.36–1.98)[1], persists during the 2–3 years following smoking cessation[10]. A link between smoking habit and UC course in intermittent smokers has been reported. Many patients note symptom exacerbation when they stop smoking, followed by symptom relief when they smoke again[10]. Moreover, smokers with UC who quit experience an increase in disease activity, hospital admissions, and need for major medical therapy (oral steroids, immunosuppressants), within the first years following the cessation of smoking[11]. The risk of colectomy in the short term is, however, not increased when compared to matched non-smokers and continuing smokers[11].

## SMOKING AND CD

### Smoking increases the risk of developing CD

The percentage of current smokers in a group of patients with CD is significantly higher than that observed in a control population matched for sex and age (45–55% vs 30–40%). The meta-analysis of Calkins yielded a pooled odds ratio of 2.0 (1.65–2.47) in current smokers when compared to lifetime non-smokers[1]. Smoking might determine disease location, most studies reporting a higher prevalence of ileal disease and a lower prevalence of colonic involvement in smokers[12,13].

### Smoking worsens the course of CD

Current smoking increases by more than 50% the risk of flare-up as compared to never-smokers[14], and is associated with a lower quality of life[15]. A prospective 10-year follow-up of our patients demonstrated that, whatever the treatment, smokers experienced more years with active disease than non-smokers (Figure 1). The need for steroids and immunosuppressants is increased in smokers compared to non-smokers[13,16]. The risk of being operated on at least once during the disease course is also usually increased in smokers vs non-smokers[12,17]. In patients who have been operated on, Cottone et al. have shown that macroscopic lesions on the ileal site of the anastomosis were observed 1 year after surgery in 70% of smokers, vs 35% of non-smokers and 27% of ex-smokers[18]. The cumulative rates of clinical and surgical recurrence, respectively, were found consistently more elevated in smokers than in non-smokers[18–22].

### Smoking cessation improves CD

Although former smokers have an increased risk of developing CD, this risk decreases 4 years after quitting[1]. After surgery the risk of endoscopic and clinical recurrence in former smokers who had quit at least 1 year before are similar to that of non-smokers[18]. Similarly, CD activity in ex-smokers is not different to that of non-smokers, and less marked than in current smokers[14]. A prospective study performed in a selected group of 59 patients who stopped smoking, following a smoking cessation intervention[23], examined the disease course from 1 year after the quit date. Regarding the flare-up rate and therapeutic needs, disease severity was similar in patients who had never smoked and in those who stopped smoking, and both had a better course than continuing smokers. Quitters had a 65% decreased risk of flare as compared with paired continuing smokers. They were less likely to require corticosteroids, start immunosuppressive therapy, or require an increased dose of immunosuppressants. The beneficial effect of smoking cessation was long-lasting, with a proportion of active disease each year similar to non-smokers (Figure 2)

The respective effects of current smoking and smoking cessation in UC and CD are summarized in Table 1.

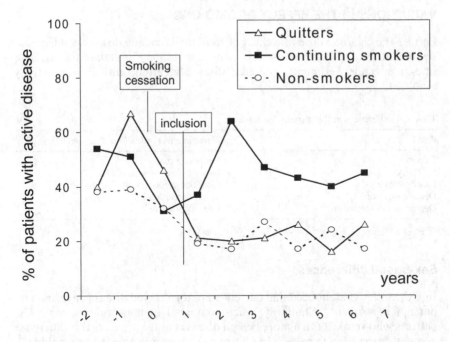

**Figure 2** Year-by-year evolution of the proportion of patients with active disease in a cohort of 59 Crohn's disease patients who quit smoking (quitters) and two matched control cohorts of 59 non-smokers and 59 continuing smokers. The first visit following the anniversary of the quit date (or matching in controls) was the date of inclusion into the prospective follow-up, while disease activity during the three previous years was assessed retrospectively (see ref. 23)

**Table 1** Respective effects of current smoking and smoking cessation in IBD

|  | Ulcerative colitis | | Crohn's disease | |
|---|---|---|---|---|
|  | Current smoking* | Smoking cessation[†] | Current smoking* | Smoking cessation[†] |
| Development | ↓ | ↑ | ↑ | ± |
| Activity (flares) | ↓ | ↑ | ↑ | ↓ |
| Surgery | ↓ | – | ↑ | ↓? |

*Current smoking compared to non-smoking.

[†]smoking cessation compared to continuing smoking.

## VARIATIONS IN THE EFFECT OF SMOKING

Besides the cigarette dose (the effect of smoking is usually dose-dependent and significant only in heavy smokers), many factors may modulate the effect of smoking: gender, ethnicity, genetics, disease location, and disease activity (Table 2).

**Table 2** Differences in the response to smoking

|  | Harmful effect of smoking |
| --- | --- |
| Gender | Females > males |
| Ethnicity | Non-Jewish > Jewish (CD)? |
| Family studies | Genetics predominant |
| Disease location | Small bowel > colon |
| Degree of inflammation | Chronic > acute |

### Sex-based differences

In UC Motley et al. noticed that current smoking delayed disease onset in men, but not in women[10]. Similarily, when compared to non-smokers, male UC patients who smoked ran a more benign disease course, whereas this difference was not observed in females[4] . In CD, women are more affected by smoking[24]. The study by Sutherland et al. first reported a doubled proportion of early postoperative recurrences in female smokers as compared to female non-smokers, whereas there was no significant difference between smokers and non-smokers in men[19]. In Crohn's colitis, smoking is clearly harmful in women, whereas colitis of men is not affected by smoking[4]. In other terms, the dual effect of smoking in colitis, beneficial in UC, and harmful in Crohn's colitis, is importantly modulated by gender, with women taking more disadvantage than men. Differences in the way of smoking (women use more filter cigarettes and more light cigarettes, and have a higher relative exposure to smoke than nicotine[25]) and the negative effect of oestrogens on proinflammatory cytokine gene regulation and T–B cell interactions[26] are possible explanations.

### Ethnicity

In Israel there is no association between smoking and CD in Jewish patients[27], whereas the opposite association does exist in UC. The reason why CD in Israeli Jews is not so sensitive to smoking as in other populations is not clear. Some studies in Spain[6] and Norway[28] also failed to identify a deleterious effect of smoking on the course of CD.

### Heritability

Tobacco consumption acts on IBD genetic predisposition to shift the phenotype from UC towards CD. Family studies have reported a high concordance rate within a family between smoking habits and the phenotype of IBD, with

CD developing in smokers and UC in non-smokers[29]. Studies of sibling pairs[30] and twins[31] also confirmed the determining role of smoking habits upon the IBD phenotype, and their significance for discordance for disease. It should also be noted that the effect of smoking was found to be independent from that of NOD2 mutant genotypes on the CD phenotype, with smoking associated with ileal disease and early surgery, whereas NOD2 mutations were associated with ileal disease and development of stricturing or internal fistulizing complications[32].

## Disease location

In both IBD there is a trend towards a more segmental disease in smokers vs non-smokers. In patients with distal UC at diagnosis, retrograde extension of the disease process is less frequent in smokers than in non-smokers[2,33]. In CD, smokers have a higher prevalence of ileal disease and a lower prevalence of colonic involvement[12,13]. Besides, patients with colonic involvement only are less sensitive to the harmful effect of smoking[14]. Moreover, after quitting, some patients develop lesions of the distal colon whereas they had more proximal CD before. Finally, development and severity of anoperineal disease seem not to be influenced by smoking status[14].

## Degree of inflammation

There are no data supporting the view that current smoking aggravates the clinical presentation of a CD flare or jeopardizes the response to treatment, although it may decrease the efficacy of infliximab[34]. Smoking is also associated with more frequent intestinal penetrating complications[7,12,35,36]; however, considering patients with an internal fistula, surgery can be postponed a longer time in smokers vs non-smokers (unpublished data).

## UNDERSTANDING THE EFFECT OF SMOKING

### Putative mechanisms of the effect of smoking in IBD

Nicotine modifies the thickness of adherent mucus to the colonic mucosa in rabbits[37] and increases mucin synthesis in UC patients[38]. However, transdermal nicotine has no effect on mucin gene transcription[39]. Nicotine abolishes the synthesis of IL-1$\beta$ and TNF-$\alpha$ by mouse colonic mucosa[40]. In humans, nicotine decreases some proinflammatory cytokines production by human mononuclear cells, such as IL-2[41], TNF-$\alpha$[42], and IL-8[43]. Smokers with IBD have a significant reduction in mucosal cytokine levels, specifically IL-1$\beta$ and IL-8 for patients with UC and IL-8 for patients with CD[44]. Beneficial effects of nicotine in active UC may result from the decrease of IL-8 expression[39].

Other pertinent effects of nicotine or tobacco smoke on the intestine are the reduction of smooth muscle tone and contractile activity through the release of nitric oxide[45], modified permeability to macromolecules[46,47], and changes in the microcirculation with transient ischaemia[47]. Smoking through increased carbon monoxide concentration may decrease the perfusion to the rectum and to acutely damaged colonic tissue, which may be beneficial in UC and flares of CD, whereas in the chronic state it may amplify the impairment in vasodilation capacity of the chronically inflamed microvessel[48], resulting in ischaemia, and perpetuating ulceration and fibrosis[49].

Moreover, cigarette smoke increases lipid peroxidation[50], while CD is characterized by a decreased total radical-trapping antioxidant potential[51]. The immune suppression and a defect in bacterial clearance or macrophage deficiency induced by smoking may also have a detrimental role[52].

The effect of nicotine and tobacco smoke has been studied in different experimental models of colitis[53–57]. In general, cigarette smoke and nicotine had a beneficial effect on the colonic damage, but this effect was reversed at high dose. The effect of nicotine was abolished by the ganglion blocker hexamethonium. The same dose of nicotine which was beneficial in colitis worsened small bowel lesions; also, carbon monoxide, while aggravating ischaemia, was found to have potent anti-inflammatory effects on macrophages of IL-10$^{-/-}$ mice[58]. Therefore the same substance, either nicotine or carbon monoxide, may exert opposite effects depending on dose, duration of exposure, and the target which is examined.

## Relevance to understanding the mechanisms of IBD

The sensitivity of IBD to smoking concerns both disease onset and disease course. Smoking (or smoking cessation) has a trigger role and, on the same predisposing genetic background, may be a major determinant of the disease phenotype, CD or UC. Smoking may also recruit patients who would not have had IBD if they had never smoked. Then, after disease onset, smoking (or smoking cessation) either protects the intestine or perpetuates the mucosal damage, probably according to many factors, such as gender (women are more susceptible to the harmful effect), disease location (the protective effect is more efficient in the distal intestine), degree of inflammation (smoking through decreased tissue perfusion aggravates chronic inflammatory state but may improve acute colitis), daily cigarette dose (harmful effects dominant above 15 cigarettes), concentration of nicotine in cigarettes (less nicotine and more damage), and probably other unidentified factors, for example genetic factors such as the apoE genotype which is known to determine the sensitivity to smoking in coronary disease[59]. Smoking also acts on different targets: mucus layer, cytokine production, macrophage function, and microvasculature, probably in different directions in one individual. The balance of these various effects is unpredictable and unstable, and this instability may explain the rapidity of the disease response to changes in smoking habits (Figure 3). Finally, one can consider three types of IBD: (a) some patients are genetically scheduled to develop CD, and they will have CD; (b) others are scheduled to develop UC; and they will have UC; and (c) others are scheduled to develop either type of IBD. In these latter patients, if there is no particular environmental factor, the disease will be UC-like, involving the distal intestine, with continuous, superficial and extensive lesions. If these patients smoke, however, they will have a different disease, which may involve the small bowel, have a

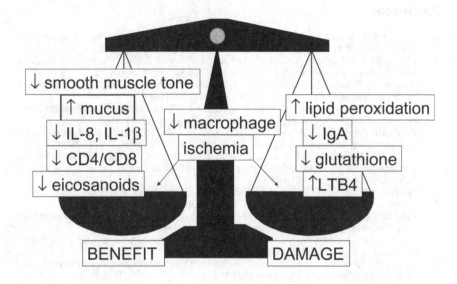

**Figure 3** Representation of the various mechanisms of the effect of smoking in IBD. The balance in one individual is unpredictable and unstable

patchy distribution, be located to one digestive segment, and develop rapid stricturing and penetrating complications; thus a CD-like phenotype. The same effect of smoking will also be observed in patients with the typical genetic background, with CD becoming more severe, and on the contrary UC less extensive and less active.

## SUMMARY AND CONCLUSION

Smoking modifies the risk of developing IBD, increasing the risk of CD and decreasing that of UC. In an individual genetically at risk for IBD, smoking may be the main factor determining the phenotype of the disease, either UC or CD. In addition, smoking changes considerably the course of the disease, improving UC and worsening CD, while smoking cessation is followed rapidly by reversal of the effect. Even though the effect of smoking appears opposite in the two IBD, smoking should be discouraged in both. In UC the increased risks of neoplasia and vascular disease in smokers largely overcome the benefits. In CD smoking is harmful, and persuading patients to stop smoking is probably the most cost-effective strategy for controlling disease activity.

# References

1.  Calkins BM. A meta-analysis of the role of smoking in inflammatory bowel disease. Dig Dis Sci. 1989;34:1841–54.
2.  Mokbel M, Carbonnel F, Beaugerie L, Gendre JP, Cosnes J. [Effect of smoking on the long-term course of ulcerative colitis]. Gastroenterol Clin Bio. 1998;22:858–62.
3.  Srivasta ED, Newcombe RG, Rhodes J, Avramidis P, Mayberry JF. Smoking and ulcerative colitis: a community study. Int J Colorectal Dis. 1993;8:71–4.
4.  Cosnes J, Nion-Larmurier I, Afchain P, Beaugerie L, Gendre JP. Gender differences in the response of colitis to smoking. Clin Gastroenterol Hepatol. 2004;2:41–8.
5.  Boyko EJ, Perera DR, Koepsell TD, Keane EM, Inui TS. Effects of cigarette smoking on the clinical course of ulcerative colitis. Scand J Gastroenterol. 1988;23:1147–52.
6.  Fraga XF, Vergara M, Medina C, Casellas F, Bermejo B, Malagelada JR. Effects of smoking on the presentation and clinical course of inflammatory bowel disease. Eur J Gastroenterol Hepatol. 1997;9:683–7.
7.  Holdstock G, Savage D, Harman M, Wright R. Should patients with inflammatory bowel disease smoke? Br Med J (Clin Res Ed). 1984;288:362.
8.  Odes HS, Fich A, Reif S et al. Effects of current cigarette smoking on clinical course of Crohn's disease and ulcerative colitis. Dig Dis Sci. 2001;46:1717–21.
9.  Woseth DMHS. Influence of smoking history on disease course in ulcerative colitis. Gastroenterology. 1998;114:A1116.
10. Motley RJ, Rhodes J, Ford GA et al. Time relationships between cessation of smoking and onset of ulcerative colitis. Digestion. 1987;37:125–7.
11. Beaugerie L, Massot N, Carbonnel F, Cattan S, Gendre JP, Cosnes J. Impact of cessation of smoking on the course of ulcerative colitis. Am J Gastroenterol. 2001;96:2113–16.
12. Lindberg E, Jarnerot G, Huitfeldt B. Smoking in Crohn's disease: effect on localisation and clinical course. Gut. 1992;33:779–82.
13. Russel MG, Volovics A, Schoon EJ et al. Inflammatory bowel disease: is there any relation between smoking status and disease presentation? European Collaborative IBD Study Group. Inflamm Bowel Dis. 1998;4:182–6.
14. Cosnes J, Carbonnel F, Carrat F, Beaugerie L, Cattan S, Gendre J. Effects of current and former cigarette smoking on the clinical course of Crohn's disease. Aliment Pharmacol Ther. 1999;13:1403–11.
15. Russel MG, Nieman FH, Bergers JM, Stockbrugger RW. Cigarette smoking and quality of life in patients with inflammatory bowel disease. South Limburg IBD Study Group. Eur J Gastroenterol Hepatol. 1996;8:1075–81.
16. Cosnes J, Carbonnel F, Beaugerie L, Le Quintrec Y, Gendre JP. Effects of cigarette smoking on the long-term course of Crohn's disease. Gastroenterology. 1996;110:424–31.
17. Breuer-Katschinski BD, Hollander N, Goebell H. Effect of cigarette smoking on the course of Crohn's disease. Eur J Gastroenterol Hepatol. 1996;8:225–8.
18. Cottone M, Rosselli M, Orlando A et al. Smoking habits and recurrence in Crohn's disease. Gastroenterology. 1994;106:643–8.
19. Sutherland LR, Ramcharan S, Bryant H, Fick G. Effect of cigarette smoking on recurrence of Crohn's disease. Gastroenterology. 1990;98:1123–8.
20. Moskovitz D, McLeod RS, Greenberg GR, Cohen Z. Operative and environmental risk factors for recurrence of Crohn's disease. Int J Colorectal Dis. 1999;14:224–6.
21. Yamamoto T, Keighley MR. The association of cigarette smoking with a high risk of recurrence after ileocolonic resection for ileocecal Crohn's disease. Surg Today. 1999;29:579–80.
22. Ryan WR, Allan RN, Yamamoto T, Keighley MR. Crohn's disease patients who quit smoking have a reduced risk of reoperation for recurrence. Am J Surg. 2004;187:219–25.
23. Cosnes J, Beaugerie L, Carbonnel F, Gendre JP. Smoking cessation and the course of Crohn's disease: an intervention study. Gastroenterology. 2001;120:1093–9.
24. Persson PG, Ahlbom A, Hellers G. Inflammatory bowel disease and tobacco smoke – a case–control study. Gut. 1990;31:1377–81.
25. Zeman MV, Hiraki L, Sellers EM. Gender differences in tobacco smoking: higher relative exposure to smoke than nicotine in women. J Womens Health Gend Based Med. 2002;11:147–53.

26. Rider V, Abdou NI. Gender differences in autoimmunity: molecular basis for estrogen effects in systemic lupus erythematosus. Int Immunopharmacol. 2001;1:1009–24.
27. Reif S, Lavy A, Keter D et al. Lack of association between smoking and Crohn's disease but the usual association with ulcerative colitis in Jewish patients in Israel: a multicenter study. Am J Gastroenterol. 2000;95:474–8.
28. Moum B, Ekbom A, Vatn MH et al. Clinical course during the 1st year after diagnosis in ulcerative colitis and Crohn's disease. Results of a large, prospective population-based study in southeastern Norway, 1990–93. Scand J Gastroenterol. 1997;32:1005–12.
29. Smith MB, Lashner BA, Hanauer SB. Smoking and inflammatory bowel disease in families. Am J Gastroenterol. 1988;83:407–9.
30. Bridger S, Lee JC, Bjarnason I, Jones JE, Macpherson AJ. In siblings with similar genetic susceptibility for inflammatory bowel disease, smokers tend to develop Crohn's disease and non-smokers develop ulcerative colitis. Gut. 2002;51:21–5.
31. Orholm M, Binder V, Sorensen TI, Rasmussen LP, Kyvik KO. Concordance of inflammatory bowel disease among Danish twins. Results of a nationwide study. Scand J Gastroenterol. 2000;35:1075–81.
32. Brant SR, Picco MF, Achkar JP et al. Defining complex contributions of NOD2/CARD15 gene mutations, age at onset, and tobacco use on Crohn's disease phenotypes. Inflamm Bowel Dis. 2003;9:281–9.
33. Samuelsson SM, Ekbom A, Zack M, Helmick CG, Adami HO. Risk factors for extensive ulcerative colitis and ulcerative proctitis: a population based case–control study. Gut. 1991; 32:1526–30.
34. Parsi MA, Achkar JP, Richardson S et al. Predictors of response to infliximab in patients with Crohn's disease. Gastroenterology. 2002;123:707–13.
35. Picco MF, Bayless TM. Tobacco consumption and disease duration are associated with fistulizing and stricturing behaviors in the first 8 years of Crohn's disease. Am J Gastroenterol. 2003;98:363–8.
36. Louis E, Michel V, Hugot JP et al. Early development of stricturing or penetrating pattern in Crohn's disease is influenced by disease location, number of flares, and smoking but not by NOD2/CARD15 genotype. Gut. 2003;52:552–7.
37. Zijlstra FJ, Srivastava ED, Rhodes M et al. Effect of nicotine on rectal mucus and mucosal eicosanoids. Gut. 1994;35:247–51.
38. Finnie IA, Campbell BJ, Taylor BA et al. Stimulation of colonic mucin synthesis by corticosteroids and nicotine. Clin Sci (Lond). 1996;91:359–64.
39. Louvet B, Buisine MP, Desreumaux P et al. Transdermal nicotine decreases mucosal IL-8 expression but has no effect on mucin gene expression in ulcerative colitis. Inflamm Bowel Dis. 1999;5:174–81.
40. Van Dijk JP, Madretsma GS, Keuskamp ZJ, Zijlstra FJ. Nicotine inhibits cytokine synthesis by mouse colonic mucosa. Eur J Pharmacol. 1995;278:R11–12.
41. van Dijk AP, Meijssen MA, Brouwer AJ et al. Transdermal nicotine inhibits interleukin 2 synthesis by mononuclear cells derived from healthy volunteers. Eur J Clin Invest. 1998; 28:664–71.
42. Madretsma GS, Donze GJ, van Dijk AP, Tak CJ, Wilson JH, Zijlstra FJ. Nicotine inhibits the *in vitro* production of interleukin 2 and tumour necrosis factor-alpha by human mononuclear cells. Immunopharmacology. 1996;35:47–51.
43. Bhatti MA, Hodgson HJF. Nicotine downregulates IL-8 production and tissue expression in inflammatory bowel disease. Gastroenterology. 1997;112:A934.
44. Sher ME, Bank S, Greenberg R et al. The influence of cigarette smoking on cytokine levels in patients with inflammatory bowel disease. Inflamm Bowel Dis. 1999;5:73–8.
45. Green JT, McKirdy HC, Rhodes J, Thomas GA, Evans BK. Intra-luminal nicotine reduces smooth muscle tone and contractile activity in the distal large bowel. Eur J Gastroenterol Hepatol. 1999;11:1299–304.
46. Suenaert P, Bulteel V, Den Hond E et al. The effects of smoking and indomethacin on small intestinal permeability. Aliment Pharmacol Ther. 2000;14:819–22.
47. Thomas GA, Rhodes J, Green J. Nicotine and gastrointestinal disease. Q J Med. 1996;89: 485–8.
48. Wakefield AJ, Sawyerr AM, Dhillon AP et al. Pathogenesis of Crohn's disease: multifocal gastrointestinal infarction. Lancet. 1989;2:1057–62.

49. Hatoum OABD, Otterson MF, Gutterman DD. Acquired microvascular dysfunction in inflammatory bowel disease: loss of nitric oxide-mediated vasodilatation. Gastroenterology. 2003;125:58–69.
50. Euler DE, Dave SJ, Guo H. Effect of cigarette smoking on pentane excretion in alveolar breath. Clin Chem. 1996;42:303–8.
51. Genser D, Kang MH, Vogelsang H, Elmadfa I. Status of lipidsoluble antioxidants and TRAP in patients with Crohn's disease and healthy controls. Eur J Clin Nutr. 1999;53:675–9.
52. Otterbein LE, Bach FH, Alam J et al. Carbon monoxide has anti-inflammatory effects involving the mitogen-activated protein kinase pathway. Nat Med. 2000;6:422–8.
53. Eliakim R, Fan QX, Babyatsky MW. Chronic nicotine administration differentially alters jejunal and colonic inflammation in interleukin-10 deficient mice. Eur J Gastroenterol Hepatol. 2002;14:607–14.
54. Eliakim R, Karmeli F, Cohen P, Heyman SN, Rachmilewitz D. Dual effect of chronic nicotine administration: augmentation of jejunitis and amelioration of colitis induced by iodoacetamide in rats. Int J Colorectal Dis. 2001;16:14–21.
55. Galeazzi F, Blennerhassett PA, Qiu B, O'Byrne PM, Collins SM. Cigarette smoke aggravates experimental colitis in rats. Gastroenterology. 1999;117:877–83.
56. Guo X, Ko JK, Mei QB, Cho CH. Aggravating effect of cigarette smoke exposure on experimental colitis is associated with leukotriene B(4) and reactive oxygen metabolites. Digestion. 2001;63:180–7.
57. Qiu BS, Deng YK, Galeazzi F, Collins SM. Neural involvement in the bivalent action of nicotine on experimental colitis in rats. Gastroenterology. 1997;112:A1065.
58. Hegazi RSR, Otterbein O, Plevy SE. Immunosuppressive effects of carbon monoxide in IL-10 deficient macrophages. Gastroenterology. 2003;124:A101.
59. Humphries SE, Talmud PJ, Hawe E, Bolla M, Day IN, Miller GJ. Apolipoprotein E4 and coronary heart disease in middle-aged men who smoke: a prospective study. Lancet. 2001; 358:115–19.

# 13
# Microparticles: a link between modern life and inflammatory bowel disease?

L. PELE and J. J. POWELL

## INTRODUCTION

In a 1939 broadcast, Winston Churchill famously referred to his consideration of the political and military ambitions of Russia as a 'riddle wrapped in a mystery inside an enigma'. The same could be said for many scientists and clinicians today in their consideration of the aetiology of Crohn's disease (CD), more than 70 years after its first convincing description.

Nonetheless, clinical, genetic, molecular and epidemiological advances have been made that provide us with a series of 'expressions' that must at least be satisfied in attempting to explain the aetio-pathogenesis of the disease. In the first place aetiological factor(s) or trigger(s) must be distinguished from factors that contribute to the pathogenesis or maintenance of disease. These are briefly outlined below, but we return in the final paragraph to this important differentiation when considering the potential role of dietary microparticles in the aetiology or the pathogenesis of disease.

Genetic susceptibility has a very clear role in CD, and the frequency of inflammatory bowel disease (IBD) in first-degree relatives of patients may be as high as 40%; this frequency tends to be higher in patients with CD than for those with ulcerative colitis[1]. IBD1, located in the pericentromeric region of chromosome 16, was the first susceptibility genetic locus linked to CD[2], and further studies led to the location of a specific gene on chromosome 16 q12, namely *CARD15* which encodes for NOD2, where mutations were reproducibly observed in more of the patients with CD than controls in Western populations[3–5]. Other loci have also been associated with IBD, namely: IBD2 locus on chromosome 12[6,7], where linkage analyses are shown to be greater in ulcerative colitis compared with CD; IBD3 locus on chromosome 6p encompassing the major histocompatibility complex[6,8] and the tumour necrosis factor gene; and IBD5 locus on chromosome 5q31[9], coding for a number of immunoregulatory cytokines. By resequencing genes in the latter, Peltekova et al. recently found that functional variants in the gene encoding the organic cation transporter gene (OCTN) (missense substitution in SLC22A4 (C→T) and a transversion in SLC22A5 (G→C)), were associated with CD[10]. Although translation of these genetic alterations into a full understanding of cellular dysfunction is still in progress, hypotheses on the cause of CD need to heed these findings.

The incomplete concordance rate for CD within monozygotic twins ($<50\%$) and the variation in risk in different geographical locations supports a role for environmental factors in the pathogenesis and occurrence of CD. In acute manifestations of IBD the intestinal pathology of CD is similar to that of infectious *Yersinia enterocolitica* or *Mycobacterium tuberculosis*. Moreover, granulomas are a feature of around 65% of CD cases, and given that *Mycobacterium paratuberculosis* causes a granulomatous intestinal disease in ruminants (Johne's disease) similar to CD, an association of CD with mycobacteria has long been suggested. However, in spite of numerous studies, findings have failed to show any consistent association between mycobacterial infections and disease. Moreover, if infection is the trigger in CD then 'atypical' exposure has to be invoked due to the modern Western incidence of CD.

It is generally accepted that mycobacteria and other infectious agents gain entry to intestinal tissue mainly via the M cell-rich mucosa of intestinal lymphoid aggregates (i.e. follicle-associated epithelium) although, even then, in the absence of invasive enteric organisms, translocation is inefficient. Interestingly, lymphoid aggregates appear to be the first site of early CD and, classically, the pathology presents as skip lesions and predominates in the terminal ileum and colon, both of which are features that are compatible with disease origins located to intestinal lymphoid aggregates. However, while bacteria clearly can access the follicle-associated epithelium of the gut (e.g. Peyer's patches), they are in fact rarely observed in uninflamed CD or healthy mucosa, either at the site of intestinal lymphoid aggregates or elsewhere[11]. This contrasts to dietary microparticles which are bacterial-sized but of non-biological origin, and are readily observed at the base of human Peyer's patches (Figure 1). Here we develop the thesis that such microparticles, in the presence of luminal bacterial products, may trigger inflammation in susceptible individuals but that commensal, luminal organisms are then sufficient to maintain disease activity.

The intestinal flora contains $10^{12}$–$10^{14}$ bacteria per gram of colonic tissue and between 400 and 500 bacterial species[12]. The bacterial distribution varies according to the different regions of the gastrointestinal tract and bacteria are predominantly present in the colon[12], caecum and, less so, the terminal ileum.

Lactobacilli, which are the main constituents of the Gram-positive bacteria of the intestinal flora, have been shown to prevent Gram-negative translocation and to regulate GALT humoral and cellular responses. Furthermore, lipopoly-saccharide (LPS), a component of the Gram-negative bacteria cell wall, is highly proinflammatory. Interestingly, the concentration of lactobacilli in colonic biopsy specimens from ulcerative colitis patients and in faeces from CD patients is decreased[13], which may therefore enable better LPS transloca-tion. In addition, it is possible that products of the commensal flora promote inflammation in the presence of an impaired mucosal barrier and increased intestinal permeability[1,14]. Moreover, Swidsinski et al. showed that IBD

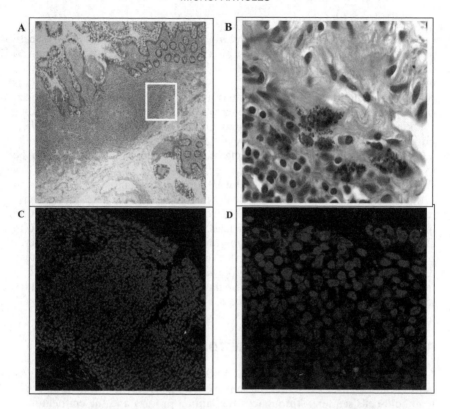

**Figure 1** Exogenous and endogenous microparticles within Peyer's patches. **A, B**: Haematoxylin and eosin staining of human Peyer's patch at low and high magnifications (original × 100 and × 1000, respectively) showing pigment cells (macrophages) containing exogenous dietary microparticles (granular pigment easily observed in **B**). Note the large majority of such pigment cells are located at the base of Peyer's patches, as indicated by the white box. **C, D**: Mouse Peyer's patch stained with calcein (original green but shown as bright spots in **C** or brighter areas of cytoplasm in **D**), demonstrating the presence of endogenous calcium phosphate particles in lysosomes of phagocytic cells, and also counterstained with propidium iodide (dark nuclei) at low and high magnifications (original × 400 and × 2560, respectively). Note that such endogenous microparticles are found throughout the patch in both human and murine intestine. All micrographs were kindly provided by Vinay Thoree, Gastrointestinal Laboratory, St Thomas' Hospital

patients have a higher concentration of mucosal bacteria, which progressively increases with the severity of disease[15]. A possible role for bacteria in the pathogenesis of IBD is further emphasized by the finding that diversion of the faecal stream may determine recurrence of CD[16] and from the numerous animal studies which indicate that bacterial flora are required for the onset and maintenance of colitis in susceptible animals. Finally it is important to point out that LPMNC from healthy people are non-responsive to autologous flora, but do respond when challenged with heterologous intestinal bacteria, showing that tolerance specifically exists towards the self-bacterial flora[17].

However, this tolerance is lost in patients with active IBD[17].

Among environmental factors, smoking and diet have clear effects on disease pathogenesis. Interestingly, smoking affects the host differently according to the type of IBD or the part of bowel involved. In simple terms smoking has a negative effect on the course of CD[18] but seems 'protective' in ulcerative colitis (UC)[19], although these observations may, in part, result from smoking-induced skewing of disease phenotype in the early stages of IBD development[19]. Like smoking, diet seems to affect CD and UC differently, since elemental diets improve CD but not UC. We have noted previously that cigarette smoke contains high levels of microparticles while elemental diets appear to be microparticle-free[20,21].

Overall, the increasing prevalence of IBD parallels an increase in atopy and asthma, all of which are immunologically mediated[12]. In the lung, atmospheric fine and ultrafine particles may play a role in modulating and exacerbating immune responses. Similar-sized particles, derived from the diet, are found in abundance in Peyer's patches probably due to the widespread use of micro-particulate food additives and anti-caking agents in the modern Western diet[21,22] (Figure 1). A brief review of Peyer's patches, and their critical role in immune tolerance, is given prior to discussion of the potential impact of dietary microparticles in this regulatory process.

## PEYER'S PATCHES AND IMMUNE TOLERANCE

Gut-associated lymphoid tissue (GALT) comprises intestinal lymphoid aggregates (notably Peyer's patches), mesenteric-lymph nodes and large numbers of lymphoid cells scattered throughout the lamina propria and the epithelium of the small and large intestine[23,24]. Recently described intestinal villous M cell regions may also be recognized as part of the GALT[25].

In the human intestine there are approximately 100 Peyer's patches (containing five follicles or more) at birth, and the number increases to around 300 by late adolescence[26,27]. Peyer's patches are macroscopic lymphoid aggregates found in the submucosa along the length of the small intestine and contain large B cell follicles surrounded by T cell areas and macrophages/dendritic cells[24,26,27]. These lymphoid cells are separated from the intestinal lumen by a single layer of columnar epithelial cells called the follicle-associated epithelium, and by a more diffuse area called the subepithelial dome[24]. As noted above, the follicle-associated epithelium contains specialized enterocytes termed M cells; their apical membranes are generally deprived of brush border and instead display microfolds[24,26–28], whereas the invaginated basolateral membranes form a pocket enabling partial engulfment of intraepithelial lymphocytes and macrophages[29]. M cells provide a port of entry for bacteria, other particles, macromolecules and viruses[30–33]. Endocytosed material is then transported transepithelially, avoiding digestion, and is passed to the underlying close-contact lymphoid cells. Since the dome area itself contains a wide network of dendritic cells and macrophages, Peyer's patches are considered to be an immune-inductive site where most antigen presentation occurs in the dome area. Following uptake and processing of the antigen, primed lymphocytes will exit Peyer's patches through draining lymphatics to the mesenteric lymph nodes where they reside and undergo maturation and then migrate into the blood stream through the thoracic duct, finally to home back into the mucosa[23,34]. By the same token this lymphoid site is critical for the induction of immune tolerance, ensuring that the gut mucosa does not continually respond to the myriad of luminal antigens (e.g. food proteins or commensal organisms) to which it is exposed. This is termed oral tolerance, and a key question is what processes drive such adaptive tolerance versus immune induction? T cell co-stimulation, engagement of pathogen-associated molecular patterns, down-regulation of APC co-stimulatory molecules, induction of apoptosis or activation of the inflammasome are some of the signals that help to determine whether there is induction of immune tolerance or activation. Nonetheless, such observations do not explain how the gut can differentiate invasive pathogens from commensal organisms, or why the lymphoid cells that do still express co-stimulatory molecules do not elicit immune responses to soluble antigen. We propose that constitutive 'antigen-transfection' with naturally occurring calcium phosphate particles is a major effector of immune tolerance in the gastrointestinal tract.

## DIETARY MICROPARTICLES

### Calcium phosphate: the endogenous microparticle

Both calcium and phosphate ions are actively secreted into the gastrointestinal lumen. Net secretion of calcium occurs mainly in the proximal jejunum[35], but also takes place in the ileum[36]. This is a non-saturable process where secretion of calcium occurs in a sodiumdependent paracellular fashion[36]. Although the range of luminal calcium levels varies greatly, the mean concentration of luminal calcium has been suggested[37] as 4.2 mM. Phosphate ions that are also actively secreted in the ileum co-precipitate with calcium ions to form calcium phosphate microparticles[35,36,38,39], typically of 0.5–1 μm diameter[40]. In dogs it has been shown that the precipitate comprises about one-third of the secreted calcium[35]. The luminal precipitation of calcium phosphate is both specific and surprising since luminal precipitation of other metals ions such as iron and aluminium is prevented by secretion of low-molecular-weight ligands, proteins and mucins[40,41]. Luminal calcium phosphate microparticles appear to be a form of hydroxy-apatite[40] and their occurrence and function have been considered on a number of occasions. The prevailing view is that these ions are secreted as part of the normal homeostatic mechanism for calcium and phosphate, and that co-precipitation is an inevitable consequence of saturation of the luminal contents (with respect to calcium phosphate). Since calcium phosphate avidly adsorbs organic molecules, it has further been proposed that these particles may have a role in luminal detoxification through adsorption of faecal toxins. Indeed, the binding of small quantities of the luminal components (bacterial fragments, toxins, etc.) almost certainly does occur[21,38,39,42] but, on quantitative grounds, a detoxification role seems unlikely. Instead, we have considered that calcium and phosphate ions may co-precipitate with luminal antigen forming a microparticle-sized hydroxy-apatite–antigen complex which is then scavenged by M cells of Peyer's patches and passed to underlying phagocytic lymphoid cells for processing (Figure 2). Indeed, we have recently shown that microparticles of calcium phosphate are clearly abundant in Peyer's patch lymphoid cells, morphologically typical of macrophages (unpublished observations; Figure 1). Next we considered what the consequence of these findings may be.

Numerous observations have been made suggesting that dietary calcium has a positive role in protecting against colonic carcinogenesis. This protective effect could be due to calcium precipitation with luminal fatty acids and bile acids, thus reducing their cytotoxic effects as many authors have suggested. However, luminal chemistry is complicated and events such as these cannot be assumed without *in-situ* demonstration: indeed, the high affinity of calcium ions for phosphate ions is likely to greatly reduce the free calcium that is available for binding to bile acids or fatty acids. Moreover, recent data indicate that luminal calcium also substantially modulates enteric infection and inflammation[43–45], which is not easily explained through the binding of bile acids or fatty acids. A more convincing explanation for such findings comes from the data of Penman et al. and of Hambly et al. with both studies showing that higher levels of dietary calcium increased apoptosis of the mid and distal regions of the murine colon[46,47]. Mechanisms are not clear, but previous data indicate that luminal calcium levels of the gut are proportional to dietary levels[21] and that the uptake of calcium phosphate microparticles by intestinal mononuclear cells can directly trigger cell death[48]. The pathway of cell death has not been further investigated, but may follow dissolution of calcium phosphate within acidic phagolysosomes ($\sim$pH 5), leakage of calcium ions into the cytoplasm and initiation of caspase-9-dependent cell death[48,49]. Certainly the role of intracellular calcium in cell death has been well discussed elsewhere[50–52]; however, given that antigen-bearing, luminal calcium phosphate particles appear to be taken up by phagocytes of intestinal lymphoid aggregates, consequent apoptotic induction may be important.

Apoptotic cells containing antigen, which are then taken up by dendritic cells, offer an effective means of inducing tolerance. While investigating the effects of uptake by dendritic cells of apoptotic or necrotic cells, Sauter et al. demonstrated that phagocytosis of necrotic cells induced maturation (as shown by up-regulation of CD83, CD40 and CD86), whereas phagocytosis of apoptotic cells did not[53]. Furthermore, Sauter et al. also showed that cross-presentation of antigenic peptides derived from necrotic cells resulted in an active immune response, while cross-presentation of those derived from apoptotic cells led to tolerance. A number of other investigators have demonstrated similar findings with presentation of antigen from apoptotic cells leading to tolerance[54–56].

The state of the dendritic cell (i.e. immature versus mature) is an important factor in directing the type of response towards an antigen (i.e. immunity versus tolerance). To confirm this, Mahnke et al. recently found that targeting immature dendritic cells with antigen resulted in antigen presentation in a tolerizing fashion[57], whereas addition of an 'activating factor' to induce dendritic cell maturation led to activation of T cells. This tolerizing effect seems to be largely driven through the generation of regulatory T cells ($T_{reg}$ cells). Interestingly, innate immune recognition, by Toll-like receptors (TLR) for example, can influence the activation of adaptive, regulatory immune responses through activation of co-stimulatory molecules on dendritic cells, alterations in the cytokine milieu, or changes in $T_{reg}$ survival and proliferation. Depending on the environment, such signalling can either enhance or suppress the activity of $T_{reg}$ cells[58,59]. Whether the NOD family of proteins, which are intracellular receptors for certain bacterial products and one of which is mutated in some patients with CD (NOD2), can similarly influence adaptive regulatory responses remains to be determined.

Adaptive $T_{reg}$ cells can be induced from mature T cells under certain conditions of antigenic stimulation. They express variable levels of CD25 and, unlike natural $T_{reg}$ cells, exert their suppressive functions in a cytokine-dependent manner. The first subcategory of inducible $T_{reg}$ cells is Tr1. These cells are characterized by their ability to produce IL-10 and sometimes TGF-β, and can be obtained *in vitro* by stimulating naive $CD4^+$ T cells with IL-10[60]. This T cell subset may be important for gastrointestinal immune regulation and the prevention of colitis *in vivo*[61]. The second population of inducible $T_{reg}$ cells is Th3. These cells are preferentially generated following oral antigen and are triggered in an antigen-specific manner. Nevertheless they are able to mediate bystander suppression upon encounter of antigen in the target organ. Unlike Tr1, they only produce TGF-β and provide help for IgA[62,63]. Peyer's patches appear to be the major site of induction of $T_{reg}$ cells with respect to oral tolerance and whether, as we propose (Figure 2), this is mediated through calcium phosphate-induced 'antigen transfection' and apoptosis remains to be determined. For now, this hypothesis provides a framework for the thesis that non-calcium-bearing microparticles could similarly promote 'antigen transfection' but without delivering the requisite signals for tolerance.

## Exogenous dietary microparticles

Studies on air pollution particles (i.e. $PM_{10}/PM_{2.5}$, meaning particulate matter of less than 10 μm diameter or less than 2.5 μm diameter respectively) may help to gain a better understanding of how microparticles may be involved in the aetio-pathogenesis of CD. Indeed, PM are associated with an increased rate of mortality and morbidity[64] and are increasingly linked to cardiovascular and respiratory diseases[65,66]. PM have been shown to induce the release of proinflammatory cytokines (e.g. IL-6, IL-8 and TNF-α)[67-69], to up-regulate the expression of adhesion molecules[70], to generate oxidative stress[70-72] and induce inflammatory-cell infiltration[69,70]. The proinflammatory effects associated with the fine (i.e. microparticles) or coarse fraction of particulate matter may be related to chemical and biological contaminants[69], which is analogous to the situation described here for luminal antigen–microparticle interactions.

Exogenous dietary microparticles, found as natural dietary contaminants (dust and soil) and as man-made food additives or pharmaceutical excipients, are bacterial-sized inorganic microparticles. In the main they comprise titanium dioxide and aluminosilicates although, interestingly, chromium-containing particles have also been reported in human intestinal tissue[73]. Titanium

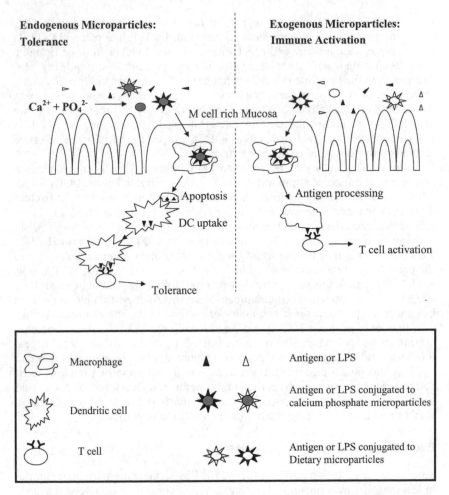

**Figure 2** Schematic diagram of proposed endogenous and exogenous microparticle uptake in Peyer's patches and immunological outcomes.

*Endogenous microparticles: tolerance.* Calcium and phosphate ions are secreted by the intestinal mucosa and co-precipitate in the intestinal lumen forming bacterial-sized 'endogenous microparticles'. Luminal antigens and/or other bacterial fragments and toxins, such as LPS, can co-precipitate with or adsorb to the calcium phosphate microparticles. The resulting conjugate is taken up through M cells overlying the Peyer's patch and passed to underlying phagocytes. Following uptake, antigen-associated calcium phosphate microparticles induce apoptosis in macrophages. Clearance of these apoptotic macrophages by dendritic cells (DC) results in the cross-presentation of antigen to T cells leading to tolerance rather than an active immune response.

*Exogenous microparticles: immune activation.* Following the ingestion of exogenous dietary microparticles (e.g. aluminosilicates or titanium dioxide), they may adsorb luminal antigens and/or other bacterial fragments and toxins, such as LPS. The resulting conjugate is taken up through M cells overlying the Peyer's patch and scavenged by underlying phagocytes. Antigen-conjugated dietary microparticles are then processed and the antigen presented to T cells, potentially resulting in T cell activation. Unrestrained responsiveness, presumably driven by genetic susceptibility, could lead to inflammation dioxide (E171), used as an intense whitening and brightening agent, is mainly found in confectionery, white sauces/dressings, non-dairy creamers or as an opacity agent in the pharmaceutical industry[21,22]. On the other hand, silicates and aluminosilicates (E554, E556 and E559) are used in the food industry to prevent poor flow or caking, and are present in cheese, sugar, powdered milk and cream powder substitute[21]. Although the actual content of these additives is not always clearly stated, it has been estimated that, in the UK, $10^{12}$–$10^{13}$ microparticles of titanium dioxide are ingested per person/day[21].

## Intestinal pigment cells and particle types

Like other particles or macromolecules, dietary microparticles, which are very resistant to degradation, are likely to be taken up via the M cell-rich mucosa overlying Peyer's patches and presented to underlying macrophages[30]. Shepherd et al. previously described the presence of macrophage-like cells located at the base of Peyer's patches, which contain dark granules or pigments[74]. These cells have consequently been named pigment cells, and ensuing analysis revealed that the pigment contained titanium, silicon and aluminium. In the mid-1990s our laboratory undertook a similar investigation[75], and confirmed the findings of Shepherd et al. However, we also demonstrated that each micron-sized granule was in fact a phagolysosome, itself containing many particles of sub-micron size ($<0.1$–$0.7$ μm). Furthermore, we demonstrated that these pigment cells contained three distinct types of microparticle: *type I* – spheres of titanium dioxide, 100–200 nm diameter, characterized as the synthetic food-additive polymorph anatase; *type II* – aluminosilicates, $<100$–400 nm in length, often with adsorbed surface iron and mostly characteristic of the natural clay mineral kaolinite; and finally *type III* – mixed environmental silicates without aluminium, 100–700 nm in length and of variable morphology.

Interestingly, these inorganic microparticles have been linked to podoconiosis, or non-filarial elephantiasis, with clinical features that include swelling and deformity of the legs due to occlusion of the lymphatic system[76–78]. Furthermore, progression of early disease can be prevented by appropriate footwear preventing uptake of soil-derived microparticles. This prompted our laboratory to consider a similar 'particle withdrawal' trial in CD (see below).

## Microparticles and CD

As noted above, the earliest lesions of CD occur at sites of intestinal lymphoid aggregates (i.e. Peyer's patches) where microparticles are taken up, phagocytosed and retained. In addition, the increased use of additives such as titanium dioxide in the Western diet has a striking relationship with the alarming rise in the incidence of CD[79]. Based on these observations, and the knowledge that this group of microparticles can induce latent, granulomatous inflammation in susceptible individuals[80], we have proposed that these microparticles could be involved in the pathogenesis of CD. However, since there are apparently equal numbers of pigment cells in healthy and disease intestine, it seems unlikely that these microparticles *per se* could be responsible for causing pathology without an underlying susceptibility[74,75].

Another characteristic of these microparticles, especially the man-made food additives (e.g. titanium dioxide), is their large negatively charged surfaces. As a result these surfaces are likely to become modified by adsorption of biomolecules[81], which are present in large numbers in the gastrointestinal tract. Bacterial LPS is abundant in the gut lumen and is a potent activator of monocytes. Once recruited into the intestinal mucosa, monocytes undergo phenotypic and functional changes during maturation to become resident macrophages. Although the latter are LPS-resistant, tolerance may be overcome when LPS is presented in a particulate form. Work within our laboratory therefore focused on the effects of microparticle–toxin conjugates on intestinal phagocyte function[82]. We have investigated the effect of titanium dioxide conjugated to LPS and found that this complex has marked proinflammatory properties, as demonstrated by the release of IL-1β, not only from PBMNC but also from LPMNC; and secondly that LPS was more potent in a particulate form than in a soluble form. Other investigators demonstrated similar findings, showing that even low levels of bacterial endotoxin in the environment influenced the production of cytokines by alveolar macrophages exposed to air particulate pollution[83,84]. These and many associated studies have underlined the importance of physical form in determining cellular responses to exogenous molecules. Along these lines, but potentially even more important, is the processing of particulate antigen which has been shown to be qualitatively and quantitatively different from that of soluble antigen.

## Processing of particulate antigen

Particulate antigen can be processed and presented by macrophages and B cells (although macrophages process a greater size range of particulate antigens compared to B cells), and very probably dendritic cells, and can stimulate class II-restricted T cells more efficiently than soluble antigen[85]. Cytochalasin D, an inhibitor of phagocytosis, prevents the processing of particulate antigen[86,87], indicating that phagocytic internalization and phagolysosomal processing are essential for presentation of particulate antigen, although small conjugates (i.e. ultrafine particles of <60–100 nm) seem less likely to be dependent on actin-driven phagocytosis and are probably taken up by constitutive pinocytosis[86]. Unlike soluble antigen, particulate antigen may gain entry to an alternative class I MHC pathway, especially in macrophages[86–88]. During particulate antigen solubilization, leakage of antigen from the phagosome may occur, resulting in classical class I pathway antigen-processing[87] or, on the other hand, the conjugation of class I molecules following antigen processing may take place intravesicularly or at the cell surface during regurgitation[86,88]. This will, either way, result in the presentation of particulate antigen in the context of MHC class I as well as class II. It is possible, however, that such cross-presentation of exogenous antigen occurs only as a result of 'particle gorging' by phagocytes in a fashion that is unlikely to occur under normal physiological and ambient conditions. More importantly, in repeated studies, particulate antigen appears to markedly promote T cell proliferation compared to soluble antigen regardless of whether these T cells are CD4+ (i.e. MHC class II-stimulated) or CD4+ and CD8+ (i.e. MHC class II- and class I-stimulated).

The effects of particulate antigen have also been investigated *in vivo*, and studies tend to show that oral or even systemic delivery of antigens entrapped into microparticles increases the immunogenicity of the antigen alone, either by increasing the production of specific immunoglobulins and/or by triggering a cell-mediated immune response[89–92]. Although immunization with particle bound-antigen results in a mixed cytokine population (both Th1 and Th2), the presence of certain co-factors such as IL-12 is able to shift the immune response towards a Th1 response[93].

Overall these findings clearly show that the influence of the physical form of antigen might be of great importance in dictating cellular uptake and processing in the gastrointestinal tract, and that the immunological state of the host (or inflamed versus non-inflamed) might direct the immune response. The implications are obvious: non-calcium-bearing, exogenous (i.e. dietary) particles may bind luminal antigens and be taken up into Peyer's patch macrophages. In the absence of a calcium signal the macrophage will not undergo apoptosis but, instead, will present the antigen in a potentially immune-activating fashion (Figure 1). If this mechanism proceeds unchecked (i.e in susceptible individuals) then inflammation could ensue.

## A final note: aetiology or pathogenesis?

Following the success of Ernest Price in preventing the advancement of podoconiosis by negating further microparticle exposure we established studies to address the effect of dietary microparticle withdrawal in active CD. In spite of a successful pilot study we have been unable to reproduce the effects in a much larger and properly powered multi-centre trial[94], leading us to believe that, once the disease is active, the removal of microparticles is ineffective; the mucosal barrier is breeched and uptake of luminal bacteria is probably sufficient to maintain disease activity. However, such findings should not be confused with the potential role of microparticles in the initiation or triggering of disease. If further laboratory studies can identify a dysfunctional response in microparticle handling in cells from subjects with CD then the final trial would be one of long-term withdrawal of exogenous microparticles in subjects with quiescent disease.

## Acknowledgement

We would like to thank Mr Vinay Thoree for the provision of micrographs and for his kind help during the preparation of this manuscript. We also give special thanks to the Sir Halley Stewart Trust for funding the studentship of L.P. and V.T., along side the MRC, for supporting much of the work described within.

## References

1.  Fiocchi C. Inflammatory bowel disease: etiology and pathogenesis. Gastroenterology. 1998;115:182–205.
2.  Hugot JP, Laurent-Puig P, Gower-Rousseau C et al. Mapping of a susceptibility locus for Crohn's disease on chromosome 16. Nature. 1996;379:821–3.

3.  Hampe J, Cuthbert A, Croucher PJ et al. Association between insertion mutation in NOD2 gene and Crohn's disease in German and British populations. Lancet. 2001;357:1925–8.
4.  Hugot JP, Chamaillard M, Zouali H et al. Association of NOD2 leucine-rich repeat variants with susceptibility to Crohn's disease. Nature. 2001;411:599–603.
5.  Ogura Y, Bonen DK, Inohara N et al. A frameshift mutation in NOD2 associated with susceptibility to Crohn's disease. Nature. 2001;411:603–6.
6.  Satsangi J, Welsh KI, Bunce M et al. Contribution of genes of the major histocompatibility complex to susceptibility and disease phenotype in inflammatory bowel disease. Lancet. 1996;347:1212–17.
7.  Yang H, Ohmen JD, Ma Y et al. Additional evidence of linkage between Crohn's disease and a putative locus on chromosome 12. Genet Med. 1999;1:194–8.
8.  Hampe J, Shaw SH, Saiz R et al. Linkage of inflammatory bowel disease to human chromosome 6p. Am J Hum Genet. 1999;65:1647–55.
9.  Rioux JD, Daly MJ, Silverberg MS et al. Genetic variation in the 5q31 cytokine gene cluster confers susceptibility to Crohn disease. Nat Genet. 2001;29:223–8.
10. Peltekova VD, Wintle RF, Rubin LA et al. Functional variants of OCTN cation transporter genes are associated with Crohn disease. Nat Genet. 2004;36:471–5.
11. Rayment N, Mylonaki M, Hudspith B, Brostoff J, Rampton DS. Co-localisation of *Escherichia coli* with macrophages in lamina propria in patients with active inflammatory bowel disease (IBD). Gut. 2003;52 (abstract).
12. Hart AL, Stagg AJ, Frame M et al.. The role of the gut flora in health and disease, and its modification as therapy. Aliment Pharmacol Ther. 2002;16:1383–93.
13. Caradonna L, Amati L, Magrone T, Pellegrino NM, Jirillo E, Caccavo D. Enteric bacteria, lipopolysaccharides and related cytokines in inflammatory bowel disease: biological and clinical significance. J Endotoxin Res. 2000;6:205–14.
14. Merger M, Croitoru K. Infections in the immunopathogenesis of chronic inflammatory bowel disease. Semin Immunol. 1998;10:69–78.
15. Swidsinski A, Ladhoff A, Pernthaler A et al. Mucosal flora in inflammatory bowel disease. Gastroenterology. 2002;122:44–54.
16. Rutgeerts P, Goboes K, Peeters M et al. Effect of faecal stream diversion on recurrence of Crohn's disease in the neoterminal ileum. Lancet. 1991;338:771–4.
17. Duchmann R, Kaiser I, Hermann E, Mayet W, Ewe K, Meyer zum Buschenfelde KH. Tolerance exists towards resident intestinal flora but is broken in active inflammatory bowel disease (IBD). Clin Exp Immunol. 1995;102:448–55.
18. Cosnes J, Carbonnel F, Carrat F, Beaugerie L, Cattan S, Gendre J. Effects of current and former cigarette smoking on the clinical course of Crohn's disease. Aliment Pharmacol Ther. 1999;13:1403–11.
19. Bridger S, Lee JC, Bjarnason I, Jones JE, Macpherson AJ. In siblings with similar genetic susceptibility for inflammatory bowel disease, smokers tend to develop Crohn's disease and non-smokers develop ulcerative colitis. Gut. 2002;51:21–5.
20. Hiller FC, Anderson PJ, Mazumder MK. Deposition of sidestream cigarette smoke in the human respiratory tract. II. Deposition of ultrafine smoke particles. Toxicol Lett. 1987;35:95–9.
21. Lomer MC, Thompson RP, Powell JJ. Fine and ultrafine particles of the diet: influence on the mucosal immune response and association with Crohn's disease. Proc Nutr Soc. 2002;61:123–30.
22. Lomer MC, Thompson RP, Commisso J, Keen CL, Powell JJ. Determination of titanium dioxide in foods using inductively coupled plasma optical emission spectrometry. Analyst. 2000;125:2339–43.
23. Mowat AM, Viney JL. The anatomical basis of intestinal immunity. Immunol Rev. 1997;156:145–66.
24. Mowat AM. Anatomical basis of tolerance and immunity to intestinal antigens. Nat Rev Immunol. 2003;3:331–41.
25. Jang MH, Kweon MN, Iwatani K et al. Intestinal villous M cells: an antigen entry site in the mucosal epithelium. Proc Natl Acad Sci USA. 2004;101:6110–15.
26. MacDonald TT, Spencer J. Gut-associated lymphoid tissue. Handbook of Mucosal Immunology 1994. Academic Press; 415–24.
27. MacDonald TT. The mucosal immune system. Parasite Immunol. 2003;25:235–46.

28. Neutra MR, Mantis NJ, Kraehenbuhl JP. Collaboration of epithelial cells with organized mucosal lymphoid tissues. Nat Immunol. 2001;2:1004–9.

29. Gebert A, Rothkotter HJ, Pabst R. M cells in Peyer's patches of the intestine. Int Rev Cytol. 1996;167:91–159.

30. Owen RL. Uptake and transport of intestinal macromolecules and microorganisms by M cells in Peyer's patches – a personal and historical perspective. Semin Immunol. 1999;11:157–63.

31. Pappo J, Ermak TH. Uptake and translocation of fluorescent latex particles by rabbit Peyer's patch follicle epithelium: a quantitative model for M cell uptake. Clin Exp Immunol. 1989;76:144–8.

32. Sass W, Dreyer HP, Seifert J. Rapid insorption of small particles in the gut. Am J Gastroenterol. 1990;85:255–60.

33. Smith MW, Thomas NW, Jenkins PG, Miller NG, Cremaschi D, Porta C. Selective transport of microparticles across Peyer's patch follicle-associated M cells from mice and rats. Exp Physiol. 1995;80:735–43.

34. MacDermott RP. The immune system in inflammatory bowel disease. Mt Sinai J Med. 1990;57:273–8.

35. Schedl HP, Osbaldiston GW, Mills IH. Absorption, secretion, and precipitation of calcium in the small intestine of the dog. Am J Physiol. 1968;214:814–19.

36. Favus MJ. Factors that influence absorption and secretion of calcium in the small intestine and colon. Am J Physiol. 1985;248:G147–57.

37. Lentner C. 1981. Geigy Scientific Tables, Volume 1. Units of Measurements, Body Fluids, Composition of the Body Nutrition, 8th edn, (Ciba-Geigy Ltd, Basle)

38. Lapre JA, De Vries HT, Koeman JH, Van der MR. The antiproliferative effect of dietary calcium on colonic epithelium is mediated by luminal surfactants and dependent on the type of dietary fat. Cancer Res. 1993;53:784–9.

39. Van der MR, Welberg JW, Kuipers F et al. Effects of supplemental dietary calcium on the intestinal association of calcium, phosphate, and bile acids. Gastroenterology. 1990;99:1653–9.

40. Powell JJ, Whitehead MW, Ainley CC, Kendall MD, Nicholson JK, Thompson RP. Dietary minerals in the gastrointestinal tract: hydroxypolymerisation of aluminium is regulated by luminal mucins. J Inorg Biochem. 1999;75:167–80.

41. Powell JJ, Jugdaohsingh R, Thompson RP. The regulation of mineral absorption in the gastrointestinal tract. Proc Nutr Soc. 1999;58:147–53.

42. Govers MJ, Termont DS, Van Aken GA, Van der MR. Characterization of the adsorption of conjugated and unconjugated bile acids to insoluble, amorphous calcium phosphate. J Lipid Res. 1994;35:741–8.

43. Bovee-Oudenhoven IM, Termont DS, Weerkamp AH, Faassen-Peters MA, Van der MR. Dietary calcium inhibits the intestinal colonization and translocation of salmonella in rats. Gastroenterology. 1997;113:550–7.

44. Bovee-Oudenhoven IM, Wissink ML, Wouters JT, Van der MR. Dietary calcium phosphate stimulates intestinal lactobacilli and decreases the severity of a salmonella infection in rats. J Nutr. 1999;129:607–12.

45. Bovee-Oudenhoven IM, Lettink-Wissink ML, Van Doesburg W, Witteman BJ, Van der MR. Diarrhea caused by enterotoxigenic *Escherichia coli* infection of humans is inhibited by dietary calcium. Gastroenterolog. 2003;125:469–76.

46. Hambly RJ, Saunders M, Rijken PJ, Rowland IR. Influence of dietary components associated with high or low risk of colon cancer on apoptosis in the rat colon. Food Chem Toxicol. 2002;40:801–8.

47. Penman ID, Liang QL, Bode J, Eastwood MA, Arends MJ. Dietary calcium supplementation increases apoptosis in the distal murine colonic epithelium. J Clin Pathol. 2000;53:302–7.

48. Evans SM, Ashwood P, Warley A, Berisha F, Thompson RP, Powell JJ. The role of dietary microparticles and calcium in apoptosis and interleukin-1beta release of intestinal macrophages. Gastroenterology. 2002;123:1543–53.

49. Tantral L, Malathi K, Kohyama S, Silane M, Berenstein A, Jayaraman T. Intracellular calcium release is required for caspase-3 and -9 activation. Cell Biochem Funct. 2004;22:35–40.

50. Boehning D, Patterson RL, Snyder SH. Apoptosis and calcium: new roles for cytochrome c and inositol 1,4,5-trisphosphate. Cell Cycle. 2004;3:252–4.
51. Orrenius S, Zhivotovsky B, Nicotera P. Regulation of cell death: the calcium–apoptosis link. Nat Rev Mol Cell Biol. 2003;4:552–65.
52. Yu SP, Canzoniero LM, Choi DW. Ion homeostasis and apoptosis. Curr Opin Cell Biol. 2001;13:405–11.
53. Sauter B, Albert ML, Francisco L, Larsson M, Somersan S, Bhardwaj N. Consequences of cell death: exposure to necrotic tumor cells, but not primary tissue cells or apoptotic cells, induces the maturation of immunostimulatory dendritic cells. J Exp Med. 2000;191:423–34.
54. Chernysheva AD, Kirou KA, Crow MK. T cell proliferation induced by autologous non-T cells is a response to apoptotic cells processed by dendritic cells. J Immunol. 2002;169:1241–50.
55. Ferguson TA, Herndon J, Elzey B, Griffith TS, Schoenberger S, Green DR. Uptake of apoptotic antigen-coupled cells by lymphoid dendritic cells and cross-priming of CD8(+) T cells produce active immune unresponsiveness. J Immunol. 2002;168:5589–95.
56. Liu K, Iyoda T, Saternus M, Kimura Y, Inaba K, Steinman RM. Immune tolerance after delivery of dying cells to dendritic cells in situ. J Exp Med. 2002;196:1091–7.
57. Mahnke K, Qian Y, Knop J, Enk AH. Induction of CD4+/CD25+ regulatory T cells by targeting of antigens to immature dendritic cells. Blood. 2003;101:4862–9.
58. Caramalho I, Lopes-Carvalho T, Ostler D, Zelenay S, Haury M, Demengeot J. Regulatory T cells selectively express toll-like receptors and are activated by lipopolysaccharide. J Exp Med. 2003;197:403–11.
59. Pasare C, Medzhitov R. Toll pathway-dependent blockade of CD4+CD25+ T cell-mediated suppression by dendritic cells. Scienc. 2003;299:1033–6.
60. Roncarolo MG, Bacchetta R, Bordignon C, Narula S, Levings MK. Type 1 T regulatory cells. Immunol Rev. 2001;182:68–79.
61. Groux H, O'Garra A, Bigler M et al. A CD4+ T-cell subset inhibits antigen-specific T-cell responses and prevents colitis. Nature. 1997;389:737–42.
62. Levings MK, Sangregorio R, Sartirana C et al. Human CD25+CD4+ T suppressor cell clones produce transforming growth factor beta, but not interleukin 10, and are distinct from type 1 T regulatory cells. J Exp Med. 2002;196:1335–46.
63. Weiner HL. Induction and mechanism of action of transforming growth factor-beta-secreting Th3 regulatory cells. Immunol Rev. 2001;182:207–14.
64. Samet JM, Dominici F, Curriero FC, Coursac I, Zeger SL. Fine particulate air pollution and mortality in 20 US cities, 1987–1994. N Engl J Med. 2000;343:1742–9.
65. Nemmar A, Hoylaerts MF, Hoet PH, Nemery B. Possible mechanisms of the cardiovascular effects of inhaled particles: systemic translocation and prothrombotic effects. Toxicol Lett. 2004;149:243–53.
66. Wong GW, Lai CK. Outdoor air pollution and asthma. Curr Opin Pulm Med. 2004;10:62–6.
67. Hetland RB, Cassee FR, Refsnes M et al. Release of inflammatory cytokines, cell toxicity and apoptosis in epithelial lung cells after exposure to ambient air particles of different size fractions. Toxicol In Vitro. 2004;18:203–12.
68. Monn C, Becker S. Cytotoxicity and induction of proinflammatory cytokines from human monocytes exposed to fine (PM2.5) and coarse particles (PM10-2.5) in outdoor and indoor air. Toxicol Appl Pharmacol. 1999;155:245–52.
69. Schins RP, Lightbody JH, Borm PJ, Shi T, Donaldson K, Stone V. Inflammatory effects of coarse and fine particulate matter in relation to chemical and biological constituents. Toxicol Appl Pharmacol. 2004;195:1–11.
70. Salvi S, Blomberg A, Rudell B, Kelly F, Sandstrom T, Holgate ST, Frew A. Acute inflammatory responses in the airways and peripheral blood after short-term exposure to diesel exhaust in healthy human volunteers. Am J Respir Crit Care Med. 1999;159:702–9.
71. Dick CA, Brown DM, Donaldson K, Stone V. The role of free radicals in the toxic and inflammatory effects of four different ultrafine particle types. Inhal Toxicol. 2003;15:39–52.
72. Wilson MR, Lightbody JH, Donaldson K, Sales J, Stone V. Interactions between ultrafine particles and transition metals in vivo and in vitro. Toxicol Appl Pharmacol. 2002;184:172–9.

73. Mazumder R.N, Bode J, Lewin J McIntyre MM, Ghosh S. Detection of chromium microparticles in inflammatory bowel disease (IBD) tissues by energy dispersive analysis of X-rays. Gut. 2002;50 (abstract).

74. Shepherd NA, Crocker PR, Smith AP, Levison DA. Exogenous pigment in Peyer's patches. Hum Pathol. 1987;18:50–4.

75. Powell JJ, Ainley CC, Harvey RS et al. Characterisation of inorganic microparticles in pigment cells of human gut associated lymphoid tissue. Gut. 1996;38:390–5.

76. Evans SM, Powell JJ, Thompson RPH. Podoconiosis. In: DA Warrell, TM Cox, JD Firth, ED Benz Jr, editors. Oxford Textbook of Medicine, 4th edn. Oxford: Oxford University Press, 2003:989–91.

77. Harvey RSJ, Powell JJ, Thompson RPH. A review of the geochemical factors linked to podoconiosis. Environ Geochem Health. 1996;113:255–60.

78. Price E.W. Podoconiosis: Non-filarial Elephantiasis. Oxford: Oxford University Press, 1990.

79. Mahmud N, Weir DG. The urban diet and Crohn's disease: is there a relationship? Eur J Gastroenterol Hepatol. 2001;13:93–5.

80. Roge J, Fabre M, Levillain P, Dubois P. Unusual particles and crystals in Crohn's disease granulomas. Lancet. 1991;337:502–3.

81. Ellingsen JE. A study on the mechanism of protein adsorption to $TiO_2$. Biomaterials. 1991;12:593–6.

82. Powell JJ, Harvey RS, Ashwood P, Wolstencroft R, Gershwin ME, Thompson RP. Immune potentiation of ultrafine dietary particles in normal subjects and patients with inflammatory bowel disease. J Autoimmun. 2000;14:99–105.

83. Becker S, Fenton MJ, Soukup JM. Involvement of microbial components and toll-like receptors 2 and 4 in cytokine responses to air pollution particles. Am J Respir Cell Mol Biol. 2002;27:611–18.

84. Ning Y, Imrich A, Goldsmith CA, Qin G, Kobzik L. Alveolar macrophage cytokine production in response to air particles *in vitro*: role of endotoxin. J Toxicol Environ Health A. 2000;59:165–80.

85. Vidard L, Kovacsovics-Bankowski M, Kraeft SK, Chen LB, Benacerraf B, Rock KL. Analysis of MHC class II presentation of particulate antigens of B lymphocytes. J Immunol. 1996;156:2809–18.

86. Harding CV, Song R. Phagocytic processing of exogenous particulate antigens by macrophages for presentation by class I MHC molecules. J Immunol. 1994;153:4925–33.

87. Kovacsovics-Bankowski M, Clark K, Benacerraf B, Rock KL. Efficient major histocompatibility complex class I presentation of exogenous antigen upon phagocytosis by macrophages. Proc Natl Acad Sci USA. 1993;90:4942–6.

88. Song R, Harding CV. Roles of proteasomes, transporter for antigen presentation (TAP), and beta 2-microglobulin in the processing of bacterial or particulate antigens via an alternate class I MHC processing pathway. J Immunol. 1996;156:4182–90.

89. Alonso MJ, Gupta RK, Min C, Siber GR, Langer R. Biodegradable microspheres as controlled-release tetanus toxoid delivery systems. Vaccine. 1994;12:299–306.

90. Delgado A, Lavelle EC, Hartshorne M, Davis SS. PLG microparticles stabilised using enteric coating polymers as oral vaccine delivery systems. Vaccine. 1999;17:2927–38.

91. Heritage PL, Loomes LM, Jianxiong J, Brook MA, Underdown BJ, McDermott MR. Novel polymer-grafted starch microparticles for mucosal delivery of vaccines. Immunology. 1996;88:162–8.

92. Maloy KJ, Donachie AM, O'Hagan DT, Mowat AM. Induction of mucosal and systemic immune responses by immunization with ovalbumin entrapped in poly(lactide-co-glycolide) microparticles. Immunology. 1994;81:661–7.

93. Sedlik C, Deriaud E, Leclerc C. Lack of Th1 or Th2 polarization of CD4+ T cell response induced by particulate antigen targeted to phagocytic cells. Int Immunol. 1997;9:91–103.

94. Lomer MC, Grainger SL, Ede R et al. Lack of efficacy of a reduced microparticle diet in a multi-centred trial of patients with active Crohn's disease. Eur J Gastroenterol Hepatol. 2004.

# 14
# Are we able to treat inflammatory bowel disease with diet?

T. NIC SUIBHNE, C. O'MORAIN and M. O'SULLIVAN

## INTRODUCTION

Environmental factors play an important role in the manifestation, course and prognosis of inflammatory bowel disease (IBD), with observations pointing to risk factors associated with the Western lifestyle. Shoda et al.[1] studied the correlation between the increasing incidence of Crohn's disease (CD) and dietary change in a homogeneous Japanese population from 1966 to 1985 and found that the increased incidence was strongly correlated with increased dietary intake of animal protein and an increased ratio of n-6 to n-3 polyunsaturated fatty acids. This, together with increasing evidence, suggests that diet may play an important role in the development and subsequent course of IBD, but is not to date proven.

The use of diet in treating IBD can be viewed in two ways: as a nutritional role as adjunct to medical therapy and as a primary therapeutic role in the form of enteral diet therapy.

## DIET IN IBD: NUTRITIONAL ROLE

### Malnutrition

Malnutrition and weight loss are common and important features associated with IBD, appearing at all stages of the disease. Undernutrition is most overt during active disease but is also common at time of diagnosis and during remission[2]. A high prevalence (40%) of malnutrition has been observed in hospitalized patients[3]. The causes and physiological manifestations of under-nutrition in both health and disease are well reported (Table 1), with weight loss, growth failure and hypoalbuminaemia more prevalent in CD than in ulcerative colitis (UC). These, together with other problems such as vitamin deficiencies, are likely to have a negative impact on disease outcome and recovery. Malnourished patients are known to have a higher risk of post-operative complications, longer hospital stays, increased risk of new infections and, furthermore, the long-term mortality of malnourished patients is much

**Table 1** Causes and manifestations of undernutrition in IBD

| Causes of undernutrition in IBD | Possible adverse effects |
| --- | --- |
| Poor dietary intake | Weight loss |
| Poor appetite | Hypoalbuminaemia |
| Increased nutrient requirements | Anaemia |
| Reduced nutrient absorption | Negative nitrogen balance |
| Nutrient loss through GI tract | Vitamin D deficiency |
| Drug-nutrient interactions | Osteoporosis |
| Intestinal obstruction | Impaired muscle, respiratory and immune function |
|  | Impaired recovery from illness and longer hospital stay |
|  | Impaired growth in children |
|  | Psychological effects on mood and well-being |

higher than that of comparable well-nourished patients. Systematic nutritional screening, intervention and monitoring of patients should ideally be an integrated component of a multi-disciplinary management approach for IBD. Treating malnutrition early, especially in those patients who have chronic IBD but with clinically inactive or mildly active disease, could, in turn, reduce other health events such as osteoporosis, anaemia, infections, delayed recovery from illness/surgery. Targeting nutritionally at-risk groups of IBD (such as those at increased risk of osteoporosis and children) patients for appropriate intervention would improve their nutritional status and may, accordingly, help optimize immune and clinical responses.

## Osteoporosis

Patients with IBD have an increased fracture risk and are at increased risk of developing osteoporosis. The overall relative risk of fractures is 40% greater in IBD than that of the general population[4]. The prevalence of osteoporosis in IBD is 15%[4], which is due to a number of factors including corticosteroid use, increasing age, malabsorption, malnutrition and systemic inflammation. In response to this increased risk, the British Society of Gastroenterology has published detailed guidelines[5] for the prevention and treatment of osteoporosis in IBD. The guidelines are primarily aimed at maintaining optimal bone mass and ensuring that patients receive optimal nutritional bone-sparing therapies and advice. All IBD patients should consume an overall daily intake of 1500 mg of calcium, either in the diet, or by adding calcium tablets if necessary. Vitamin D levels should be monitored and treated if deficiency occurs. Patients are advised to increase weight-bearing exercise, avoid smoking and alcohol excess and to reduce use of corticosteroids. If and when steroids are prescribed, patients should be given the lowest dose for as short a time as possible, with vitamin D (800 units) prescribed concurrently. Bone mineral density should be measured and repeated in each year in which steroids are prescribed, and if the *T*-score is less than 1.5, bisphosphonates should be offered in addition to vitamin D. The American Gastroenterology Association provide similar advice; ensure that patients receive adequate calcium and be especially vigilant when dealing with at-risk subgroups[4].

## Children

The problem of undernutrition is particularly important in children and adolescents as linear growth and development can be impaired. Preliminary data from a 12-year prospective study of paediatric IBD in northern France found that weight loss was present in 70% of children with CD and 34% of those with UC[6]. Nutritional deficiencies are also commonly seen in children with IBD, with iron deficiency (70%), hypoalbuminaemia (60%), folic acid deficiency (60%), and vitamin D deficiency (50%) the most common[7]. Minor deficiencies in copper, magnesium, zinc, vitamin $B_{12}$ and vitamin A have also been reported[7]. A study by Sentongo et al.[8] found that growth in males was more impaired than in female children despite similarities in disease activity, duration and lifetime steroid exposure. Gender, therefore, may confer risk for impaired growth in CD in childhood and adolescence. The conventional therapy for IBD, namely corticosteroids, may further compromise nutritional status and linear growth in this high-risk group. Maintaining good nutritional status, limiting the use of corticosteroids and controlling the inflammatory response are sensible strategies for promoting optimal growth and development in children with IBD. This provides strong grounds for the use of enteral nutrition as primary therapy in paediatric CD, as enteral diet therapy has been shown to improve linear growth[9] in children with CD, as well as inducing and maintaining remission. Wilschanski et al.[9] observed that patients who continued nasogastric supplementary feeding after resumption of an otherwise normal diet remained well longer than those who discontinued nocturnal supplements completely and, furthermore, continued use of nasogastric supplements before completion of puberty was associated with improved linear growth.

## DIET IN IBD: PRIMARY THERAPEUTIC ROLE

Nutritional intervention (in the form of elemental diet) was first shown to have a primary therapeutic effect in CD, but not in UC, in the early 1970s. In 1984 the first controlled study confirmed that an elemental diet was as effective as steroids in inducing remission in acute CD[10] (for reviews see refs 11–13).

Subsequent studies supported this primary therapeutic effect for both elemental and polymeric enteral diets, although published meta-analyses in the 1990s concluded that, while enteral diet therapy was more effective than placebo at maintaining remission, corticosteroids were more effective than enteral diet therapy[14,15]. In contrast, Teahon et al.[16], in a single centre, reported similar remission rates for diet therapy and steroids during their 10-year experience with enteral diets. Although conflicting results exist, enteral diet therapy should be considered as primary therapy in Crohn's patients, especially in certain subgroups including paediatric or pregnant patients, those with strictures due to oedema or those who are intolerant to steroids and/or other medical treatments.

## Elemental diet

Elemental diets used as sole source of nutrition have been shown to induce remission rates similar to that of steroids[10]. Elemental diets provide nutrients in their simplest form – protein as free amino acids, carbohydrate as glucose or short-chain maltodextrins and fat as short-chain triglycerides – and are commercially available in a powder format requiring reconstitution, or as a ready-to-use liquid feed. The diet is administered either orally or by nasogastric feeding and the volume consumed is calculated based on the nutritional requirements of the individual patient. The diet is used as sole source of nutrition for 2 weeks, although if clinical improvement is seen, continuance of the diet for a period of 3 weeks is encouraged. Patients who fail to show an improvement in disease parameters within 7–10 days may be assigned to alternative therapy.

The most common problems encountered with elemental diets include intolerance, poor compliance and unpalatability of the diet. Adherence is critical to the success, as with any therapeutic approach poor patient cooperation usually produces poor outcome. Practical strategies can be used for most of the problems encountered, such as use of flavouring for oral administration, nasogastric feeding if oral administration proves difficult along with careful monitoring and encouragement. Although the total replacement of normal diet for 2 weeks may seem harsh to most healthy people, most patients do not see it as a problem as they frequently have a poor appetite, weight loss, nausea and are often fasting for tests. As they begin to improve, they are encouraged and motivated to continue with the regime. It has been reported that 65% of patients treated with an enteral diet would opt for the diet in the event of a further relapse[17].

The elemental diet has been considered an expensive therapy, at least compared to the standard corticosteroid treatment, but in the light of today's biologic therapies, this is probably no longer the case.

## Semi-elemental diet

Peptide-based diets, also known as oligopeptide, protein hydrolyses, oligomeric or semi-elemental diets, contain a nitrogen source of peptides of varying chain lengths, thought to be better absorbed than free amino acids. These diets, in theory, offer an advantage over elemental diets in terms of lower osmolarity, better palatability, less expense and ease of use. Complications should therefore be less frequent than with elemental diets, although in practice similar rates of withdrawal and intolerance have been noted for both peptide and elemental diets[18].

In the largest controlled trial[19] only 53% of the oligopeptide diet group achieved remission compared to 79% of the steroid group, and a meta-analysis suggested that oligopeptide diets may be therapeutically inferior to elemental and polymeric diets[14].

## Polymeric diet

The standard routine whole protein enteral feeds are administered orally or by nasogastric feeding, and have been shown to result in remission rates comparable to steroids and elemental diets[20]. These feeds are the commercial standard feeds used in hospitals for the provision of routine enteral nutritional support. Polymeric diets are cheaper than both elemental and peptide feeds, are widely available and have reported fewer problems with tolerance and palatability.

## Side-effects

One of the main advantages of enteral feeding is that it is free from side-effects (apart from possible short-term effects) compared to other drug treatments. Steroids and other Crohn's drugs have many side-effects, both short- and long-term (including growth suppression, bone disease and hypertension), making enteral feeding an attractive alternative. The main side-effects associated with enteral feeding are nausea and headaches, but these usually resolve after the first few days. Diarrhoea can also occur but should be limited if the feed is gradually introduced during the initial 3–4 days of the feeding regime. Weight loss, abdominal cramps and vomiting can also occur but they usually resolve as the patient adapts to the diet.

## ENTERAL DIET THERAPY: MODE OF ACTION

The mechanism underlying a therapeutic response to enteral diets remains unclear. Initially it was hypothesized that the absence of whole protein in elemental feeds was responsible for their efficacy in inducing remission in CD, although it has subsequently been shown that polymeric feeds are as effective as elemental diets[20]. Several other theories have been proposed regarding the possible mode of action (Figure 1), with the most exciting of these relating to the potential anti-inflammatory effect of an enteral diet on the gastrointestinal mucosa[21].

## Mucosal inflammation

For many years it has been known that enteral diet therapy is associated with decreased intestinal permeability[16]. More recently the effects of a specific oral polymeric diet, CT3211, on the mucosal inflammatory process in active CD have been studied[21]. Twenty-nine children received an 8-week course of CT3211, which is rich in transforming growth factor $\beta_2$, as sole source of nutrition and were assessed clinically and endoscopically before and after treatment. After the 8-week treatment period 79% of patients achieved clinical remission. C-reactive protein (CRP) levels significantly declined from 19 mg/L to 13.5 mg/L with treatment while serum tumour necrosis factor $\alpha$ levels fell by a mean of 4.7 pg/ml. Weight and body mass index increased. Macroscopic healing occurred in 16 cases (eight in the colon and eight in the terminal ileum). Interleukin-1$\beta$ mRNA, interferon-$\gamma$ mRNA and interleukin-8 mRNA were

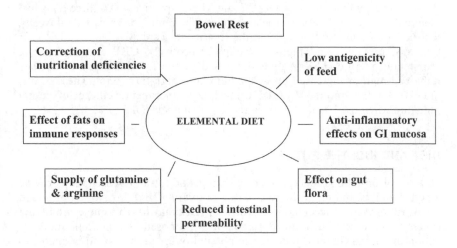

**Figure 1** Elemental diet: mode of action

elevated in patients at the start of treatment compared to controls; these levels decreased in response to treatment while ileal transforming growth factor $\beta_1$ mRNA rose with treatment. This study provides evidence of mucosal healing together with a down-regulation of the proinflammatory cytokines following treatment with enteral diet therapy. The fall in mucosal interleukin-1$\beta$ in response to treatment is consistent with the concept that enteral nutrition influences the underlying mucosal inflammatory process in CD.

In an *in-vitro* study, Meister et al.[22] studied the anti-inflammatory effects of enteral diet components on CD-affected tissues. Biopsies were taken from 39 IBD patients and controls, and incubated for 24 h with enteral diets. Incubation of Crohn's tissues in both elemental diet and casein resulted in an increase in the IL-1 receptor antagonist/IL-10$\beta$ ratio compared to controls. The authors concluded that enteral diet incubation increased the anti-inflammatory to proinflammatory cytokine ratio in CD and that the anti-inflammatory effect was not specifically due to amino acid composition, as both elemental and casein diets had similar anti-inflammatory effects.

## Role of fat

In the past the underlying mechanism of action focused on the presentation of nitrogen in the feed, but recent evidence suggests that the fat content of feeds may be more important[23]. Evidence is accumulating that additional long-chain triglyceride impairs the response rate in active CD, whereas no deleterious effects of medium-chain triglyceride have been identified[24]. In a recent study Bamba et al.[24] randomized 36 patients to receive elemental diets with very low, medium or high amounts of additional long-chain triglyceride (LCT) so that the total energy from fat was 1%, 6% and 11% respectively. All three diets had identical total calories and nitrogen sources. The remission rates after 4 weeks, for the 28 patients who completed the treatment, were 80%, 40% and 25% in the low-, medium- and high-fat groups respectively. CRP and erythrocyte sedimentation rate (ESR) values decreased in the low-fat group while they remained high or increased in the medium- and high-fat groups. These results indicate that the amount of LCT is critical to the therapeutic efficacy of enteral nutrition when used as the primary therapy in active CD.

## DIET AND IBD: THE FUTURE

New nutrition-related therapies are emerging, and future research may provide a better understanding of the role of diet in treating IBD. Probiotics, although not as yet proven or accepted as therapy for IBD, may have immune-modifying effects[25]. As alteration in gut flora has been proposed as a possible mode of action in enteral therapy, there is the possibility for probiotics to be incorporated into dietary supplements and enteral feeds. There is also evidence that n-3 polyunsaturated fatty acids (PUFA) may play a role in reducing post-surgical recurrence of Crohn's[26] and in treating paediatric UC[27], along with the evidence that the fat content of enteral feeds may be important. Short-chain fatty acids (SCFA) may also offer potential therapeutic strategies for IBD as butyrate, a SCFA derived from colonic bacterial fermentation of dietary fibre, has gained attention for its role as a key fuel source for the colonocyte and its possible ability to down-regulate inflammatory cytokines[28].

## CONCLUSION

In IBD, nutrition plays an important role in the prevention and treatment of malnutrition, in the promotion of growth and development in children and in the prevention of osteoporosis. In its role as a primary therapy, enteral diets seem to have lost their attractiveness in favour of drug treatment, molecular genetic intervention and other high-tech therapies. However, these diets offer a safe and alternative treatment to specific subgroups of patients and should not be overlooked, as they continue to be used with success as primary therapy for CD in Japan. While the debate surrounding the role of enteral nutrition as primary therapy in CD continues, it should be remembered that enteral diet therapy is free from significant side-effects, is not associated with known long-term risks and improves, rather than impairs, nutritional status.

## References

1.  Shoda R, Matsueda K, Yamato S, Umeda N. Epidemiologic analysis of Crohn disease in Japan: increased dietary intake of n-6 polyunsaturated fatty acids and animal protein relates to the increased incidence of Crohn disease in Japan. Am J Clin Nutr. 1996;63:741–5.

2. Geerling BJ, Badart-Smook A, Stockbrugger RW, Brummer RJ. Comprehensive nutritional status in patients with long-standing Crohn disease currently in remission. Am J Clin Nutr. 1998;67:919–26.
3. Pirlich M, Schutz T, Kemps M et al Prevalence of malnutrition in hospitalized medical patients: impact of underlying disease. Dig Dis. 2003;21:245–51.
4. Bernstein CN, Leslie WD, Leboff MS. American Gastroenterological Association medical position statement: guidelines on osteoporosis in gastrointestinal diseases. Gastroenterology. 2003;124:791–4.
5. Scott EM, Gaywood I, Scott BB. Guidelines for osteoporosis in coeliac disease and inflammatory bowel disease. British Society of Gastroenterology. Gut. 2000;46:i1–8.
6. Auvin S, Molinie C, Gower-Rousseau C et al. Incidence, clinical presentation and localisation of pediatric inflammatory bowel disease at diagnosis in northern France (1998–99). Gut (Suppl. II). 2003;35:A53.
7. Reif S, Klein I, Lubin F, Farbstein M, Hallak A, Gilat T. Pre-illness dietary factors in inflammatory bowel disease. Gut. 1997;40:754–60.
8. Sentongo TA, Semeao EJ, Piccoli DA, Stallings VA, Zemel BS. Growth, body composition, and nutritional status in children and adolescents with Crohn's disease. J Pediatr Gastroenterol Nutr. 2000;31:33–40.
9. Wilschanski M, Sherman P, Pencharz P, Davis L, Corey M, Griffiths A. Supplementary enteral nutrition maintains remission in paediatric Crohn's disease. Gut. 1996;38:543–8.
10. O' Morain C, Segal AW, Levi AJ. Elemental diet as primary treatment of acute Crohn's disease: a controlled trial. Br Med J. 1984;288:1859–62.
11. O'Sullivan M, O'Morain C. Liquid diets for Crohn's disease. Gut. 2001;48:757.
12. O' Sullivan M, O' Morain C. Nutritional therapy in inflammatory bowel disease. Current treatment options. Gastroenterology. 2004;7:191–8.
13. Gassull MA. Nutrition and inflammatory bowel disease: its relation to pathophysiology, outcome and therapy. Dig Dis. 2003;21:220–7.
14. Fernandez-Banares F, Cabre E, Esteve-Comas M, Gassull MA. How effective is enteral nutrition in inducing clinical remission in active Crohn's disease? A meta-analysis of the randomized clinical trials. J Parent Ent Nutr. 1995;19:356–64.
15. Messori A, Trallori G, D'Albasio G, Milla M, Vannozzi G, Pacini F. Defined-formula diets versus steroids in the treatment of active Crohn's disease: a meta-analysis. Scand J Gastroenterol. 1996;31:267–72.
16. Teahon K, Bjarnason I, Pearson M, Levi AJ. Ten years' experience with an elemental diet in the management of Crohn's disease. Gut. 1990;31:1133–7.
17. Teahon K, Pearson M, Levi AJ, Bjarnason I. Practical aspects of enteral nutrition in the management of Crohn's disease. J Parent Ent Nutr. 1995;19:365–8.
18. Mansfield JC, Giaffer MH, Holdsworth CD. Controlled trial of oligopeptide versus amino acid diet in treatment of active Crohn's disease. Gut. 1995;36:60–6.
19. Lochs H, Steinhardt HJ, Klaus-Wentz B et al. Comparison of enteral nutrition and drug treatment in active Crohn's disease. Results of the European Cooperative Crohn's Disease Study. IV. Gastroenterology. 1991;101:881–8.
20. Verma S, Brown S, Kirkwood B, Giaffer MH. Polymeric versus elemental diet as primary treatment in active Crohn's disease: a randomized, double-blind trial. Am J Gastroenterol. 2000;95:735–9.
21. Fell JM, Paintin M, Arnaud-Battandier F et al. Mucosal healing and a fall in mucosal pro-inflammatory cytokine mRNA induced by a specific oral polymeric diet in paediatric Crohn's disease. Aliment Pharm Ther. 2000;14:281–9.
22. Meister D, Bode J, Shand A, Ghosh S. Anti-inflammatory effects of enteral diet components on Crohn's disease-affected tissues *in vitro*. Dig Liver Dis. 2002;34:430–8.
23. Gassull MA, Fernandez-Banares F, Cabre E et al. Fat composition may be a clue to explain the primary therapeutic effect of enteral nutrition in Crohn's disease: results of a double blind randomised multicentre European trial. Gut. 2002;51:164–8.
24. Bamba T, Shimoyama T, Sasaki M et al. Dietary fat attenuates the benefits of an elemental diet in active Crohn's disease: a randomized, controlled trial. Eur J Gastroenterol Hepatol. 2003;15:151–7.
25. Tamboli CP, Caucheteux C, Cortot A, Colombel JF, Desreumaux P. Probiotics in inflammatory bowel disease: a critical review. Best Pract Res Clin Gastroenterol. 2003;17:805–20.

26. Belluzzi A, Brignola C, Campieri M, Pera A, Boschi S, Miglioli M. Effect of an enteric-coated fish-oil preparation on relapses in Crohn's disease. N Engl J Med. 1996;334:1557–60.
27. Shimizu T, Fujii T, Suzuki R et al. Effects of highly purified eicosapentaenoic acid on erythrocyte fatty acid composition and leukocyte and colonic mucosa leukotriene B4 production in children with ulcerative colitis. J Pediatr Gastroenterol Nutr. 2003;37:581–5.
28. Segain JP, de La Bletiere DR, Bourreille A et al. Butyrate inhibits inflammatory responses through NFkappaB inhibition: implications for Crohn's disease. Gut. 2000;47:397–403.

# Section V
# New diagnostic tools

**Chair: R.W. STOCKBRÜGGER and Z. KRASTEV**

# 15
# Novel endoscopic diagnostic developments – bioendoscopy

**G. N. J. TYTGAT**

## INTRODUCTION

A plethora of novel endoscopy methods have been developed[1-4]. An important area relates to bioendoscopy, the molecular characterization of tissue by endoscopic means. All methods have a common aim:

1. To enhance the resolution (the optical distinction between two points) and by doing so the detailed analysis of tissue microarchitecture.

2. To characterize the biochemical characteristics of tissue in order to improve neoplastic alteration.

This overview will summarize some of those novel developments. Spectroscopic analysis of tissue will not be discussed, except fluorescence spectroscopy. The ultimate objective of all novel developments is to eliminate the need for conventional biopsy; yet a more realistic goal at this stage is still largely to identify suspicious areas likely to harber neoplastic change for targeted biopsy.

## HIGH-RESOLUTION–HIGH-MAGNIFICATION ENDOSCOPY

High-resolution endoscopes provide more detailed images of the intestinal lining. The image resolution is further enhanced by optical/electronic magnification at the expense of reducing the surface area that is visualized[5-7]. Super-CCD chip technology further advances the detection of minute structural detail through increase of the light-receiving photo diode area[8].

High-resolution–high-magnification is often combined with chromoscopy to allow detailed inspection of villous pattern and especially pit patterns in the colon. The latter application is rapidly expanding worldwide to facilitate the detection of flat (adenomatous) neoplastic change[9].

## NARROW-BAND IMAGING

Narrow-band imaging has recently been developed to improve the quality of endoscopic images and to enhance the microvasculature and mucosal villous or pit patterns. The sequential lighting method in video endoscopes has a rotation disc with RGB (red, green, blue) filters in front of a white light source (xenon lamp). Narrow-band imaging (NBI) is a novel technique that changes these optical filters to spectral narrow-band filters. The NBI system enables one to narrow the bandwith of the spectral transmittance of the optical filters, used in the light source. The penetration depth of light is dependent on its wavelength. Visible blue light penetrates only the very superficial areas. The narrow-band blue image is somewhat enhanced compared to the narrow-band red and green signal. NBI accentuates the relief pattern of the mucosa without the need for chromoscopy. In addition, the image of the size and distribution of capillaries is accentuated[10–12].

## LASER-SCANNING CONFOCAL MICROSCOPY (LCM)

The image from LCM is obtained by counting the reflective light of 405 nm wavelength diode laser. Virtual histological image can be obtained successfully from unstained fresh tissue[13–15]. The special resolution of LCM is around 1 μm. The light penetration depth is limited to about 200 μm. Staining with 2% cresyl violet facilitates identification of nuclei and cell membranes[16].

## ENDOCYTOSCOPY

Inoue and colleagues (Inoue, personal communication) developed a catheter-type contact endomicroscope which has 1100 times magnifying power at maximum. After 1% methylene blue vital staining of the mucosa, living normal and cancer cells can be successfully observed in high-quality images, comparable to conventional cytology. The cellular architecture and composition can be analysed at micron scale resolution. Movement of erythrocytes in capillaries is readily identified. Neoplastic cells have blurred and irregularly enlarged nuclei. This technology approaches virtual biopsy and virtual histology.

## FLUORESCENCE SPECTROSCOPY

Fluorescence spectroscopy can be used to extract biochemical properties from tissue. Fluorescence photons are scattered and absorbed during their path to the tissue surface where they are collected via an optical fibre probe. Fluorescence occurs when tissue emits light of longer wavelengths following excitation by a shorter wavelength light source. One can exploit either fluorescence generated by naturally occurring tissue fluorophores such as collagen and porphyrins (autofluorescence) or fluorescence emitted by an exogenously administered drug (5-aminolevulinic acid or haematoporphyrin derivate). In autofluorescence the detection of dysplastic lesions depends on changes in tissue microarchitecture and in the concentration, spatial distribution and metabolic activity of endogenous fluorophores which ultimately influence the spectral line-shape. Using excitation wavelength in the violet-blue, Barrett's epithelium generally emits a fluorescence spectrum characterized by a relative increase in the red-to-green fluorescence intensity ratio. This ratio is accentuated as the mucosa evolves via dysplasia (neoplasia) to carcinoma. Thus neoplastic tissue exhibits slightly weaker red fluorescence but much weaker green fluorescence than normal tissues when illuminated by blue light[17-21].

The latest video-autofluorescence imaging prototype has one CCD chip for white light endoscopy and another chip for integration of total auto fluorescence after blue light excitation (395–475 nm) plus green (550 nm) and red (610 nm) reflectance into a real-time pseudo-colour image.

## BIOENDOSCOPY

The use of optically detectable affinity ligands to target molecules of interest allows more precise characterization of tissue. The identification of targets that are highly specific to the disease process, together with high-affinity attachment of the labelling substance to the target, is obviously critical. Such fluorophores typically emit light in the near-infrared zone, often rather powerful fluorescence. There are now several examples of such an approach. The use of fluorescently labelled antibodies against carcinoembryonic antigen may facilitate the detection of gastric tumours[22]. Whole-body imaging of green fluorescent protein can be used to test the efficiency of gene carriers for *in-vivo* transduction[23]. The use of reporter molecules or molecular beacons such as fluorescent indocyanine fluorochromes, linked to a peptide backbone carried by a pegylated polymer, accumulates in neoplastic tissue and strongly fluoresces after proteolysis by cathepsin B when stimulated with excitation light[24].

Beacon chemicals may perhaps be sprayed endoscopically on suspected targets instead of being infused systemically, in a way similar to antibodies to various lumenally exposed epitopes carrying a fluorescent tag. Another example of bioendoscopy is the use of FISH to characterize chromosomal changes and mutations *in vivo*(?) or in exfoliated cells.

## DISCUSSION

In searching for (neoplastic) lesions without overt suspicious morphological changes, random tissue sampling remains a 'hit-or-miss' process. If histological confirmation of a lesion is required before deciding to treat it topically, the time lag associated with tissue processing necessitates a second procedure. There is obviously a need to improve targeting surveillance biopsies to dysplastic areas such as in patients with long-standing ulcerative colitis or Barrett's oesophagus.

Several novel modalities have been developed to improve the detailed endoscopic inspection and analysis of the lining of the gastrointestinal tract. The main purpose of these novel developments is to facilitate the detection of microarchitectural change suspicious of neoplastic alteration, allowing targeted biopsy or local ablation, preferably with mucosal resection to allow ultimate histopathological analysis.

High-resolution, high-magnification endoscopy, usually combined with chromoscopy (lugol, indigo carmine, methylene blue) allows high-resolution inspection of mucosal microarchitecture (pit patterns, villous patterns). Common indications for its use are the detection of villous areas in Barrett's oesophagus or abnormal pit patterns in flat adenoma/neoplasia.

Narrow-band imaging is an emerging method of highlighting features of the mucosa that are invisible with standard white light video endoscopy. NBI accurately depicts villous patterns and pit patterns and accentuates vascular microstructures.

In fluorescence endoscopy, tissue fluorophores are activated with bluish light to emit minute quantities of fluorescent light (autofluorescence). Fluorescent spectra from normal and neoplastic mucosa differ, allowing computer-generated false colouring of abnormal areas for targeted biopsy.

Probe-based laser-scanning confocal microscopy imaging may characterize suspicious areas as malignant cells and malignant tissue architecture is readily identifiable.

The above-mentioned novel promising endoscopic imaging techniques represent a significant improvement over standard endoscopy for diagnosing neoplastic change. Other novel technology is based on the interaction with light and tissue (bioendoscopy). Bioendoscopy may be defined as any endoscopic technique that either provides biological information about the target tissue or exploits unique biological features of the target tissue. The potential of bioendoscopy is dazzling. The possibility to monitor the function of a variety of proteins *in vivo* by designing appropriate beacons is on the horizon.

New optical techniques will continue to emerge, such as multiphoton, second harmonic and magnetic resonance microscopy, which can peer deeper and less destructively into living tissue. Green fluorescence protein, a reporter of gene expression and protein localization in living cells, unifies digitized fluorescence microscopy/endoscopy to molecular biology. By using blue, cyan, yellow and red fluorescent protein tags, confocal fluorescence microscopy and related imaging technologies will allow tracking of multiple proteins within living tissue.

It is readily obvious from the bewildering array of novel developments that the role of imaging in future biomedical research will only be increasing at an unprecedented rate.

## References

1.  Bruno MJ. Magnification endoscopy, high resolution endoscopy, and chromoscopy; towards a better optical diagnosis. Gut. 2003;5(Suppl. 4):iv7–11.
2.  Tajiri H, Matsuda K, Fukisaki J. What can we see with the endoscope? Present status and future perspectives. Dig Endsc. 2002;14:131–7.
3.  Dacosta RS, Wilson BC, Marcon NE. New optical technologies for earlier endoscopic diagnosis of premalignant gastrointestinal lesions. J Gastroenterol Hepatol. 2002;17 (Suppl.):S85–104.
4.  Dacosta RS, Wilson BC, Marcon NE. Photodiagnostic techniques for the endoscopic detection of premignant gastrointestinal lesions. Dig Endosc. 2003;15:153–73.

5.  Otsuka Y, Niwa Y, Ohmiya N et al. Usefulness of magnifying endoscopy in the diagnosis of early gastric cancer. Endoscopy. 2004;36:165–9.
6.  Hurlstone DP, Cross SS, Adam I et al. Efficacy of high magnification chromoscopic colonoscopy for the diagnosis of neoplasia in flat and depressed lesions of the colorectum: a prospective analysis. Gut. 2004;53:284–90.
7.  Kiesslich R, Jung M. Magnification endoscopy: does it improve mucosal surface analysis for the diagnosis of gastrointestinal neoplasias? Endoscopy. 2002;34:819–22.
8.  Udagawa T, Amano M, Okada F. Development of magnifying video endoscopes with high resolution. Dig Endosc. 2001; 13:163–9.
9.  Participants in the Paris Workshop. The Paris endoscopic classification of superficial neoplastic lesions: esophagus, stomach, and colon. Gastrointest Endosc. 2003;58:S3–50.
10. Sharma P, McGregor D, Cherian R, Weston A. Use of narrow band imaging, a novel imaging technique, to detect intestinal metaplasia and high-grade dysplasia in patient with Barrett's esophagus. Gastrointest Endosc. 2003;57:AB77.
11. Kara MA, Bergman JJ, Fockens P et al. Narrow Band imaging for improved mucosal pattern recognition in Barrett's esophagus. Gastrointest Endosc. 2003;57:AB176.
12. Fujisaki J, Saito N, Matsuda K et al. Clinicopathological analysis of superficial type Barrett's esophageal cancer and efficacy of magnifying endoscopy with narrow band image system for specialized columnar epithelium. Gastrointest Endosc. 2003;57:AB179.
13. Inoue H, Igari T, Nishikage T, Ami K, Yoshida T, Iwai T. A novel method of virtual histopathology using laser-scanning confocal microscopy *in-vitro* with untreated fresh specimens from the gastrointestinal mucosa. Endoscopy. 2000;32:439–43.
14. Inoue H, Cho JY, Satodate H et al. Development of virtual histology and virtual biopsy using laser-scanning confocal microscopy. Scand J Gastroenterol. 2003;(Suppl. 237):37–9.
15. Sakashita M, Inoue H, Kashida H et al. Virtual histology of colorectal lesions using laser-scanning confocal microscopy. Endoscopy. 2003;35:1033–8.
16. George M, Meining A. Cresyl violet as a fluorophore in confocal laser scanning microscopy for future *in-vivo* histopathology. Endoscopy. 2003;35:585-9.
17. Haringsma J, Tytgat GN, Yano H et al. Autofluorescence endoscopy: feasibility of detection of GI neoplasms unapparent to white light endoscopy with an evolving technology. Gastrointest Endosc. 2001;53:642–50.
18. Tajiri H, Kobayashi M, Izuishi K, Yoshida S. Fluorescence endoscopy in the gastrointestinal tract. Dig Endosc. 2000;12:S28–31.
19. Prosst RL, Gahlen J. Fluorescence diagnosis of colorectal neoplasms: a review of clinical applications. Int J Colorectal Dis. 2002;17:1–10.
20. Bhunchet E, Hatakawa H, Sakai Y, Shibata T. Fluorescein electronic endoscopy: a novel method for detection of early stage gastric cancer not evident to routine endoscopy. Gastrointest Endosc. 2002;55:562–71.
21. Mayinger B, Jordan M, Horbach T et al. Evaluation of *in vivo* endoscopic autofluorescence spectroscopy in gastric cancer. Gastrointest Endosc. 2004;59:191–8.
22. Keller R, Winde G, Terpe HJ, Foerster EC, Domschke W. Fluorescence endoscopy using a fluorescein-labeled monoclonal antibody against carcinoembryonic antigen in patients with colorectal carcinoma and adenoma. Endoscopy. 2002;34:801–7.
23. Wack S, Hajri A, Heisel F et al. Feasibility, sensitivity, and reliability of laser-induced fluorescence imaging of green fluorescent protein-expressing tumors *in vivo*. Mol Ther. 2003;7:765–73.
24. Marten K, Bremer C, Khazaie K et al. Detection of dysplastic intestinal adenomas using enzyme-sensing molecular beacons in mice. Gastroenterology. 2002;122:406–14

# 16
# Cross-sectional imaging in inflammatory bowel disease

**K. TURETSCHEK**

## INTRODUCTION

Patients suffering from inflammatory bowel disease (IBD) require prompt and accurate treatment in order to relieve their symptoms and to minimize potential complications. Conventional enteroclysis (CE) has been considered to be the imaging technique of choice for evaluating the small bowel. Adequate distension is mandatory because collapsed bowel loops may hide or even simulate small bowel disease. However, CE provides no information on extralumenal disease extension and does not allow any determination of disease activity.

The modalities for diagnostic imaging in IBD have dramatically changed over the past decade. Large bowel enemas have lost their importance compared to colonoscopy, and conventional enteroclysis has been widely replaced by CT- and MR enteroclysis (CTE, MRE).

CT has emerged as one of the most important imaging techniques for the evaluation of patients with gastrointestinal disorders for the following reasons: it is a non-invasive and fast imaging modality, it is well tolerated and widely accepted by patients, and the accuracy for detecting intramural and intraperitoneal extension of various diseases, including IBD, is excellent. However, one essential disadvantage of conventional CT is the fact that inflammatory altered bowel walls can be differentiated reliably from normal bowel only as long as they are adequately distended. Collapsed or non-distended inflamed bowel segments may mimic normal bowel loops; thus, bowel wall thickening and bowel stenosis – as seen in many patients suffering from Crohn's disease (CD) – might potentially be overlooked.

## CT ENTEROCLYSIS

CTE has been introduced to overcome the individual deficiencies of conventional CT and conventional enteroclysis, and to combine the advantages of both in a single technique[1–3]. This modality enables detailed morphological characterization of inflamed bowel segments and visualizes mural and extramural manifestations of CD. Entities such as stenoses, fistulas, abscesses or skip lesions, which might emphasize the need for elective gastrointestinal surgery, can easily be detected.

In order to obtain adequate lumen distension, up to 2 L of a water-like contrast medium (i.e. PEG, mannitol, sorbitol, methylcellulose) is usually administered via a nasoenteric tube. CT examinations are performed after intravenous injection of up to 150 ml non-ionic contrast material. The new generation of multi-detector-row scanner (MDCT) permits images of perfect quality in axial, coronal, sagittal or even oblique planes. Reformated coronal planes enable a better and faster overview of the entire abdomen and facilitate the orientation for clinicians, especially in difficult cases where gastrointestinal surgery is considered a therapeutic option. CTE has become an accepted method for evaluating the small bowel in IBD and, at least in our department, has almost completely replaced conventional enteroclysis.

## MR ENTEROCLYSIS

MRE is an emerging technique for evaluating patients suffering from IBD. Recent technical advances have significantly improved the image quality of MRI, and offer novel diagnostic perspectives in imaging IBD[3,4].

As for CTE, optimal lumen distension is a key requisite in imaging the small bowel with MRE: different intestinal contrast agents are available on the market which are classified as positive (gadolinium-based), negative (iron-based) or biphasic (water-like). Depending on the contrast agent used, the lumen appears dark (negative) or bright (positive) on T1 and T2 weighted sequences. Biphasic contrast agents (i.e. PEG, mannitol, methylcellulose) are either bright or dark depending on the applied pulse sequence.

There are two main ways in which to administer the intestinal contrast agent: (a) via a nasoenteric tube or (b) ingested orally[5–7]. Nasoenteric tubes are placed under fluoroscopic guidance beyond the duodenojejunal junction and offer best small bowel distension. In addition, the filling of the small bowel may be monitored with MR (= MR fluoroscopy), thereby giving information on small bowel motility and potential adhesions or fixations of small bowel loops. Thus the administration of the distension medium is monitored and might be stopped at any time, or when adequate distension of the entire small bowl is achieved.

Another quite innovative approach is to administer the distension medium orally. Naturally this method is well accepted by patients because they do not suffer any discomfort from the nasoenteric tube; moreover, no radiation is applied.

Although different contrast agents and administration techniques have been proposed, and give sufficient distension in many cases, the quality of bowel distension is poorer compared to the nasoenteric tube approach, the degree of distension is quite unpredictable in some patients and sometimes the whole procedure becomes rather time-consuming (up to 4 h)[6].

MRE can demonstrate a wide spectrum of abnormalities including mucosal changes, transmural and extramural manifestations and, of course, all kinds of complications of the disease. Only early changes of the mucosa (i.e. aphthous lesions, fine nodularity of the mucosa) are often missed because of insufficient resolution.

Another advantage of MRE is its excellent soft-tissue contrast and the ability to differentiate between active inflammatory and chronic-fibrotic bowel wall segments.

### Assessment of disease activity

Activity assessment is of high importance for patients suffering from CD. Patients' symptoms may be non-specific and due to active inflammation, fibrotic scarring or stricture formation. **MRE** has the potential to differentiate between these entities and thereby offer crucial information on further patient management (drug versus surgical therapy).

Many MR parameters have been proposed to assess disease activity throughout the literature[8,9]; however, there is some agreement on some parameters being of high value: bowel wall thickening and bowel wall enhancement after intravenous contrast administration, high signal of the bowel wall and oedema of the perivisceral fatty tissue on T2 weighted sequences.

The key points favouring the use of MRE are the lack of radiation exposure, the multiplanar capabilities and the high intrinsic contrast. The use of different MR sequences enables detailed characterization of the thickened bowel wall, thereby giving information on disease activity (inflammation versus fibrosis). MRE will play an important role in monitoring therapeutic effects (i.e. anti-TNF-$\alpha$) and will play a crucial role in patient management throughout the course of the disease.

## CT AND MR COLONOGRAPHY (VIRTUAL COLONOSCOPY)

Colonoscopy still remains the mainstay of imaging the large bowel in patients with IBD. CT or MR colonography represents complementary imaging methods which might be used in cases of incomplete colonoscopy or severe pain. However, the resolution of both radiological methods is still too low to detect minimal changes of the mucosa; therefore the indications are limited to patients suffering from stenoses and suspected extraintestinal complications.

As with all cross-sectional imaging methods both techniques need colonic distension which can be achieved either with fluid (i.e. water) or air. The use of air (room air or $CO_2$) is generally preferred because it is easy, fast and inexpensive.

Various post-processing methods create different ways of looking at the colon, including fly-through, 3D or multiplanar views.

Another interesting approach to image the colon is so-called barium-tagged dark lumen colonography[10,11]. This new method eliminates some of the disadvantages of colonoscopy and is well tolerated and accepted by patients. Faecal tagging makes oral cleansing of the bowel unnecessary, and in fact increases the acceptance and tolerance of patients for the evaluation of their large bowel. The most common tagging agent is barium sulphate, because it is widely available, safe and inexpensive, and this method of bowel cleansing might be used in conjunction with CT or MR colonography.

So far, barium-tagged dark lumen colonography is considered an alternative or additional method in evaluating the large bowel in patients with IBD, especially in patients who have severe pain and faecal residue.

## CONCLUSION

Cross-sectional imaging expands the diagnostic armentarium, and has enormous potential in the diagnostic work-up of patients suffering from IBD. Traditional imaging methods, in particular conventional enteroclysis, are becoming more and more outdated, and will be replaced completely by these new imaging approaches. Nevertheless, these upcoming techniques should always be considered as complementary to the clinical assessment of patients.

## References

1. Rollandi GA, Curone PF, Biscaldi E et al. Spiral CT of the abdomen after distension of small bowel loops with transparent enema in patients with Crohn's disease. Abdom Imaging. 1999;24:544–9.
2. Turetschek K, Schober E, Wunderbaldinger P et al. Findings at helical CT-enteroclysis in symptomatic patients with Crohn disease: correlation with endoscopic and surgical findings. J Comput Assist Tomogr. 2002;26:488–92.
3. Schreyer AG, Seitz J, Feuerbach S, Rogler G, Herfarth H. Modern imaging using computer tomography and magnetic resonance imaging for inflammatory bowel disease (IBD). Inflamm Bowel Dis. 2004;10:45–54.
4. Umschaden HW, Szolar D, Gasser J, Umschaden M, Haselbach H. Small-bowel disease: comparison of MR enteroclysis images with conventional enteroclysis and surgical findings. Radiology. 2000;215:717–25.
5. Lauenstein TC, Schneemann H, Vogt FM, Herborn CU, Ruhm SG, Debatin JF. Optimization of oral contrast agents for MR imaging of the small bowel. Radiology. 2003;228:279–83.
6. Patak MA, Froehlich JM, von Weymarn C, Ritz MA, Zollikofer CL, Wentz K. Non-invasive distension of the small bowel for magnetic-resonance imaging. Lancet. 2001;358: 987–8.
7. Ajaj W, Goehde SC, Schneemann H, Ruehm SG, Debatin JF, Lauenstein TC. Oral contrast agents for small bowel MRI: comparison of different additives to optimize bowel distension. Eur Radiol. 2004;14:458–64.
8. Koh DM, Miao Y, Chinn RJ et al. MR imaging evaluation of the activity of Crohn's disease. Am J Roentgenol. 2001;177:1325–32.
9. Maccioni F, Viscido A, Broglia L et al. Evaluation of Crohn disease activity with magnetic resonance imaging. Abdom Imaging. 2000;25:219–28.
10. Lauenstein TC, Goehde SC, Ruehm SG, Holtmann G, Debatin JF. MR colonography with barium-based fecal tagging: initial clinical experience. Radiology. 2002;223:248–54.
11. Ajaj W, Lauenstein TC, Pelster G, Goehde SC, Debatin JF, Ruehm SG. MR Colonography: how does air compare to water for colonic distension? J Magn Reson Imaging. 2004;19: 216–21.

# 17
# State of the art: the transparent abdomen

**P. ASBACH, P. ROGALLA and B. HAMM**

## INTRODUCTION

Radiological evaluation of patients with suspected or proven inflammatory bowel disease is a key element in managing the patient with special respect to surgical or non-surgical treatment options. The radiological imaging techniques based on projection radiography have almost completely been replaced by the two state-of-the-art cross-sectional imaging modalities, computed tomography (CT) and magnetic resonance imaging (MRI). Cross-sectional imaging has several important advantages over the classic radiographic procedures (e.g. Sellink enteroklysma). It is a very fast and investigator-independent modality that is suitable for evaluation of the gastro-intestinal tract as well as extra-intestinal complications (abscess, fistula, lymph node enlargement). Images can be obtained in any plane and can be reconstructed to generate three-dimensional images that are more appealing to clinicians (Figure 1a–c).

The choice of the optimal cross-sectional imaging modality (CT or MRI) depends on several aspects. First, imaging protocols for the small bowel are substantially different from large bowel imaging protocols. Secondly, the decision depends on the patient's age and health condition. Thirdly, the choice of the imaging protocol is influenced by the clinical question to be answered.

## SMALL BOWEL

For the small bowel, two major imaging options are available in terms of oral contrast agent administration which is the basis for small bowel evaluation. Distension of the bowel lumen is required to increase diagnostic sensitivity in the detection of wall thickening and stenotic bowel segments. Duodenal intubation ensures optimal intraluminal contrast agent administration and maximum distension of the small bowel lumen. Alternatively, the contrast agent can be given orally, but this mode of administration produces less lumenal distension compared to duodenal intubation. In this case the contrast agent is supplemented with methycellulose and/or mannitol (in both CT and MRI ) and ingested by the patient over a period of approximately 1 h prior to the examination after 1 day of bowel cleansing.

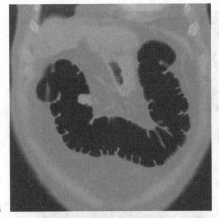

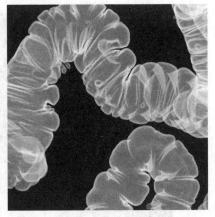

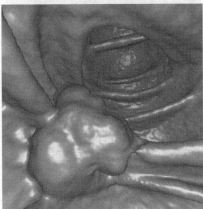

**Figure 1** (a) MSCT; coronal reformation; polype in transverse colon. (b) MSCT; TTP (transparent tissue projection). (c) MSCT; surface reconstruction; endoluminal view; virtual colonoscopy

## COMPUTED TOMOGRAPHY

Since multisclice computed tomography (MSCT) was introduced in the late 1990s the scanning time for abdominal imaging has significantly decreased to an average of 10–15 s at present. This makes the images very robust to artefacts caused by bowel motion or inability of the patient to tolerate a breath-hold examination (Figure 2a,b). Secondly, the exposure to radiation has consequently been reduced to a dimension that is comparable, with the latest CT-scanner technology even less than the classic projection radiological modalities (Sellink enteroklysma, colon double-contrast examination). Biphasic image acquisition is warrantable to distinguish transient spasms from stenotic bowel segments; the second phase is acquired in low-dose technique in order to further decrease exposure to radiation. Near isotropic imaging allows three-dimensional-reformation and surface and volume rendering techniques to generate virtual endoscopy images[1–4].

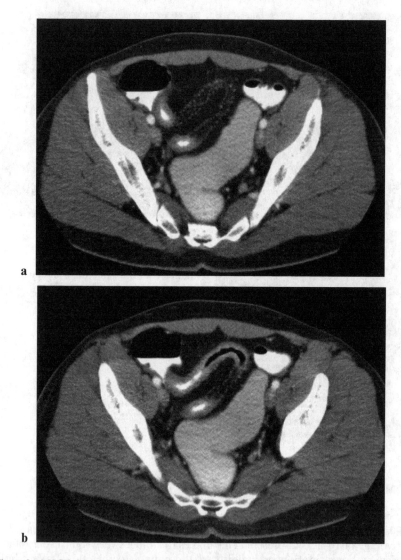

**Figure 2** MSCT; 24-year-old patient with Crohn's disease. Wall thickening of the terminal ileum with consecutive stenosis of the bowel segment. (a) Inflammation of the bowel segment marked by accentuated mesenteric blood vessels. (b) Wall thickening of the bowel segment, narrowing of the bowel lumen

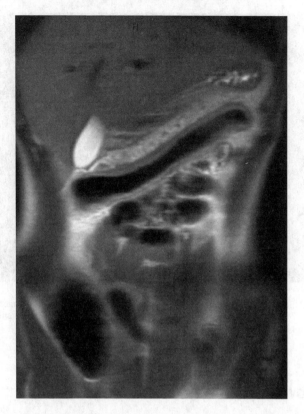

**Figure 3** MRI; coronal T2w half-Fourier single-shot turbo spin-echo sequence; 22-year-old patient with Crohn's disease with colonic manifestation. Wall thickening of the transverse colon

## MAGNETIC RESONANCE IMAGING

In comparison to CT, MRI has a lower spatial resolution and longer image acquisition time, making the images more susceptible to bowel motion artefacts. Image quality can be significantly increased by the administration of anti-spasmolytic drugs. The absence of radiation exposure makes MRI especially suitable for use in very young patients. In addition, the latest MRI scanner technology allows real-time image acquisition with a decent spatial resolution. It is thus possible to visualize peristalsis of the bowel, which is of specific interest in grading stenotic bowel segments and monitoring anti-inflammatory therapy[5].

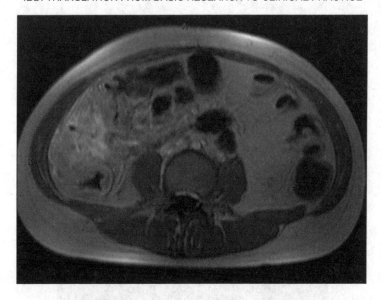

**Figure 4**  MRI; axial T1w gradient echo sequence after intravenous contrast agent application (gadolinium); 27-year-old patient with Crohn's disease. Thickening and massive enhancement of the inflamed bowel wall of terminal ileum and ilio-caecal valve.

## LARGE BOWEL

As in small bowel imaging, the lumenal contrast agent is also the key factor in the radiological evaluation of the large bowel. Two options are available, administration of a contrast agent solution (saline, positive oral contrast) or insufflation of carbon dioxide, which is more widely used because it is more suitable for use in daily clinical routine. However, the air–tissue interface leads to significant susceptibility artefacts in MRI, making CT the predominant imaging modality for the large bowel. MRI is currently only a second-line imaging modality in the evaluation of inflammatory large bowel disease (Figures 3 and 4). In local cancer staging, in particular of rectal tumours, MRI has major advantages, but this goes beyond the scope of this review. Faecal tagging (administration of a positive oral contrast agent in combination with the cleansing procedure) increases specificity in large bowel evaluation[6–8].

## CONCLUSION

State-of-the-art cross-sectional imaging modalities yield high-resolution images of the small and large bowel and are highly sensitive and specific in evaluating patients with inflammatory bowel disease. Virtual endoscopy has the potential to replace flexible endoscopies with diagnostic intent.

# References

1.  Rust GF, Eisele O, Hoffmann JN, Kopp R, Furst H, Reiser M. Virtual coloscopy with multi-slice computerized tomography. Radiologe. 2000;40:27–82.
2.  Horton KM, Corl FM, Fishman EK. CT of nonneoplastic diseases of the small bowel: spectrum of disease. J Comput Assist Tomogr. 1999;23:417–28.
3.  Rogalla P, Werner Rustner M et al. Virtual endoscopy of the small bowel: phantom study and preliminary clinical results. Eur Radiol. 1998;8:563–7.
4.  Scholmerich J. Inflammatory bowel disease. Endoscopy. 1999;31:66–73.
5.  Aschoff AJ, Zeitler H, Merkle EM, Reinshagen M, Brambs HJ, Rieber A. MR enteroclysis for nuclear spin tomographic diagnosis of inflammatory bowel diseases with contrast enhancement. Fortschr Röntgenstr. 1997;167:387–91.
6.  Bielen D, Thomeer M, Vanbeckevoort D et al. Dry preparation for virtual CT colonography with fecal tagging using water-soluble contrast medium: initial results. Eur Radiol. 2003;13:453–8.
7.  Lefere P, Gryspeerdt S, Dewyspelaere J, Baekelandt M, Van HB. Dietary fecal tagging as a cleansing method before CT colonography: initial results polyp detection and patient acceptance. Radiology. 2002;224:393–403.
8.  Lauenstein TC, Goehde SC, Debatin JF. Fecal tagging: MR colonography without colonic cleansing. Abdom Imaging. 2002;27:410–17.

# Section VI
# Mechanism of action of established drugs

Chair: R.N. FEDORAK and P. MICHETTI

# 18
# Apoptosis as a target

M. F. NEURATH

## INTRODUCTION

Ulcerative colitis (UC) and Crohn's disease (CD) are prototypes of inflammatory bowel diseases (IBD) in humans. Recent evidence suggests that genetic
factors, environmental factors and bacterial antigens play a key role in the
pathogenesis of IBD. Specifically, bacterial antigens from the commensal flora
seem to cause an unbalanced activation of the mucosal immune system in
genetically predisposed patients leading to chronic intestinal inflammation. In
this context, lamina propria T lymphocytes have emerged as key effector cells in
the chronic phase of IBD and experimental animal models of colitis[1-9]. Whereas
in the normal gut T cells are more susceptible to undergo programmed cell death
(apoptosis) as compared to peripheral blood T cells, T cells in IBD patients
(particularly in CD) are resistant against apoptosis. This observation probably
contributes to an expansion of effector T cells in the lamina propria of IBD
patients. This chapter will focus first on the activation of lamina propria T cells in
IBD patients and subsequently on the role of T cell apoptosis in IBD patients. In
particular, the molecular mechanism of action of azathioprine, which is based
on the induction of apoptosis, will be described in detail.

## ACTIVATION OF LAMINA PROPRIA T CELLS IN IBD

T cell activation in IBD appears to be driven by luminal antigens and antigen-
presenting cells that are capable of producing cytokines inducing T cell
activation and differentiation, such as interleukins IL-12, IL-23 and
IL-27[10-12]. Whereas lamina propria T cells in CD produce a Th1-type cytokine
profile characterized by IFN-$\gamma$ and TNF production, the cytokine profile in UC
is characterized by the increased production of the Th2 cytokines IL-5 and IL-
13[9]. Such CD4+ Th1 or Th2 cells seem to play a key pathogenic role in IBD
patients. For instance, antibodies to TNF that are successfully used in treating
CD patients (e.g. infliximab) have been shown to induce rapid mucosal T cell
apoptosis within several days, indicating that the therapeutic efficacy of these
antibodies could be due to the induction of T cell apoptosis.

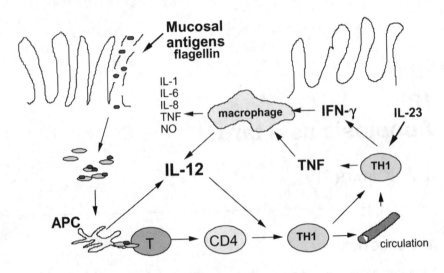

**Figure 1** Antigens from the bacterial flora such as flagellin cause an activation of the mucosal immune system such as antigen-presenting cells (e.g. dendritic cells). Subsequently, mucosal T cell activation and IL-12-dependent Th1 differentiation occurs, leading to production of both IFN-γ and TNF

In recent years there has been a growing interest in understanding cytokine signalling events. In CD much progress has been made in understanding the molecular pathways leading to Th1 T cell differentiation. The IL-12 hetero-dimer produced by lamina propria dendritic cells or macrophages is one key cytokine that induces Th1 T cell differentiation in CD. This mechanism involves the activation and phosphorylation of the transcription factor STAT4 (signal transducer and activator of transcription 4). This concept is supported by the observation that Th1-mediated animal models of IBD are associated with increased IL-12 production and neutralizing antibodies to IL-12 have been shown to suppress Th1-mediated chronic intestinal inflammation, pre-sumably by the prevention of Th1 T cell development and the induction of Fas-mediated T cell apoptosis. Furthermore, STAT4 deficient T cells failed to induce Th1-mediated colitis in an adoptive transfer system, whereas STAT4 transgenic mice can develop Th1-mediated colitis. However, it is not clear whether the effects of STAT4 *in vivo* can be entirely attributed to IL-12 (p35/p40), since IL-23 (p19/p40) has been recently shown to activate STAT4 in T cells and may activate Th1 effector cells in CD.

At the molecular level, recent studies in CD have analysed the expression and function of Th1-associated transcription factors. One such transcription factor, denoted T-bet (T box protein expressed in T cells) has recently been identified as a factor that controls IFN-γ production and Th1 development in CD. The potential clinical importance of this transcription factor is shown by the fact that T-bet-deficient T cells fail to induce Th1-mediated experimental colitis in the $CD4^+$ $CD62L^+$ adoptive transfer model of colitis. Since CD in humans is associated with strongly increased expression of T-bet in lamina propria T cells, these data provide a rationale for selective targeting of this transcription factor in this disease.

## ON THE REGULATION OF APOPTOSIS OF LAMINA PROPRIA T CELLS IN IBD

The intestine is one of the largest reservoirs of lymphocytes in the human body. Since unchecked proliferation of such lamina propria lymphocytes may provoke the risk of developing chronic inflammatory diseases, the immune system controls efficient elimination of activated lymphocytes in the gut in a process known as apoptosis or programmed cell death. Apoptosis appears to be critical for the design of novel therapeutic approaches in IBD. For instance, antibodies against IL-12 or the IL-6 receptor (IL-6R) have been successfully used for treatment of experimental mucosal inflammation, and these antibodies have been shown to act via the induction of apoptosis of lamina propria T lymphocytes[13-15]. Administration of anti-IL-6R antibodies to mice with experimental colitis suppressed experimental colitis via the induction of T cell apoptosis[7]. The mechanism underlying this anti-apoptotic effect is due to complexes of soluble IL-6R (sIL-6R) and IL-6 that interact with gp130 on the membrane of $CD4^+$ T cells and initiate an anti-apoptotic IL-6 signal. The anti-apoptotic effect mediated by such IL-6 *trans*-signalling involves Bcl-2 and bcl-xL. Indeed, these proteins are elevated in lamina propria T cells from patients with CD, leading to an increase of the Bcl-2/Bax ratio, and are known to be induced by STAT-3[7].

In addition to anti-IL-6R antibodies, the administration of neutralizing anti-TNF antibodies to patients with CD has recently been shown to rapidly induce apoptosis of lamina propria T cells (reviewed in ref. 16). This finding could explain the rapid beneficial effects of anti-TNF therapies in clinical trials. Recent data suggest that the interaction between mTNF and TNFR2 is critical for the regulation of colitis activity. Consistent with this hypothesis, it has recently been shown that etenercept, a protein that blocks soluble but not membrane-bound TNF, does not appear to be effective in the treatment of active CD in two prospective studies.

## ON THE MOLECULAR MECHANISM OF ACTION OF AZATHIOPRINE

Apoptosis is particularly important for the mucosal immune system, since a resistance of lamina propria T cells to apoptosis may lead to chronic inflammatory responses in the gut. We have recently provided evidence that azathioprine, a drug that has been used for four decades for successful treatment of IBD[17], induces T cell apoptosis of CD45RO memory T cells. Since CD45RO T lymphocytes are considered to be key effector cells in the lamina propria of IBD patients, it appears that the excellent therapeutic efficacy of azathioprine in IBD could be due to the induction of local T cell

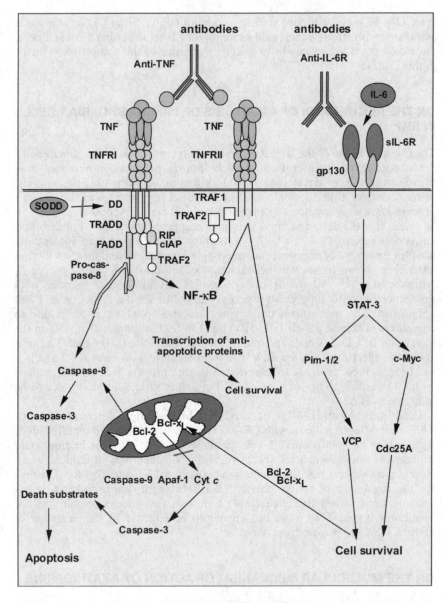

**Figure 2** Mechanisms used by anti-IL-6R and anti-TNF antibodies to induce T cell apoptosis. Anti-TNF antibodies block the mTNF/TNFRII interaction and actively induce T cell apoptosis in IBD patients[23]. Furthermore, anti-IL-6R antibodies block the bcl-xL activation in IBD lamina propria T cells (modified according to ref. 14).

apoptosis. This hypothesis is supported by the finding that successful azathioprine treatment in IBD patients leads to an increased number of apoptotic T cells in the peripheral blood and the lamina propria, whereas no apoptosis is seen in patients unresponsive to azathioprine therapy. The induction of apoptosis by azathioprine was critically dependent on T cell costimulation via the costimulatory molecule CD28 that is known to inhibit TCR-induced apoptosis during a primary T cell response by activation of the antiapoptotic bcl-xL protein. Interestingly, azathioprine down-regulates bcl-xL expression at the mRNA and protein level, strongly suggesting that this drug blocks a key regulatory pathway in CD28 signalling.

The CD28 signalling pathway is important not only for the initial activation of T cells, but also for maintaining their viability and responsiveness during a persistent immune response[17-20]. Additional studies have shown that azathioprine-induced suppression of CD28 signalling is due to suppression of the activation of Rac1; a Rac GTPase that is known to play a key role in CD28 signalling. Rac proteins play a major role in T cell development, differentiation and proliferation[21,22]. For instance, whereas dominant positive Rac mutations have been associated with increased cell proliferation and tumours, functionally inactive Rac mutations are associated with immunodeficiencies in humans. Azathioprine-mediated suppression of Rac1 activation in T cells was mediated by direct binding of an azathioprine metabolite (6-thioGTP) to Rac1. The observed up-regulation of vav expression in azathioprine-treated T cells is most likely an insufficient compensatory mechanism to achieve Rac1 activation by up-regulation of the corresponding guanosine exchange factor.

Thus, the molecular mechanism of action of azathioprine is based on the suppression of Rac1 activation via 6-thioGTP. Azathioprine-induced suppression of Rac1 activation leads to suppression of bcl-xL expression (via blockade of NFκB and STAT-3 activation) and a mitochondrial pathway of T cell apoptosis. Our data thus suggest that azathioprine-induced immunosuppression is mediated by suppression of Rac1 activation and the consecutive induction of T cell apoptosis. It should be noted, however, that azathioprine-induced apoptosis affects mainly CD45RO effector T cells upon costimulation via CD28, suggesting that azathioprine may be particularly effective in eliminating pathogenic memory T cells in autoimmune and chronic inflammatory diseases. These data have important implications for the design of novel and more specific diagnostic tools to monitor azathioprine therapies, and for the development of novel therapeutic approaches for autoimmune diseases.

Taken together, considerable progress has been made in understanding the immunopathogenesis of IBD in humans. Various studies point towards an important role of T cell apoptosis in experimental colitis and IBD patients. These data provide a rationale for the design of novel therapeutic strategies in IBD in the near future that are based on the specific induction of T cell apoptosis.

# References

1. Blumberg RS, Saubermann LJ, Strober W. Animal models of mucosal inflammation and their relation to human inflammatory bowel disease. Curr Opin Immunol. 1999;11:648–56.
2. Powrie F, Leach MW, Mauze S, Menon S, Caddle LB, Coffman RL. Inhibition of Th1 responses prevents inflammatory bowel disease in scid mice reconstituted with CD45RBhi CD4+ T cells. Immunity. 1994;2:553–62.
3. Mombaerts P, Mizoguchi E, Grusby MJ, Glimcher LH, Bahn AK, Tonegawa S. Spontaneous development of inflammatory bowel disease in T cell receptor mutant mice. Cell. 1993;75:275–82.
4. Mizoguchi A, Mizoguchi E, Bhan AK. The critical role for interleukin-4 but not interferon-gamma in the pathogenesis of colitis in T-cell receptor alpha mutant mice. Gastroenterology. 1999;116:320–6.
5. Wirtz S, Finotto S, Kanzler S et al. Cutting edge: chronic intestinal inflammation in STAT-4 transgenic mice: characterization of disease and adoptive transfer by TNF- plus IFN-gamma producing CD4+ T cells that respond to bacterial antigens. J Immunol. 1999;162:1884–8.
6. Boirivant M, Fuss IJ, Chu A, Strober W. Oxazolone colitis: a murine model of T helper cell type 2 colitis treatable with antibodies to interleukin-4. J Exp Med. 1998;188:1929–39.
7. Atreya R, Mudter J, Finotto S et al. Blockade of IL-6 trans-signaling suppresses T cell resistance against apoptosis in chronic intestinal inflammation: evidence in Crohn's disease and experimental colitis *in vivo*. Nature Med. 2000;6:583–8.
8. Simpson SJ, Shah S, Comiskey M et al. T cell-mediated pathology in two models of experimental colitis depends predominantly on the interleukin 12/signal transducer and activator of transcription (Stat)-4 pathway, but is not conditional on interferon gamma expression by T cells. J Exp Med. 1998;187:1225–34.
9. Fuss I, Neurath MF, Boirivant M et al. Disparate CD4+ lamina propria (LP) lymphocyte secretion profiles in inflammatory bowel disease. J Immunol. 1996;157:1261–70.
10. Rengarajan J, Szabo SJ. Transcriptional regulation of Th1/Th2 polarization. Immunol Today. 2000;21:479–83.
11. Asnagli H, Murphy KM. Stability and commitment in T helper cell development. Curr Opin Immunol. 2001;13:242–7.
12. Moser M, Murphy KM. Dendritic cell regulation of TH1-TH2 development. Nat Immunol. 2000;1:199-205.
13. Neurath MF, Fuss I, Kelsall BL, Stuber E, Strober W. Antibodies to IL-12 abrogate established experimental colitis in mice. J Exp Med. 1995;182:1280–9.
14. Neurath MF, Finotto S, Fuss I, Boirivant M, Galle PR, Strober W. Regulation of T-cell apoptosis in inflammatory bowel disease: to die or not to die, that is the mucosal question. Trends Immunol. 2001;22:21–6.
15. Fuss IJ, Marth T, Neurath MF, Pearlstein GR, Jain A, Strober W. Anti-interleukin 12 treatment regulates apoptosis of Th1 T cells in experimental colitis in mice. Gastroenterology. 1999;117:1078–88.
16. Neurath MF, Finotto S, Strober W. Immunology of inflammatory bowel disease. In: Clinical Immunology, 2nd edn, (Mosby, St Louis), 2001:1282–97.
17. Mueller DL, Seiffert S, Fang W, Behrens TW. Differential regulation of bcl-2 and bcl-x by CD3, CD28, and the IL-2 receptor in cloned CD4+ helper T cells. A model for the long-term survival of memory cells. J Immunol. 1996;156:1764–71.
18. Boise LH, Noel PJ, Thompson CB. CD28 and apoptosis. Curr Opin Immunol. 1995;7:620–5.
19. Radvanyi LG, Shi Y, Vaziri H et al. CD28 costimulation inhibits TCR-induced apoptosis during a primary T cell response. J Immunol. 1996;156:1788–98.
20. Noel PJ, Boise LH, Green JM, Thompson CB. CD28 costimulation prevents cell death during primary T cell activation. J Immunol. 1996;157:636–47.
21. Li B, Yu H, Zheng W.-P. et al. Role of the guanosine triphosphate Rac2 in T helper 1 cell differentiation. Science. 2000;288:2219–22.
22. Gomez M, Tybulewicz V, Cantrell DA. Control of pre T cell proliferation and differentiation by the GTPase Rac1. Nat Immunol. 2000;1:348–52.
23. Van den Brande JMH, Braat H, van den Brink GR et al. Infliximab but not etanercept induces apoptosis in lamina propria T-lymphocytes from patients with Crohn's disease. Gastroenterology. 2003;124:1774–85.

# 19
# Mechanisms of action of conventional drugs in inflammatory bowel disease

L. J. EGAN

## INTRODUCTION

Effective therapy of inflammatory bowel disease (IBD) often requires the use of more than one drug. This observation underscores the complexity of the pathogenesis of IBD and the fact that a single aetiological factor has yet to be identified. Many of the drugs used widely in IBD were initially selected for further evaluation based on the empirical observation of clinical improvement of IBD when those drugs were administered for other diseases. This is the case for 5-aminosalicylic acid drugs, corticosteroids, and infliximab. Other immune-modifying drugs, such as cyclosporine, methotrexate, and thiopurines, were initially tested in IBD because of established benefit in other inflammatory diseases, principally rheumatoid arthritis. Studies of the mechanism of action of all these drugs in IBD have not only provided the means to optimize their formulation and use, but have also established the usefulness of therapeutics as an experimental means to probe the pathophysiology of IBD. In this chapter I will set forth some of what is known concerning the mechanisms of action of drugs with established benefits in IBD. This discussion will include cyclosporine, tacrolimus, methotrexate, 5-aminosalicylic acid (5-ASA) drugs, and corticosteroids.

## CYCLOSPORINE AND TACROLIMUS

Cyclosporine is a cyclic polypeptide derived from a fungus, and tacrolimus is a macrolide antibiotic[1] These chemically unrelated entities share a very similar mechanism of immunosuppression. Cyclosporine binds to cyclophilin, an immunophilin protein in cytoplasm, whereas tacrolimus binds a related immunophilin, FK 506 binding protein (FKBP)[2]. Cyclosporine or tacrolimus–immunophilin complexes inhibit the phosphatase calcineurin, which prevents activation of the transcription factor, nuclear factor of activated T cells. Nuclear factor of activated T cells is an essential component of the T lymphocyte activation pathway after stimulation of those cells through the T cell receptor or through intracellular calcium signaling. Through this mechanism, T cell activation, as assessed by expression and secretion of interleukin-2, is inhibited potently by both cyclosporine and tacrolimus. Other putative molecular mechanisms of action of cyclosporine and tacrolimus include blockade of p38 mitogen-activated protein kinase pathways and inhibition of c-Rel containing nuclear factor κB (NFκB) complexes[3].

Cyclosporine has been observed to be highly effective when administered intravenously at a dose of either 2 or 4 mg/kg per day in severely active ulcerative colitis patients[4]. This observation highlights the importance of T lymphocytes in the pathogenesis of ulcerative colitis. However, the beneficial effects of cyclosporine are rarely realized without significant adverse effects[1,5]. In Crohn's disease cyclosporine has not been effective at non-toxic doses.

Tacrolimus has been studied in fistulizing Crohn's disease. In this clinical situation tacrolimus has been observed to improve fistula drainage, but at non-toxic doses does not appear to be all that effective in inducing fistula closure[6]. Several small uncontrolled series have also suggested benefit of tacrolimus in ulcerative colitis[7].

## METHOTREXATE

Methotrexate is an antimetabolite, anticancer drug that structurally is an analogue of folic acid. Methotrexate competes with folic acid for the enzyme dihydrofolate reductase. In the presence of methotrexate, folic acid cannot be reduced to its tetrahydrofolate form. As a consequence of this, purine nucleotide biosynthesis and thymidylate synthesis are both blocked. This leads to ineffective DNA synthesis during the S phase of the cell cycle and consequently cell death.

The above-described classic mechanism of action of methotrexate is widely believed to be the relevant anticancer mechanism of this drug. However, several observations regarding the use of methotrexate in chronic inflammatory diseases have suggested that inhibition of dihydrofolate reductase may not be the relevant mechanism in these diseases. Studies in rheumatoid arthritis, psoriasis and more recently IBD have indicated that doses significantly lower than those used during cancer chemotherapy are sufficient for a beneficial effect of methotrexate in inflammatory diseases[8,9]. The lower doses of methotrexate used in inflammatory diseases usually do not result in significant cytopenias; in fact the development of a cytopenia during methotrexate therapy from inflammatory disease is considered a toxicity[10]. Furthermore, the side-effects of methotrexate during chronic use in inflammatory disease can be somewhat mitigated by the co-administration of folic acid[11]. Importantly, this therapeutic manoeuvre does not appear to abrogate the therapeutic effect of the drug. Together, these observations have suggested that the anti-inflammatory action of methotrexate might be unrelated to inhibition of dihyrofolate reductase.

A number of studies invoked increased release of the anti-inflammatory autocoid, adenosine, as a potential anti-inflammatory mechanism of methotrexate. However, adenosine has been measured in the plasma and in the rectum of methotrexate-treated IBD patients, and no effects of this drug on adenosine concentrations were found[12]. This indicates that alterations of adenosine metabolism are probably not the relevant anti-inflammatory mechanism of action of methotrexate either.

A more likely explanation for the therapeutic benefit of methotrexate lies in its effects on T lymphocyte survival. One study convincingly showed that lymphocytes, both *in vitro* and after isolation from methotrexate-treated rheumatoid arthritis patients, underwent methotrexate-induced apoptosis when activated[13]. This effect was specific as it did not occur in the absence of methotrexate treatment or T lymphocyte activation.

Clinical studies have indicated that the therapeutic benefit of methotrexate takes at least 4 and perhaps up to 8 weeks before occurring[9,14]. A study of the clinical pharmacology of methotrexate in IBD patients showed that methotrexate concentrations, measured in erythrocytes, increased gradually over this period and plateaued after about 6–8 weeks[14,15]. Thus, it appears that in IBD patients, as has been observed in cancer patients, methotrexate does accumulate during chronic therapy in cells, probably as polyglutamated derivatives. These polyglutamated forms of methotrexate are likely to be important for the therapeutic benefit of this drug, and to explain the adequacy of once-weekly dosing.

## 5-AMINOSALICYLIC ACID DRUGS

Currently, there are four 5-ASA drugs available to treat IBD. Mesalamine (unconjugated 5-ASA) is one drug of this class. Mesalamine is rapidly and extensively absorbed in the upper small intestine and so, for use in IBD, has been formulated as a delayed or timed-release preparation. This results in effective delivery of the active 5-ASA to the lower small bowel and colon. Sulphasalazine, olsalazine, and balsalazide are all azo-conjugated formulations of 5-ASA. These drugs are believed to be inactive parent molecules that become activated when bacteria in the terminal ileum or colon cleave the azo bond which liberates free molecules of 5-ASA from the parent compound. As a result of the delayed-release properties of all the clinically available 5-ASA preparations, systemic absorption is low, ranging from 10% to 40%. Drug concentrations in the colon are also high, and range from 10 to 30 mmol/L.

The mechanisms of action of 5-ASA drugs have been extensively studied over the past 30–40 years. Observations in rectal biopsies or rectal dialysate have shown that 5-ASA drugs appear to decrease the production of prostaglandins, thromboxanes, and leukotrienes[16–19]. However, it is somewhat unlikely that inhibition of prostaglandin production is the relevant anti-inflammatory action of 5-ASA drugs, because non-steroidal anti-inflammatory drugs, such as aspirin, are more effective at this, although they aggravate IBD. In more recent years a series of *in-vitro* studies have examined the effect of 5-ASA drugs on intracellular signalling pathways that are believed to be important in inflammation. Both mesalamine and sulphazalazine have been shown to block the activation of a transcription factor, NFκB. In one study of young adult mouse colon cells, 5-ASA was shown to block degradation of IκBα induced by proinflammatory cytokines[20]. In another study an important post-transcriptional modification of the NFκB polypeptide RelA was shown to be blocked by mesalamine[21]. Yet another study showed that sulphasalazine blocked the activity of the NFκB activating kinase, IKK[22,23]. It is thus possible that the relevant *in-vivo* therapeutic mechanism of 5-ASA drugs might relate to blocking the activation of NFκB, perhaps in intestinal epithelial cells. However, the fact that 5-ASA drugs need to be present in quite high (mmol/L) concentrations to be effective suggests that perhaps these agents do not possess a single specific therapeutic target, but rather have a multitude of molecular effects, whose aggregate tends to be beneficial in IBD.

## CORTICOSTEROIDS

Corticosteroids were the first drugs proven to be beneficial in IBD. Indeed, the effective use of corticosteroids probably explains the dramatic reduction in mortality associated with ulcerative colitis in the middle of the last century. A variety of different naturally occurring and synthetic corticosteroids are used to treat IBD. The anti-inflammatory effects of corticosteroids in IBD and other inflammatory diseases is due to the glucocorticoid effects of these steroids. Mineralocorticoid effects are unrelated to inhibition of inflammation. It is important to note that the anti-inflammatory activity of corticosteroids is tightly coupled to their metabolic effects. This fact underlies the important side-effects that are observed when using corticosteroids for prolonged periods.

The classic mechanism of action of corticosteroids involves the diffusion of the steroid across the plasma membrane into the cytoplasm. Here the steroid encounters its endogenous receptor. Glucocorticoid receptors, once bound by their steroid ligand, dissociate from heat-shock proteins in the cytoplasm and migrate to the nucleus. In the nucleus the steroid receptor pair binds to glucocorticoid response elements in the promoter regions of many genes. Usually this leads to up-regulated transcription of those genes, although in certain cases binding of the steroid receptor pair can repress transcription.

Alternative non-transcriptional mechanisms of action of corticosteroids have been sought to explain their anti-inflammatory effects. In one model the corticosteroid–receptor pair physically interacts with other transcription factors in the nucleus through leucine zippers. Such an interaction impairs the localization of the other transcription factors to their binding sites in the promoter regions of target genes. Transcriptional factors known to be blocked by corticosteroids include NFκB and activator protein 1. Inhibition of the activity of these proinflammatory transcriptions factors might in part underlie the benefit of corticosteroids in inflammatory diseases[24,25]. In another model the corticosteroid receptor pair, when present in molar excess over other transcription factors such as NFκB, results in the sequestration of transcriptional co-activators by the corticosteroid receptor pair[26]. For example, cyclic AMP response element binding protein or another related transcriptional co-activator, p300, are required for the transcriptional activities of both corticosteroid receptors and NFκB. If the corticosteroid receptors are activated and present in excess compared to NFκB, then NFκB is not able to up-regulate the transcription of the target genes because a functional transcription-activating complex cannot be assembled on the target gene promoters. The importance of the non-classical mechanisms of action of corticosteroids as anti-inflammatory drugs lies in the potential for designing corticosteroid analogues that will cause glucocorticoid receptors to functionally interact with transcriptional co-activators or other transcription factors as described in the models above, but which will be unable to initiate gene transcription. This offers the promise of uncoupling the metabolic side-effects of corticosteroids from their anti-inflammatory effects.

The effects of corticosteroids on markers of inflammatory activity in IBD patients have been studied. A variety of studies have observed decreased expression of activation markers on circulating peripheral lymphocytes, decreased activation of NFκB in biopsies of rectal mucosa, and decreased migration of leucocytes from the vasculature into inflamed tissues[27,28]. At the cellular level any or all of these mechanisms are likely to be important.


## CONCLUSIONS

The therapy of IBD has greatly improved in recent years, but there is still a long way to go. Knowledge of the underlying molecular mechanisms of existing drugs in IBD has permitted, and will permit in the future, the design of superior novel therapeutic agents. The benefit will derive from the development of novel, more targeted therapies. For example, the observation that cyclosporine blocked interleukin-2 production led in part to the development of novel interleukin-2 inhibiting drugs, such as daclizumab[29]. The anti-leucocyte trafficking effects of corticosteroids also led in part to the development of anti-integrin molecules, such as natalizumab[30]. Finally, knowledge of underlying molecular mechanisms of action of drugs can also lead to the development of safer and better-tolerated drugs. This is best exemplified in the development of mesalamine from sulphasalazine.

## References

1.  Loftus CG, Loftus EV Jr, Sandborn WJ. Cyclosporin for refractory ulcerative colitis. Gut. 2003;52:172–3.
2.  Liu J, Farmer JD Jr, Lane WS, Friedman J, Weissman I, Schreiber SL. Calcineurin is a common target of cyclophilin–cyclosporin A and FKBP–FK506 complexes. Cell. 1991;66: 807–15.
3.  Kalli K, Huntoon C, Bell M, McKean DJ. Mechanism responsible for T-cell antigen receptor- and CD28- or interleukin 1 (IL-1) receptor-initiated regulation of IL-2 gene expression by NF-kappaB. Mol Cell Biol. 1998;18:3140–8.
4.  Van Assche G, D'Haens G, Noman M et al. Randomized, double-blind comparison of 4 mg/kg versus 2 mg/kg intravenous cyclosporine in severe ulcerative colitis. Gastroenterology. 2003;125:1025–31.
5.  Sandborn WJ. Cyclosporine in ulcerative colitis: state of the art. Acta Gastroenterol Belg. 2001;64:201–4.
6.  Sandborn WJ, Present DH, Isaacs KL et al. Tacrolimus for the treatment of fistulas in patients with Crohn's disease: a randomized, placebo-controlled trial. Gastroenterology. 2003;125:380–8.

7.  Hogenauer C, Wenzl HH, Hinterleitner TA, Petritsch W. Effect of oral tacrolimus (FK 506) on steroid-refractory moderate/severe ulcerative colitis. Aliment Pharmacol Ther. 2003; 18:415–23.
8.  Feagan BG, Fedorak RN, Irvine EJ et al. A comparison of methotrexate with placebo for the maintenance of remission in Crohn's disease. North American Crohn's Study Group Investigators. N Engl J Med. 2000;342:1627–32.
9.  Feagan BG, Rochon J, Fedorak RN et al. Methotrexate for the treatment of Crohn's disease. The North American Crohn's Study Group Investigators. N Engl J Med. 1995;332: 292–7.
10. Egan LJ, Sandborn WJ. Methotrexate for inflammatory bowel disease: pharmacology and preliminary results. Mayo Clin Proc. 1996;71:69–80.
11. Whittle SL, Hughes RA. Folate supplementation and methotrexate treatment in rheumatoid arthritis: a review. Rheumatology (Oxford). 2004;43:267–71.
12. Egan LJ, Sandborn WJ, Mays DC, Tremaine WJ, Lipsky JJ. Plasma and rectal adenosine in inflammatory bowel disease: effect of methotrexate. Inflamm Bowel Dis. 1999;5:167–73.
13. Genestier L, Paillot R, Fournel S, Ferraro C, Miossec P, Revillard JP. Immunosuppressive properties of methotrexate: apoptosis and clonal deletion of activated peripheral T cells. J Clin Invest. 1998;102:322–8.
14. Egan LJ, Sandborn WJ, Tremaine WJ et al. A randomized dose–response and pharmacokinetic study of methotrexate for refractory inflammatory Crohn's disease and ulcerative colitis. Aliment Pharmacol Ther. 1999;13:1597–604.
15. Egan LJ, Sandborn WJ, Mays DC, Tremaine WJ, Fauq AH, Lipsky JJ. Systemic and intestinal pharmacokinetics of methotrexate in patients with inflammatory bowel disease. Clin Pharmacol Ther. 1999;65:29–39.
16. Sharon P, Ligumsky M, Rachmilewitz D, Zor U. Role of prostaglandins in ulcerative colitis. Enhanced production during active disease and inhibition by sulfasalazine. Gastroenterology. 1978;75:638–40.
17. Ligumsky M, Karmeli F, Sharon P, Zor U, Cohen F, Rachmilewitz D. Enhanced thromboxane A2 and prostacyclin production by cultured rectal mucosa in ulcerative colitis and its inhibition by steroids and sulfasalazine. Gastroenterology. 1981;81:444–9.
18. Greenfield SM, Boswell DJ, Punchard NA, Thompson RP. The effects of 5-aminosalicylic acid and acetyl-5-aminosalicylic acid on lipid peroxidation in erythrocytes and prostaglandin production by mononuclear cells. Aliment Pharmacol Ther. 1992;6:671–83.
19. Punchard NA, Boswell DJ, Greenfield SM, Thompson RP. The effects of sulphasalazine and its metabolites on prostaglandin production by human mononuclear cells. Biochem Pharmacol. 1992;43:2369–76.
20. Yan F, Polk DB. Aminosalicylic acid inhibits IkappaB kinase alpha phosphorylation of IkappaBalpha in mouse intestinal epithelial cells. J Biol Chem. 1999;274:36631–6.
21. Egan LJ, Mays DC, Huntoon CJ et al. Inhibition of interleukin-1-stimulated NF-kappaB RelA/p65 phosphorylation by mesalamine is accompanied by decreased transcriptional activity. J Biol Chem. 1999;274:26448–53.
22. Weber CK, Liptay S, Wirth T, Adler G, Schmid RM. Suppression of NF-kappaB activity by sulfasalazine is mediated by direct inhibition of IkappaB kinases alpha and beta. Gastroenterology. 2000;119:1209–18.
23. Wahl C, Liptay S, Adler G, Schmid RM. Sulfasalazine: a potent and specific inhibitor of nuclear factor kappa B. J Clin Invest. 1998;101:1163–74.
24. Scheinman RI, Gualberto A, Jewell CM, Cidlowski JA, Baldwin AS Jr. Characterization of mechanisms involved in transrepression of NF-kappa B by activated glucocorticoid receptors. Mol Cell Biol. 1995;15:943–53.
25. Ray A, Prefontaine KE. Physical association and functional antagonism between the p65 subunit of transcription factor NF-kappa B and the glucocorticoid receptor. Proc Natl Acad Sci USA. 1994;91:752–6.
26. De Bosscher K, Vanden Berghe W, Haegeman G. The interplay between the glucocorticoid receptor and nuclear factor-kappaB or activator protein-1: molecular mechanisms for gene repression. Endocrinol Rev. 2003;24:488–522.
27. Tillinger W, Gasche C, Reinisch W et al. Influence of topically and systemically active steroids on circulating leukocytes in Crohn's disease. Am J Gastroenterol. 1998;93:1848–53.

28. Bernstein CN, Sargent M, Rawsthorne P, Rector E. Peripheral blood lymphocyte beta 2 integrin and ICAM expression in inflammatory bowel disease. Dig Dis Sci. 1997;42:2338–49.
29. Van Assche G, Dalle I, Noman M et al. A pilot study on the use of the humanized anti-interleukin-2 receptor antibody daclizumab in active ulcerative colitis. Am J Gastroenterol. 2003;98:369–76.
30. Ghosh S, Goldin E, Gordon FH et al. Natalizumab for active Crohn's disease. N Engl J Med. 2003;348:24–32.

# 20
# Optimizing the use of current drugs for inflammatory bowel disease

G. D'HAENS

'Optimizing' therapy means that for every individual patient the most effective strategy is to be selected. This will depend on the location of the disease, the activity of the inflammation, the presence of intestinal and/or extraintestinal complications and specific circumstances such as pregnancy, old age, etc. Efficacy of treatments can be optimized by selecting the right doses, the correct duration of therapy and by increasing patient compliance with his/her treatment. Four categories of drugs currently being used in the management of inflammatory bowel disease (IBD) will be discussed in this chapter: aminosalicylates, corticosteroids, immunomodulators and biologicals. Antibiotics will be discussed elsewhere.

For aminosalicylates the most common error in Crohn's disease (CD) is that too-low doses are being used. For induction of remission in mild to moderate disease at least 3–4 g of mesalazine need to be given. Maintenance benefits with aminosalicylates are questionable, although their use in the prevention of postoperative recurrence is well established. Modigliani and the French GETAID demonstrated that mesalazine 4 g/day following corticosteroid induction for active CD relapse did not lead to a significant reduction of the time to relapse when compared to placebo[1]. The largest meta-analysis so far on this topic was performed by Camma et al., demonstrating a moderate maintenance benefit of aminosalicylates in CD[2]. In ulcerative colitis (UC) a clear dose/response relationship has been demonstrated both for active and quiescent disease[3,4]. Many patients with an acute attack of UC will benefit more from a combination of oral and topical therapy than from either formulation alone[5]. The bottom line is that aminosalicylate therapy in UC needs to tailored to the individual 'needs' of every single patient. If 5-ASA is used only in the setting of the prevention of dysplasia in long-standing colitis, however, doses $\geqslant 1.2$ g/day usually suffice[6]. Finally, aminosalicylate drugs are perfectly safe during pregnancy.

Corticosteroids are still the most commonly used drugs for moderate to severe relapses of both CD and UC. They are effective in 60–70% of patients, provided sufficiently high dosages are given. Abdominal sepsis/abscess should be ruled out in order to prevent infectious complications[7]. Many patients can be treated with topical instead of systemic steroids, a strategy which clearly leads to a significant reduction in the number of steroid-induced complications[8]. Maintenance therapy with corticosteroids should virtually never be accepted. Some patients who remain dependent on systemic steroids can be switched to topical agents[9]. When given as maintenance agents, however, budesonide did not significantly reduce the number of relapses at the end of 1 year follow-up when compared to placebo[10]. Osteopenia must be considered in IBD patients on steroids, and specific measures such as calcium supplements, vitamin D and/or biphosphonates should be instituted whenever necessary. In UC, steroids can often be started orally in addition to 5-ASA. In patients failing oral steroids, hospitalization for intravenous corticosteroid therapy will often improve the patient's condition. Only after 3–7 days of intravenous steroid failure does the start of intravenous cyclosporine treatment need to be considered.

Immunomodulators have undoubtedly taken an extremely important position in the modern management of IBD. In a paediatric trail in which children with newly diagnosed CD were given 6-mercaptopurine (6-MP) or placebo in addition to corticosteroids, more than 90% of the children were still in remission after 600 days, versus only 50% ($p < 0.007$) with placebo[11]. The Cochran analysis demonstrated an odds ratio benefit for azathioprine of 2.36, with a number of patients needed to treat of four, provided the therapy is continued beyond 17 weeks. In addition, azathioprine therapy was shown to induce mucosal healing in both recurrent neoterminal ileal disease and in primary colitis and ileocolitis[12,13].

It is important to understand the metabolism of azathioprine and 6-MP in order to use these drugs in an optimal fashion. The most important active metabolite is 6-thioguanine nucleotide (6-TGN), whereas another metabolite, 6-methylmercaptopurine ribonucleotides (6-MMPR), metabolized via the enzyme TPMT, is responsible for hepatic toxicity[14]. TPMT activity is genetically determined, with three groups of enzyme activity to be distinguished: the 'wild type' in 89% of patients, heterozygote deficiency in a little more than 10% of patients and homozygote deficiency in 0.3% of patients. It is currently not recommended, however, to test TPMT routinely in all patients[15]. Aminosalicylates can inhibit TPMT activity to a certain extent, leading to a decrease in 6-MMPR, but elevated levels of 6-TGN with enhanced risk of leukopenia. After hepatic non-enzymatic conversion of azathioprine into 6-MP (approximately 88% of the drug undergoes this conversion), part of this drug is metabolized by xanthine oxidase into 6-thiouric acid (6-TU). Allopurinol, a drug widely used for the treatment and prevention of gout, blocks xanthine oxidase and also leads to higher 6-TGN levels and leukopenia. Clinical response to azathioprine or 6-MP correlates with blood 6-TGN levels above 235 pmol/$8 \times 10^8$ RBC. So far, data demonstrating that routine monitoring of 6-TG levels in patients offers a better tailoring and outcome of antimetabolite therapy remain equivocal. On the other hand, dose escalation in patients failing this type of therapy can 'rescue' these patients in up to one-third of patients[16]. 6-Thioguanine (6-TG) is an alternative antimetabolite drug that has been tested in IBD. It is immediately metabolized into the effective compound 6-TGN. Herrlinger and colleagues demonstrated that, with doses of 20–40 mg/day, clinical response and steroid reduction could be achieved in the majority of patients after 6–12 months[17]. The problem with this drug, however, is the development of hepatotoxicity, most commonly nodular regenerative hyperplasia. This problem can even occur in patients with perfectly normal liver function tests. For this reason current recommendations would restrict the use of 6-TG to controlled clinical trials.

A frequently asked question is how long antimetabolite therapy should be continued. The GETAID addressed this issue in an elegant trial in which patients in remission on azathioprine were randomized to continued azathioprine or to placebo. A limited benefit in patients on continued therapy was demonstrated, most pronounced in patients with elevated CRP levels (and probably subtle ongoing inflammation)[18]. Azathioprine can safely be continued during pregnancy, but the excretion in breast milk is high, and breast-feeding should not be encouraged. Overall, clinical data demonstrate equal biological activity of azathioprine and 6-MP. The conversion factor to calculate the dose of 6-MP to be used instead of azathioprine is 2.08. Usually, 6-MP is over-dosed and azathioprine under-dosed. Recommended dosages for 6-MP are 1–1.5 mg/kg per day and for azathioprine 2–2.5 mg/kg per day.

An alternative immunomodulatory drug is methotrexate (MTX), a folate analogue which is preferentially given parenterally because of its variable intestinal resorption[19]. The dose to start with is 25 mg/week for 12 weeks, followed by a maintenance dose of 15 mg/week. MTX is clearly prohibited during pregnancy. Although the drug's efficacy has been firmly demonstrated in CD, its efficacy in UC remains questionable. MTX toxicity can be reduced with folate supplements. Pancytopenia has been observed in 1–2% of patients; liver fibrosis is rare but can occur, but routine liver biopsies do not appear to be justified. Mild liver function test (LFT) abnormalities are common and should not always lead to discontinuation of this therapy. Interstitial pneumonitis (mortality 17%) is also extremely rare but should be recognized promptly and treated accordingly[20].

Cyclosporine is a treatment that is now only rarely being used for CD, given the advent of many new and potent biological agents. For UC, however, intravenous cyclosporine leads to improvement of symptoms in 60–80% of patients after they have failed corticosteroids. It was recently demonstrated that intravenous cyclosporine doses of 2 mg/kg per day are usually sufficient[21]. Ideally, patients receiving cyclosporine should receive prophylaxis against *Pneumocystis carinii* pneumonia with cotrimoxazole.

In recent years a great deal of insight in the pathophysiology of CD has been established. The proinflammatory cytokine tumour necrosis factor (TNF) was shown to play a pivotal part in the orchestration and amplification of the inflammatory process, and has therefore received much attention as a potential target for therapeutic intervention. Several strategies to interfere with the biological effects of TNF have been developed, such as antibodies which directly bind and inactivate TNF, and recombinant soluble receptors targeting TNF. Antibodies in clinical use or trials include infliximab, an $IgG_1$ chimeric antibody, CDP 870, a pegylated humanized anti-TNF Fab-fragment, and adalimumab (Humira), a fully human $IgG_1$. TNF has two receptors which have both been cloned for therapeutic use: the p75 receptor etanercept and the p55 receptor onercept. The mechanisms of action of infliximab and etanercept were recently compared in an elegant study by Van den Brande et al.[22]. Etancercept, a TNF receptor construct (two receptors on a human $IgG_1$ Fc portion) binds both TNF-$\alpha$ and TNF-$\beta$ (lymphotoxin), whereas the chimera infliximab is highly specific in binding only TNF-$\alpha$. Binding lymphotoxin might potentially increase susceptibility to infections. The most important difference between the two molecules is its effect on TNF-expressing inflammatory cells. Infliximab binds complement and leads to apoptosis of the inflammatory cells, whereas etanercept does not induce cell death. Since CD is characterized by a certain degree of apoptosis resistance, this may be the reason why infliximab is the more potent drug in the treatment of CD. Infliximab is currently on the market under the trade name 'Remicade', and is approved in many countries and reimbursed for active CD refractory to 'standard therapy', corticosteroid dependence, relapses in spite of immunosuppression and fistulizing disease. Contraindications include obstructive stenosis, absence of inflammatory activity but rather 'fibrotic disease', active infections including tuberculosis, congestive heart failure, presence of abscesses in patients with fistulizing disease and demyelinization of the central nervous system as in multiple sclerosis.

The Accent-I trial, the largest trial in CD published so far, demonstrated that repeated administration of infliximab every 8 weeks was effective to maintain remission in patients who had an initial response to the drug[23]. Fifty-three per cent of the patients receiving 10 mg/kg of infliximab as maintenance treatment still had a clinical response at the end of the year, versus 43% with 5 mg/kg and 17% with placebo, although many patients in the placebo group had received active medication on relapse during the trial. Virtually all corticosteroid-dependent patients could taper and discontinue their steroids completely after a mean period of 22 weeks. Adverse events were relatively mild and uncommon: serious infections in 3.8% (one case of tuberculosis), delayed hypersensitivity in 2.4%, acute infusion reactions in 1%. Interestingly, more than half of the patients treated with 10 mg/kg maintenance infusions every 8 weeks were found to have complete healing of the ulcerative lesions, significantly more than in the placebo group (only 1/17). Furthermore, endoscopic healing was found to be associated with fewer hospitalizations and surgical interventions for the duration of the trial.

The Accent-II trial had a somewhat similar design but focused on maintenance therapy for fistulizing CD[24]. Initial responders to a three-dose induction regimen (5 mg/kg at weeks 0, 2 and 6) were randomized to maintenance infusions every 8 weeks of 5 mg/kg or placebo. At the end of the study (1 year), 46% of the patients with fistulizing disease still had a fistula response, versus 23% of the patients receiving placebo.

As mentioned earlier, reactivation of tuberculosis in patients previously exposed to this disease is a particular problem with infliximab therapy[25]. Several fatalities were reported before the problem was fully recognized. The human immune system indeed appears to need TNF to kill intracellular pathogens such as *Mycobacterium tuberculosis*. Of all documented tuberculosis cases, 75% were diagnosed within the first three infusions and 97% within the first six infusions. In patients with signs of earlier tuberculosis exposure based on chest X-rays and PPD skin test, prophylactic treatment with INH during the first 6–12 months of infliximab treatment is therefore warranted.

Another important issue in patients repeatedly receiving infliximab is the development of antibodies to infliximab (ATI), with concomitant loss of clinical benefit and risk of infusion reactions[26]. We measured these antibodies in more than half of our patients who were treated with at least six infusions of infliximab on an 'on-demand' basis. The presence of these antibodies correlated with the occurrence of all types of infusion reactions. Both phenomena were more common in patients not using immunosuppressives such as azathioprine or methotrexate, which was in fact the only protective factor we were able to discover. Recently, routine administration of hydrocortisone prior to infusions was also shown to diminish the risk of antibody formation and infusion reactions[27]. When patients do have an infusion reaction, however, several measures can be taken. Besides antihistamines, acetaminophen and corticosteroids are helpful. The infusion may need to be discontinued for a while, but can often be resumed later at a slower infusion rate. Prior to future infusions these patients always receive prophylactic corticosteroids at our centre.

A number of unresolved questions with regard to routine infliximab use remain. Can immunogenicity be reduced by further humanization? What about long-term safety? What is the role and benefit of mucosal healing[28]? Is prevention of recurrent inflammation ('automatic retreatment') superior to repeated treatment of relapse ('on demand')? Further trials will undoubtedly give answers to these relevant questions.

In conclusion, the most difficult aspect of optimal patient management in IBD is to estimate the individual 'patient need' for therapy. We indeed need markers to predict a patient's clinical evolution and the later likelihood of complications and/or surgery. The majority of patients 'intolerant' to certain therapies will eventually tolerate treatment with specific measures. The overall prognosis of IBD will probably further improve using more 'aggressive approaches' early on.

## References

1. Modigliani R, Colombel JF, Dupas J-L et al. Mesalamine in Crohn's disease with steroid-induced remission: effect on steroid withdrawal and remission maintenance. Gastroenterology. 1996;110:688–93.
2. Camma C, Giunta M, Roselli M et al. Mesalamine in the maintenance treatment of Crohn's disease: a meta-analysis adjusted for confounding variables. Gastroenterology. 1997;113:1465–73.
3. Schroeder KW, Tremaine WJ, Ilstrup DM. Coated oral 5-ASA therapy for mild to moderate active ulcerative colitis: a randomized study. N Engl J Med. 1987;317:1625–9.
4. Hanauer S, Schwartz J, Robinson M et al. Mesalamine capsules for treatment of active ulcerative colitis: results of a controlled trial. Am J Gastroenterol. 1993;88:1188–97.
5. d'Albasio G, Pacini F, Camarri E et al. Combined therapy with 5-aminosalicylic acid tablets and enemas for maintaining remission in ulcerative colitis: a randomized double-blind study. Am J Gastroenterol. 1997;92:1143–7.
6. Eaden J, Abrams K, Ekbom A, Jackson E, Mayberry J. Colorectal cancer prevention in ulcerative colitis: a case–control study. Aliment Pharmacol Ther. 2000;14:145–53.
7. Summers RW, Switz DM, Sessions JT Jr et al. National Cooperative Crohn's Disease Study: results of drug treatment. Gastroenterology. 1979;77:847–69.
8. Rutgeerts P, Lofberg R, Malchow H et al. A comparison of budesonide with prednisolone for active Crohn's disease. N Engl J Med. 1994;331:842–5.

9. Andus T, Gross V, Caesar I et al. Replacement of conventional glucocorticoids by oral pH-modified release budesonide in active and inactive Crohn's disease: results of an open, prospective, multicenter trial. Dig Dis Sci. 2003;48:373–8.
10. Sandborn WJ et al. Budesonide capsules prolong time to relapse in Crohn's disease patients with medically induced remissions. Gastroenterology. 2003;124(Suppl. 4):A503 (abstract).
11. Markowitz J, Grancher K, Kohn N, Lesser M, Daum F. A multicenter trial of 6-mercaptopurine and prednisone in children with newly diagnosed Crohn's disease. Gastroenterology. 2000;119:895–902.
12. D'Haens G, Geboes K, Ponette E et al. Healing of severe recurrent ileitis with azathioprine therapy in patients with Crohn's disease. Gastroenterology. 1997;112:1475–81.
13. D'Haens G, Geboes K, Rutgeerts P. Endoscopic and histologic healing of Crohn's (ileo-) colitis with azathioprine. Gastrointest Endosc. 1999;50:667–71.
14. Lowry PW, Franklin CL, Weaver AL et al. Measurement of thiopurine methyltransferase activity and azathioprine metabolites in patients with inflammatory bowel disease. Gut. 2001;49:665–70.
15. Gearry RB, Barclay ML, Burt MJ et al. Thiopurine S-methyltransferase genotype does not predict adverse drug reactions to thiopurine drugs in patients with inflammatory bowel disease. Aliment Pharmacol Ther. 2003;18:395–400.
16. Dubinsky MC, Yang H, Hassard PV et al. 6-MP metabolite profiles provide a biochemical explanation for 6-MP resistance in patients with inflammatory bowel disease. Gastroenterology. 2002;122:904–15.
17. Herrlinger KR, Deibert P, Schwab M et al. Remission maintenance by thioguanine in chronic active Crohn's disease. Aliment Pharmacol Ther. 2003;17:1459–64.
18. Lemann M, Bouhnik Y, Colombel JF et al. Randomized, double-blind, placebo-controlled, multi-center, azathioprine withdrawal trial in Crohn's disease. Gastroenterology. 2002;122:A-23 (abstract).
19. Kurnik D, Loebstein R, Fishbein E et al. Bioavailability of oral vs. subcutaneous low-dose methotrexate in patients with Crohn's disease. Aliment Pharmacol Ther. 2003;18:57–63.
20. Fraser AG, Morton D, McGovern D, Travis S, Jewell D. The efficacy of methotrexate for maintaining remission in inflammatory bowel disease. Aliment Pharmacol Ther. 2002;16:693–7.
21. Van Assche G, D'Haens G, Noman M et al. Randomized, double-blind comparison of 4 mg/kg versus 2 mg/kg intravenous cyclosporine in severe ulcerative colitis. Gastroenterology. 2003;125:1025–31.
22. Van den Brande JM, Braat H, van den Brink GR et al. Infliximab but not etanercept induces apoptosis in lamina propria T-lymphocytes from patients with Crohn's disease. Gastroenterology. 2003;124:1774–85.
23. Hanauer SB, Feagan BG, Lichtenstein GR et al. ACCENT I Study Group. Maintenance infliximab for Crohn's disease: the ACCENT I randomised trial.Lancet. 2002;359:1541–9.
24. Sands B, Van Deventer S, Bernstein C et al. Long-term treatment of fistulizing Crohn's disease: response to infliximab in the Accent II trial through 54 weeks. Gastroenterology. 2002;122:A-81.
25. Keane J, Gershon S, Wise RP, Mirabile-Levens E. Tuberculosis associated with infliximab, a tumour necrosis factor alpha-neutralizing agent. N Engl J Med. 2001;345:1098–104.
26. Baert F, Noman M, Vermeire S et al. Influence of immunogenicity on the long-term efficacy of infliximab in Crohn's disease. N Engl J Med. 2003;348:601–8.
27. Farrell RJ, Alsahli M, Jeen YT, Falchuk KR, Peppercorn MA, Michetti P. Intravenous hydrocortisone premedication reduces antibodies to infliximab in Crohn's disease: a randomized controlled trial. Gastroenterology. 2003;124:917–24.
28. D'Haens G, Van Deventer S, Van Hogezand R et al. Endoscopic and histological healing with infliximab anti-tumor necrosis factor antibodies in Crohn's disease: a European multicenter trial.Gastroenterology. 1999;116:1029–34.

# Section VII
# The difficult cases of Crohn's disease

Chair: R. CAPRILLI and S. BAR-MEIR

# 21
# Diffuse jejunoileitis in an adolescent

S. CUKOVIC-CAVKA and B. VUCELIC

## INTRODUCTION

The majority of patients with Crohn's disease (CD) present with disease involving the distal ileum which often extends into the caecal pole or the proximal ascending colon. Diffuse jejunoileitis is a less common but very complex phenotype of CD. The medical management of diffuse small bowel disease is a very challenging and demanding problem. We report a case of a young man with a diffuse jejunoileitis.

## CASE REPORT

A 12-year-old boy presented in 1997 with a history of intermittent diarrhoea, growth retardation, arthralgias and iridocyclitis. Colonoscopy at that time revealed pancolitis. Small bowel series and ileoscopy revealed terminal ileitis. Granulomas were found in ileal biopsy and the diagnosis of CD was confirmed. The patient was treated initially with mesalamine and glucocorticoids, which resulted in remission within 2 months. However, he had prompt relapses on several attempts to taper off steroids, which prompted initiation of azathioprine (2 mg/kg per day). Unfortunately, azathioprine had to be discontinued due to leukopenia, especially lymphopenia. The left colonic disease progressed into long stenosis of the distal descending and sigmoid colon (Figure 1). Growth failure (height and weight chart demonstrated the values under the third percentile) and medical therapy failure prompted his referral to surgery in December 2001. The left colon was resected, a colostomy was created in the area of splenic flexure and the rectum was left *in situ* with mucous fistula. Methotrexate was initiated in the immediate postoperative period. Following surgery the patient rapidly gained weight (+ 15 kg) and height (+ 20 cm) and the illness remained in remission for 1 year (Figures 2a and 2b). During 2003 he relapsed twice. Infliximab was not available at that point. He was admitted to our unit at the end of December 2003 with large pretibial oedema and severe colicky pain in the abdomen. The flat plane of the abdomen showed partial bowel obstruction, multislice spiral computed tomography (MSCT) demonstrated dilated bowel loops and serum albumin levels were very low (18 g/L).

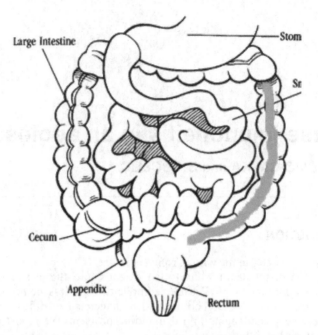

**Figure 1** The left colonic disease progressed into long stenosis of the distal descending and sigmoid colon

The patient was treated with intravenous steroids and the abdominal cramps immediately ceased. Small bowel series confirmed several jejunal short strictures with prestenotic dilations (Figure 3). While attempting to taper steroids the patient developed obstructive symptoms (colicky abdominal pain), fever and cough. MSCT revealed laminar collection around the colostomy and the inflammation of the abdominal wall. The patient was referred to surgery and multiple strictureplasties were done.

## DISCUSSION

This is the case of a very unfortunate young man with CD. The course of the disease has been difficult for a number of reasons. The extension of the disease is unusual because initial pancolitis and ileitis progressed after 5 years into diffuse jejunoileitis. Problems with medical therapy started with azathioprine, which had to be discontinued due to bone marrow toxicity. The disease was refractory to methotrexate since the patient developed diffuse jejunoileitis while on methotrexate therapy. Further, his disease obviously has a high fibrostenosing potential. According to CARD15 genetic testing he is a heterozygote for SNP13. This raises questions concerning the future course of the disease.

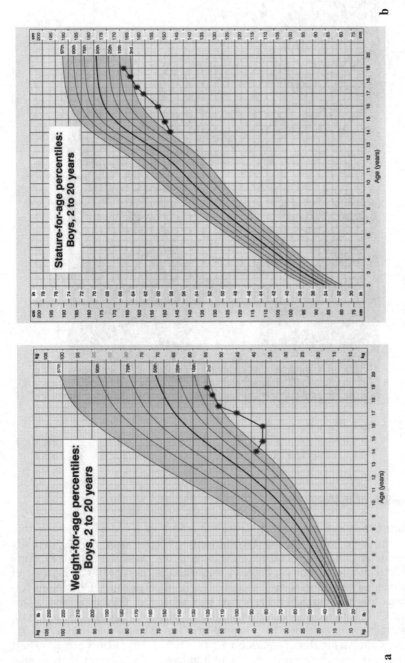

**Figure 2** Weight (a) and height (b) charts – following surgery the patient rapidly gained weight (+ 15 kg) and height (+ 20 cm)

191

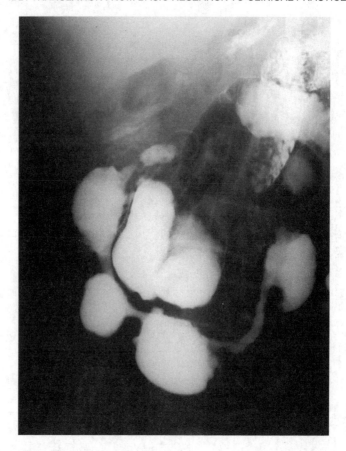

**Figure 3**   Multiple jejunal strictures

## Extension of the disease

Regarding the extension of the disease the epidemiological data indicate that ileum and colon are the most commonly affected sites in CD, usually complicated by intestinal obstruction, inflammatory mass or abscess[1,2]. Diffuse jejunoileitis is a less common but very important phenotype of CD because of the associated high morbidity and challenges in management[3]. Diffuse jejunoileitis was first recognized and reported by Crohn and Yunich in 1941[4]. This form of CD is often complicated by multifocal stenoses, bacterial overgrowth, and protein-losing enteropathy[5]. Jejunal involvement was found in 4%, 5.7% and 7% of patients in three large cohorts[3,6,7]. Diffuse jejunoileitis presents itself more often in younger patients, with children having significantly more jejunal disease than adults[8,9]. In addition, most patients with diffuse jejunoileitis present with extensive small bowel disease exclusively, though 10–15% of patients develop diffuse jejunoileitis several years after presenting initially with focal macroscopic disease, either in the distal ileum or large intestine. In nearly 25% of patients the disease extends to include the duodenum, and 30% also have colonic involvement[3,10]. Nearly all patients have significant abdominal symptoms including colicky abdominal pain, weight loss and diarrhoea[3].

## Therapy

Regarding the therapy of diffuse jejunoileitis there are still many open questions; this is a very challenging area in the field of therapy of CD. Corticosteroids and immune modulators are the primary medical therapies for jejunoileal disease. Infliximab appears effective in two-thirds of patients[11]; it is also effective in inducing remission in patients with active CD[12,13]. Unfortunately, medical management with immunosuppressants was not successful in this patient. Initially he was treated with azathioprine, which had to be discontinued because of leukopenia. Bone marrow depression leading to leukopenia, anaemia or thrombocytopenia is a dose-related toxicity of azathioprine. If severe, leukopenia can cause severe immunosuppression. Cytopenias have been reported to develop from 2 weeks to 11 years after beginning therapy, at a cumulative frequency of about 10%[14,15]. Methotrexate was initiated immediately after the left colonic resection, resulting in disease remission for 1 year. Methotrexate administered intramuscularly or subcutaneously at a dose of 25 mg per week is effective for induction of remission in patients with steroid-dependent and steroid-refractory CD[13,16]. Following several relapses during 2003 the drug was discontinued.

Surgical therapy was necessary on two occasions for this patient. The first operation occurred at the beginning of the course, indicated by the development of long stenosis in the distal descending and sigmoid colon. Growth failure and medical treatment failure prompted his referral to surgery in December 2001. The left colon was resected, a colostomy was created in the area of splenic flexure and the rectum was left *in situ* with mucous fistula. Following surgery he rapidly gained weight and the illness was in remission for 1 year. Surgery was a good decision at that point due to the boy's 'catch-up' growth. The second operation was prompted by multiple jejunal strictures.

Most patients with diffuse jejunoileitis require one or more surgical procedures because of the intestinal obstruction. The proportion of patients requiring surgery is highest in the first year after diagnosis[3,10]. Relief of obstruction is effected by strictureplasty or limited resection of areas of tight stenoses. The increasing use of strictureplasty for short strictures, and the minimal use of resection, have eliminated problems associated with the short small bowel syndrome[3]. Nowadays, strictureplasties are being used with increasing frequency in the surgical management of severe and extensive CD of the small bowel[16]. With a meticulous approach to strictureplasties the surgeon can preserve intestinal length and avoid blind or bypassed bowel loops[17,18].

The treatment of CD is very complex, especially with a phenotype described in this case report. It should be individualized due to the variable patterns seen in CD. Surgery often plays a very important role in the management of CD. The most important open question is the issue of maintenance therapy. Azathioprine, 6-mercaptopurine (6-MP) and methotrexate are all effective for maintenance of medically induced remission, particularly steroid-induced remission[13,19-22]. Infliximab is effective for maintenance of remission in patients refractory to other therapies, but concomitant immunosuppression is required[13,23]. The controversial issue is the postoperative maintenance of remission. There are no data on postoperative medical management of patients with diffuse jejunoileitis following strictureplasty or on postoperative maintenance therapy of patients undergoing resection for CD. In the absence of definitive literature data, azathioprine and 6-MP are currently the treatment of choice in patients at 'high risk' for recurrence following resection for CD[13]. Metronidazole for 3 months reduces the recurrence of severe endoscopic lesions following 3 months, but does not alter the clinical recurrence at 1 year[13].

In our patient, due to the severity and type of disease, we think that a reasonable postoperative approach is maintenance therapy with infliximab at 8-week intervals in combination with methotrexate, since azathioprine cannot be used due to the earlier-noted side-effects. We feel that biologic therapy should have been applied earlier in the course of disease, but unfortunately it was not available at that time.

The question of prognosis is still open for this patient; it will mainly depend on the efficacy of further immunosuppressive therapy. The natural history of diffuse small bowel disease remains unknown. In the literature diffuse jejunoileitis is a CD phenotype with high morbidity, particularly in the early years, commonly requiring surgical intervention. It can be discerned from recent data that the disease becomes less aggressive with time, and most patients can be restored to good health with minimal symptoms, despite the high morbidity in the early years after diagnosis[3,10].

## References

1. Sachar DB, Andrews HA, Farmer RG. Proposed classification of patient subgroups in Crohn's disease. Gastroenterol Int. 1992;5:141–54.
2. Mekhjian HS, Switz DM, Melnyk CS, Rankin GB, Brooks RK. Clinical features and natural history of Crohn's disease. Gastroenterology. 1979;77:898–906.
3. Tan WC, Allan RN. Diffuse jejunoileitis of Crohn's disease. Gut. 1993;34:1374–8.
4. Crohn BB, Yunich AM. Ileojejunitis. Ann Surg. 1941;113:371–80.
5. Touze I, Gower-Rousseau C, Grandbastien B et al. Diffuse jejuno-ileitis of Crohn's disease: a separate form of the disease? Gastroenterol Clin Biol. 1999;23:307–11.
6. Wagtmans AJG et al. Crohn's disease in the elderly: a comparison with young adults. J Clin Gastroenterol. 1998;27:129–33.
7. Goldberg HI, Caruthers SB, Nelson JA. Radiographic findings of the National Cooperative Crohn's Disease Study. Gastroenterology. 1979;77:925–37.
8. Halligan S, Nicholls S, Bartram CI, Walker-Smith JA. The distribution of small bowel Crohn's disease in children compared to adults. Clin Radiol. 1994;49:314–16.
9. Zalev AH, Prokipchuk EJ. Crohn's disease of the proximal small intestine: radiologic findings in 55 patients. Can Assoc Radiol J. 1992;43:170–8.
10. Allan RN. Crohn's disease of the small intestine diffuse jejunal ileitis. In: Allan RN, Rhodes JM, Hanauer SB, Keighley MRB, Alexander-Williams J, Fazio VW, editors. Inflammatory Bowel Diseases. New York: Churchill Livingstone, 1997:597–9.
11. Ricart E, Panaccione R, Loftus EV et al. Infliximab for Crohn's disease in clinical practice at the Mayo Clinic: the first 100 patients. Gastroenterology. 2000;118:A2967.

12. Targan SR, Hanauer SB, van Deventer SJ et al. A short-term study of chimeric monoclonal antibody cA2 to tumor necrosis factor alpha for Crohn's disease. Crohn's Disease cA2 Study Group. N Engl J Med. 1997;337:1029–35.
13. Sandborn WJ. Management of Crohn's disease. World Gastroenterol News. 2004;9:12–15.
14. Connell WR, Kamm MA, Ritchie JK, Lennard-Jones JE. Bone marrow toxicity caused by azathioprine in inflammatory bowel disease: 27 years of experience. Gut. 1993;34:1081–5.
15. Bouhnik Y, Lemann M, Mary JY et al. Long-term follow-up of patients with Crohn's disease treated with azathioprine or 6-mercaptopurine. Lancet. 1996;34:215–19.
16. Feagan BG, Rochon J, Fedorak RN et al. Methotrexate for the treatment of Crohn's disease. North American Crohn's Study Group Investigators. N Engl J Med. 1995;332:292–7.
17. Michelassi F. Side-to-side isoperistaltic strictureplasty for multiple Crohn's strictures. Dis Colon Rectum. 1996;39:345–9.
18. Dietz DW, Fazio VW, Laureti S et al. Strictureplasty in diffuse Crohn's jejunoileitis: safe and durable. Dis Colon Rectum. 2002;45:764–70.
19. Pearson DC, May GR, Fick GH, Sutherland LR. Azathioprine and 6-mercaptopurine in Crohn disease. A meta-analysis. Ann Intern Med. 1995;123:132–42.
20. Candy S, Wright J, Gerber M, Adams G, Gerig M, Goodman R. A controlled double blind study of azathioprine in the management of Crohn's disease. Gut. 1995;37:674–8.
21. Markowitz J, Grancher K, Kohn N, Lesser M, Daum F. A multicenter trial of 6-mercaptopurine and prednisone in children with newly diagnosed Crohn's disease. Gastroenterology. 2000;119:895–902.
22. Feagan BG, Fedorak RN, Irvine EJ et al. A comparison of methotrexate with placebo for the maintenance of remission in Crohn's disease. North American Crohn's Study Group Investigators. N Engl J Med. 2000;342:1627–32.
23. Hanauer SB Feagan BG, Lichtenstein GR et al. Maintenance infliximab for Crohn's disease: the ACCENT I randomised trial. Lancet. 2002;359:1541–9.

# 22
# Diffuse jejunoileitis in the adolescent patient

**M. DUBINSKY**

## INTRODUCTION

Although uncommon, diffuse jejunoilietis (JI) is one of the most difficult areas of all of Crohn's disease (CD) to treat. The reported incidence ranges from 3% to 10%[1,2] with only a small percentage of these cases having disease limited entirely to the proximal small intestine. As a result of the extensiveness of the disease process, as dictated by the underlying immune and genetic drive, this disease phenotype has traditionally been more resistant to medical management. Thus, surgical intervention has played an important part in the management of diffuse JI. Originally multiple or even massive surgical resections were the only answer for symptomatic complicating JI; however, advances in surgical techniques, most notably stricturoplasty, have changed the mortality and morbidity rates associated with diffuse JI[3]. Proximal and mid small bowel CD is more commonly seen in children as compared to adults, and thus becomes a very important diagnostic consideration in the paediatric age group This has many important implications for the diagnosis, treatment and prognosis of children afflicted with CD. The most significant factor potentially affected by diffuse JI is growth and pubertal development. The approach to a child with JI must include the recognition of the impact of JI on growth, and must be considered an important outcome as part of a successful treatment plan.

## SMALL BOWEL CD AND GROWTH

In children with diffuse small bowel disease, linear growth impairment may precede the onset of intestinal symptoms and thus may present somewhat of a diagnostic challenge to primary-care physicians. Early recognition is critical as there is a negative correlation between time to diagnosis and height. The profound effect of disease on growth can be attributable to the timing of disease onset, such that inflammatory bowel disease (IBD) often occurs at a time when the adolescent is undergoing significant physical change. Additionally, disease location has a significant impact on growth such that diffuse jejunal disease is associated with a low $z$-score for both height and weight. The three primary aspects of physical growth that are adversely affected by chronic inflammation and diminished absorptive capacity with resultant malnutrition are body weight, linear growth and sexual maturity. Although increased intestinal losses, malabsorption, increased energy requirements and psychological factors contribute to the nutritional disturbances, reduced intake remains the principal reason for continued weight loss. Colicky abdominal pain with food predominates the clinical presentation in patients with proximal small bowel disease and is the principal drive for reduced intake. In addition to weight loss a proportion of children will present with linear growth impairment and delayed sexual maturity as a result of the chronic malnutrition. Pubertal delay typically accompanies growth impairment. Growth failure is more often seen in children with CD as compared to ulcerative colitis (UC). However, approximately 6–10% of children with UC have been reported to develop growth failure over the course of their disease[4,5]. When focusing on children with CD the reported incidence of growth failure at the time of diagnosis varies between 7% and 88%[6-10]. There are three principal observed patterns of growth in CD patients: (1) those patients who experience a decrease in height velocity before onset of CD symptoms, (2) decrease in height velocity prior to diagnosis but after onset of symptoms, and (3) normal height velocity up to time of diagnosis[11]. Males may be at greater risk of growth impairment than females since they go through a growth spurt later in the stages of normal pubertal development, and this growth period tends to persist for a longer period of time than in their female counterparts.

The negative effects of therapies, corticosteroids in particular, affect the physical growth of a child and thus special attention must be paid to all these factors when approaching a child with this form of disease presentation. Hyams et al. demonstrated that, despite disease activity, the height velocity (cm/month) for children on corticosteroids was significantly reduced as compared to those patients not receiving prednisone[12]. Early identification of growth failure will enable clinicians to maximize both nutritional and medical therapies to ensure maximal catch-up growth and avoid persistent stunted growth into adulthood.

## BONES AND CROHN'S

In addition to bone growth, attention must be paid to quality of bones. Jejunal disease is associated with low bone density more so than disease at other disease locations, and early intervention to improve bone mass should be undertaken in a growing child. Other influences on bone disease include age at diagnosis, duration of disease, disease severity, nutritional status, surgical history and corticosteroid use. With regard to corticosteroid-induced bone loss, the data suggest that the greatest loss of bone occurs in young patients, with 5% occurring within the first year of therapy. Given the diffuse absorptive capacity for vitamins and minerals, calcium in particular, in the proximal small bowel special attention must be given to the effect of diffuse JI on bone health. Reports suggest that a significant proportion of CD patients have osteopenia at the time of diagnosis[13]. This is more often seen in patients with proximal small bowel disease involvement. The bone density evaluation should be considered in children with diffuse small bowel involvement as vitamin D and calcium may be necessary in addition to the anti-inflammatory medications they are prescribed. In the small percentage of patients who already have osteoporosis at presentation, biphosphonates may be considered, but the data in children are limited. The ability to measure bone density with quantitative CT scanning (QCT) has provided a more accurate assessment of bone size and quality, but normative data are not always available, limiting the interpretation of this test in children. Clinicians may want to consider bone density evaluation in patients with diffuse JI prior to initiating induction therapy. The key to bone health is controlling inflammatory mediators and non-corticosteroid induction treatment strategies will be most effective in children with existing bone disease at the time of presentation.

## MEDICAL MANAGEMENT

Given the potential long-term complications associated with diffuse small bowel disease in an adolescent, physicians should be aggressive in their choice of therapies and should consider immunomodulators as first-line therapies in addition to a more rapidly acting anti-inflammatory agent. Although corticosteroids do work well in the short term, their side-effects limit their long-term use in the paediatric age group and biologic therapies may be considered earlier on in the course of treatment. In the majority of children who present with more common CD phenotypes (i.e. ileocolonic disease), the more traditional step-up approach to treatment is an acceptable strategy. However given the potential impact of diffuse small bowel disease on bone, growth and overall health, clinicians may want to consider implementing more aggressive therapies earlier in this small subgroup of patients. The limited time window of puberty is the principal drive behind this aggressive approach, so that it is critical not to waste time with patients presenting with very active extensive small bowel disease. This becomes even more important considering the natural history of diffuse small bowel disease in the absence of aggressive medical management, is surgical intervention.

Nutritional therapy must be considered in all children presenting with diffuse small bowel disease. Whether used as supplementation or as primary therapy, adequate caloric intake must be monitored and implemented in order to maximize linear growth. Meta-analyses have supported the use of enteral feeding as primary therapy as compared to corticosteroids[14,15]. The most recent study, performed by a UK group, showed that not only does nutritional therapy induce and maintain remission, but their data suggested that nutritional therapy results in endoscopic healing[16]. Similarly cytokine production was dampened when Modulen® was introduced to children with active CD, providing evidence for its anti-inflammatory properties[17]. Another advantage for nutritional therapy is its steroid-sparing potential. Clinicians may want to consider enteral nutrition as primary therapy in lieu of corticosteroids as a bridge to long-term immunomodulators.

## SURGICAL MANAGEMENT

Prior to the introduction of immunomodulators and biologic therapies into the IBD arsenal, surgery was the mainstay of treatment for symptomatic proximal small bowel disease. The most common indication for surgery is recurrent intestinal obstruction. Conservative surgical treatment was an important development for diffuse jejunoileal disease as extensive resection was associated with increased morbidity and mortality, in particular related to short bowel syndrome. Stricturoplasty has largely replaced resection for short multiple small bowel strictures, and results in good health and small risk of short bowel syndrome. Yamamoto et al. reported on the largest series of surgical treatment outcome for diffuse jejunoileal disease[3]. After a median follow-up of 15 years there were three deaths: one from postoperative sepsis, one from small bowel carcinoma and one from bronchogenic carcinoma. The majority of patients in this study (89%) experienced a symptomatic postoperative recurrence, with the 5- and 10-year cumulative reoperation rates for recurrence at 62% and 79%, respectively. Interestingly, these rates were much higher than a reoperation rate after primary resection for localized terminal ileal disease (26% and 45%, for 5 and 10 years, respectively). Given that jejunoileal disease often burns itself out, early surgical treatment is recommended for disease that is unresponsive to medical intervention.

## NATURAL HISTORY: PREDICTORS OF COMPLICATING DISEASE BEHAVIOURS

The natural history of diffuse small bowel disease in children remains unknown. However, the implementation of potential disease-modifying therapies early in the course of disease remains an important consideration. Most of the natural history data available on all CD phenotypes essentially predate the introduction, and more regular use, of immunomodulators into clinical practice. The same can be said for the biologic therapies, infliximab in particular. The Vienna classification attempts to classify patients into more homogeneous subgroups based on age of diagnosis, disease location and disease behaviour[18]. Disease location appears to be stable over time; disease behaviour, however, does change over the course of disease. In a retrospective study the majority of adult CD patients ($\sim 70\%$) present with non-penetrating, non-stricturing disease at diagnosis[19]. Approximately 30% of these patients developed a complication every 5-year period, so that the majority of CD patients ($\sim 70\%$) after 10 years of disease have either penetrating or stricturing disease behaviour.

The currently accepted aetiopathogenic hypothesis for IBD proposes the role of genetics, immune responses and environmental factors for disease susceptibility and development. These factors and their interactions may also be important determinants of disease phenotype and disease progression. Since the discovery of NOD2, a number of studies have examined the association between genotype and disease phenotype. The most consistent associations observed were between NOD2 and ileal disease location and fibrostenosing disease behaviour[20-24]. The correlation between NOD2 and diffuse small bowel disease as a location *per se* remains unknown. Given the increased frequency of NOD2 in stricturing CD, however, and the increased frequency of stricturing disease in patients with JI the association is probably significant. There was a suggestion that NOD2 is associated with earlier age of disease onset; however, this was not replicated in all studies and these studies were conducted among predominantly adult cohorts. Our preliminary studies, however, do not suggest that NOD2 frequency is higher among children with CD[25].

Recent data suggest that immune responses may be a more direct link to disease phenotype than genotype due to the modification of genes by the environment. Landers et al. have shown that the type and amplitude of antibody responses to microbial antigens can stratify CD patients (subclinical phenotypes)[26]. The question remains as to whether these markers of immune reactivity are predictive of the development of complicating disease behaviours in a prospective manner. Furthermore, delineating the potential determinants of disease progression will lead to development of intervention studies to prevent progression of clinical disease to clinical complications. Further research is needed to understand the potential the aetiopathogenic significance of multiple antibody responses to microbial antigens in a genetically susceptible host.

## CONCLUSIONS

Although uncommon, diffuse JI remains an important CD phenotype. Intestinal obstruction is the most common complication observed among patients whose JI progresses over time. Diffuse JI in paediatric-onset CD has important implications for growth and development. Aggressive medical management with early implementation of immune-modifying agents may alter the natural history and improve the long-term outcome. It is very important to recognize diffuse JI early in the course of disease, so as to maximize physical and psychosocial growth in this very vulnerable and dynamic stage of a child's development.

## References

1. Cooke WT, Swan CHJ. Diffuse jejuno-ileitis of Crohn's disease. Q J Med. 1974;179:583–601.
2. Tan WC, Allan RN. Diffuse jejunoileitis of Crohn's disease. Gut. 1993;34:1374–8.
3. Yamamoto T, Allan RN, Keighley MR. Long-term outcome of surgical management for diffuse jejunoileal Crohn's disease. Surgery. 2001;129:96–102.
4. Griffiths AM, Nguyen P, Smith C, MacMillan JH, Sherman PM. Growth and clinical course of children with Crohn's disease. Gut. 1993; 34:939–43.
5. Markowitz, J, Grancher K, Rosa J, Aiges H, Daum F. Growth failure in pediatric inflammatory bowel disease. J Pediatr Gastroenterol Nutr. 1993;16:373–80.
6. Hildebrand H, Karlberg J, Kristiansson B. Longitudinal growth in children and adolescents with inflammatory bowel disease. J Pediatr Gastroenterol Nutr. 1994;18:165–73.
7. Kanof ME, Lake AM, Bayless TM. Decreased height velocity in children and adolescents before the diagnosis of Crohn's disease. Gastroenterology. 1988;95:1423–527.
8. Kirschner BS. Growth and development in chronic inflammatory bowel disease. Acta Pediatr Scand (Suppl). 1990;366:98–104.

9.   Lindquist BL, Jarnerot G, Wickbom G. Clinical and epidemiological aspects of Crohn's disease in children and adolescents. Scand J Gastroenterol. 1984;19:502–6.
10.  Gryboski J, Hillemeier C. Inflammatory bowel disease in children. Med Clin North Am. 1980;64:1185–202.
11.  Kanof ME, Lake AM, Bayless TM. Decreased height velocity in children and adolescents before the diagnosis of Crohn's disease. Gastroenterology. 1988;95:1523–7.
12.  Hyams JS, Carey DE. Corticosteroids and growth. J Pediatr. 1988;112:249–54.
13.  Lamb EJ, Wong T, Smith DJ et al. Metabolic bone disease is present at diagnosis in patients with inflammatory bowel disease. Aliment Pharmacol Ther. 2002;16:1895–902.
14.  Heuschkel RB. Enteral nutrition in children with Crohn's disease. J Pediatr Gastroenterol Nutr. 2000; 31:575.
15.  Zachos M, Tondeur M, Griffiths AM. Enteral nutritional therapy for inducing remission of Crohn's disease. Cochrane Database of Systematic Reviews 2001;3:CD000542.
16.  Sandhu B, Knoght C, Matary W. Longterm outcome of elemental diet as primary therapy for pediatric Crohn's disease. Gastroenterology 2002;122:A12.
17.  Fell JM, Paintin M, Arnaud-Battandier F et al. Mucosal healing and a fall in mucosal pro-inflammatory cytokine mRNA induced by a specific oral polymeric diet in paediatric Crohn's disease. Aliment Pharmacol Ther. 2000;14:281–9.
18.  Gasche C, Scholmerich J, Brynskov J et al. A simple classification of Crohn's disease: report of the working party for the World congress of Gastroenterology, Vienna, 1998. Inflamm Bowel Dis. 2000;6:8–15.
19.  Louis E, Collard A, Oger AF, Degroote E, Aboul Nasr El Yafi FA, Belaiche J. Behaviour of Crohn's disease according to the Vienna classification: changing pattern over the course of the disease. Gut. 2001;49:777–82.
20.  Hampe J, Cuthbert A, Croucher PJ et al. Association between insertion mutation in NOD2 gene and Crohn's disease in German and British populations. Lancet. 2001;357:1925–8.
21.  Cuthbert AP, Fisher SA, Mirza MM et al. The contribution of NOD2 gene mutations to the risk and site of disease in inflammatory bowel disease. Gastroenterology. 2002;122:867–74.
22.  Ahmad T, Armuzzi A, Bunce M et al. The molecular classification of the clinical manifestations of Crohn's disease. Gastroenterology. 2002;122:854–66.
23.  Lesage S, Zouali H, Cezard JP et al. EPWG-IBD Group. EPIMAD Group. GETAID Group. CARD15/NOD2 mutational analysis and genotype–phenotype correlation in 612 patients with inflammatory bowel disease. Am J Hum Genet. 2002;70:845–57.
24.  Abreu MT, Taylor KD, Lin YC et al. Mutations in NOD2 are associated with fibrostenosing disease in patients with Crohn's disease. Gastroenterology. 2002;123:679–88.
25.  Dubinsky MC, Lee-Uy N, Lin YC et al. Synergism of NOD2 and ASCA contribute to disease behavior in pediatric Crohn's disease patients. Gastroenterology. 2003;124:A372.
26.  Landers CJ, Cohavy O, Misra R et al. Selected loss of tolerance evidenced by Crohn's disease-associated immune responses to auto- and microbial antigens. Gastroenterology. 2002; 123:689–99.

# 23
# Current methods of bowel-sparing surgery in Crohn's disease*

## A. FICHERA, R. D. HURST and F. MICHELASSI

## INTRODUCTION

The recurrent nature of Crohn's disease (CD) may require repeated surgical procedures and, potentially, intestinal resections over time. Up to 30% of patients require at least two surgical procedures in their lifetime[1,2]. Patients subjected to multiple intestinal resections are at risk of diarrhoea, chronic malnutrition, electrolyte derangements, vitamin $B_{12}$ and folate deficiencies, chronic anaemia, and short gut syndrome severe enough to require temporary or permanent parenteral nutrition[3]. In an effort to minimize the occurrence of short gut syndrome, several bowel-sparing surgical techniques (i.e. stricture-plasty) have been described in the past 20 years. They all address Crohn's-related complication without, as the name implies, sacrificing bowel[4-6].

Data accrued during the course of the past two decades suggest that, with appropriate selection of patients, bowel-sparing surgical techniques are safe and effective[7-9]. Morbidity is low, and reoperative rates seem to be comparable to those obtained after resection and anastomosis. In addition, several recent studies have provided compelling evidence that CD regresses to quiescent disease at the site of a strictureplasty[10-13]. These observations provide further support for bowel-sparing procedures in CD and offer hope that regression from active to quiescent disease may translate in return of intestinal absorptive function. This chapter reviews current indications and contraindications, preoperative evaluation, operative techniques, postoperative complications, and long-term results of strictureplasty in CD.

*Reprinted with permission from Elsevier, from Advances in Surgery. 2003;37:231–50.

## INDICATIONS AND CONTRAINDICATIONS FOR BOWEL-SPARING SURGERY IN CD

Bowel-sparing techniques are safer to perform in elective rather than emergent procedures. The ability to operate after a full bowel preparation and in the absence of sepsis decreases postoperative morbidity. With these considerations in mind, complete bowel obstructions should be initially treated with nasogastric decompression and intravenous fluid administration with the hope to resolve the mechanical blockage. The obstruction is usually caused by undigested food which, with time, is pushed through the stenosis or dissolves through maceration. By the same token, easily accessible abscesses should be initially managed with image-guided drainage and with antibiotics. Percutaneous drainage is feasible and successful in avoiding early surgery in approximately 50% of cases[14]. If these strategies are followed, an otherwise emergent procedure can be converted into an elective one, at which time the complication rate after strictureplasty is markedly lower and a significant length of bowel can be spared.

Strictureplasties are indicated in patients who have duodenal enteric, and anastomotic strictures. Strictureplasties are contraindicated in the presence of active intra-abdominal sepsis (abscess or phlegmon) associated with the diseased bowel segment; free perforation of the affected bowel or generalized abdominal sepsis; long, tight strictures with a thick, unyielding intestinal wall; malignant strictures; and in patients with severe weight loss and marked hypoalbuminaemia[15]. In addition to the listed indications and contraindications, at the time of surgery the surgeon must consider factors in favour of and against their performance. Factors in favour of performing a strictureplasty include extensive jejunoileitis, previous extensive small bowel resections, existing short bowel syndrome, short strictures for which strictureplasty offers a simpler alternative to resection, and rapid symptomatic recurrence within 6–12 months of previous resection. A factor against the performance of strictureplasty is a single short stricture in close proximity to an intestinal segment to be resected. The presence of a fistula is not a contraindication to performing strictureplasty if the opening of the fistula can be debrided and then incorporated into the strictureplasty. However, a strictureplasty may not be feasible if the intestinal wall is inflamed, thickened, and rigid; if the debridement of the edges of the fistula results in a large defect that does not lend itself to a strictureplasty; and if the opening of the fistula is on the mesenteric side of the intestinal loop, making it difficult to incorporate into the strictureplasty.

## PREOPERATIVE EVALUATION AND PREPARATION

Elective surgery for abdominal CD should be preceded by a complete evaluation of the gastrointestinal tract. Conventional double-contrast enteroclysis or computed tomography enteroclysis are best to study the small bowel; colonoscopy affords the best view for the large bowel. Computed tomography scanning of the abdomen and pelvis may be necessary to validate a clinical suspicion of an abdominal abscess, inflammatory mass, or obstructive uropathy. Meticulous mechanical preparation of the small and large bowel should be undertaken in all patients before abdominal surgery for CD. Even in cases thought to be limited to the small bowel, the surgeon must always be prepared to perform surgery on the large bowel because of secondary involvement of the colon by fistulas or by an adherent inflammatory mass or abscess.

## OPERATIVE STRATEGY AND TECHNIQUE

At laparatomy the contents of the abdominal cavity are explored thoroughly. Particular care is devoted to examining the duodenum, the entire small bowel from the ligament of Treitz to the ileocaecal valve, and the colorectum, taking note of the precise location and extent of the disease. Obvious signs of disease include creeping fat on the antimesenteric border of the intestine, thickening of the mesentery, corkscrew vessels on the serosa of the bowel, and a thickened, fibrotic intestinal wall.

In obstructing disease the degree of intestinal narrowing needs to be assessed to decide whether it requires surgical correction. The presence of prestenotic dilation indicates a critical stricture in need of surgical correction; yet the presence of prestenotic dilation is not an indispensable condition for a stricture to be judged in need of correction. A proximal stenosis may be so tight that it limits the rate of intestinal contents passing through, to the point that distal strictures, although equally critical, may never be overwhelmed by the rate of intestinal contents and may never cause proximal intestinal dilation. However, if untreated, they will become future sites of chronic obstruction once the proximal obstruction has been released.

If the stenosis is short, the intestinal lumen can be assessed by pinching the stricture between the thumb and index finger. Alternatively, balloon catheters, Hager dilators, or marble stones can be used to gauge the size of the intestinal lumen. The measuring device can be conveniently introduced in the intestine through an enterotomy otherwise planned for a concurrent strictureplasty or resection. We prefer to advance a Foley catheter and then withdraw it with the balloon inflated to a diameter of 2.0 cm. The ability to inflate and deflate the Foley catheter balloon offers the advantage of measuring the size of multiple sequential strictures.

Once the intestinal disease has been mapped out, the surgeon must devise a strategy that relieves all symptoms, addresses all complications, and spares as much intestine as possible[16,17]. The surgeon can rely on resections, stricture-plasties, bypass procedures (used today almost exclusively in severe gastro-duodenal disease), or a combination of these. Although strictureplasty is an excellent option for many cases of abdominal CD, intestinal resections are still required in the majority of patients in isolation or in association with bowel-sparing procedures.

Lee and Papaioannou[6] and Alexander-Williams and Fornaro[5] are credited with the first descriptions of the use of strictureplasty in CD, although the technique had been previously described in India to correct tubercular strictures of the terminal ileum[18]. Many different strictureplasty techniques have been described since[19-21]. We describe here the most common techniques used now. In all kinds of strictureplasty, meticulous haemostasis of the intestinal edges must be obtained before and during the operation to minimize postoperative haemorrhage from the strictureplasty suture line, the most common postoperative complication[22,23]. In addition, once the bowel is opened, the stricture is inspected and a biopsy is performed on any suspicious area to rule out an occult malignancy.

## Heineke–Mikulicz strictureplasty

The Heineke–Mikulicz strictureplasty is the most commonly performed strictureplasty technique. This technique is appropriate for strictures less than 10 cm long. A longitudinal incision is made along the antimesenteric border of the stricture, extending at least 2 cm into the relatively normal proximal and distal non-strictured intestine (Figure 1). The Heineke–Milkulicz strictureplasty is then fashioned by closing the longitudinal enteromy in a transverse fashion. This is best accomplished by pulling on two seromuscular traction sutures placed at the midpoint of the entertomy. The enterotomy is best closed with a two-layer technique; alternatively, a single-layer technique can be used.

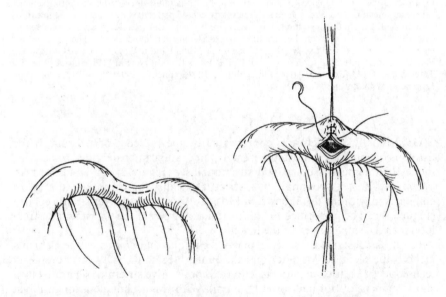

**Figure 1** Heineke–Mikulicz strictureplasty. A longitudinal incision is made along the anti-mesenteric border of the stricture, extending at least 2 cm into the normal, proximal and distal intestine. The longitudinal enterotomy is closed in a transverse fashion. (Courtesy of Mulholland MW, Zelenock GB, Michelassi F, Atlas of small intestinal surgery. In: Bell RH, Rikkers LF, Mulholland MW, editors. Digestive Tract Surgery: A Text and Atlas. Philadelphia, J.B. Lippincott, 1996:1263–311. Copyright Lippincott Williams and Wilkins)

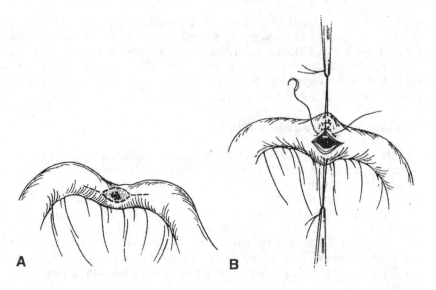

**Figure 2** Judd strictureplasty. **A**: In the presence of a fistulous opening on the antimesenteric side of an intestinal stricture, a Judd strictureplasty can be performed by debriding the intestinal wall to healthy, modestly inflamed tissue, prior to performing a longitudinal enterotomy extending on both sides of the stricture to grossly normal intestine. **B**: The longitudinal enterotomy is then closed transversely. (Courtesy of Mulholland MW, Zelenock GB, Michelassi F, Atlas of small intestinal surgery. In: Bell RH, Rikkers LF, Mulholland MW, editors. Digestive Tract Surgery: A Text and Atlas. Philadelphia, JB Lippincott, 1996:1263–1311. Copyright Lippincott Williams and Wilkins)

Occasionally, a stricture is complicated by the presence of a fistula. If the fistulous opening is on the antimesenteric side, and can be debrided to healthy, modestly inflamed tissue, a Judd strictureplasty (Figure 2) can be performed. When the bowel proximal to a stricture is disproportionally dilated in comparison with the distal bowel, a Moskel–Walske–Neumayer strictureplasty (Figure 3) can be performed to tailor the large lumen of the proximal intestinal loop into the small lumen of the distal bowel.

Fazio has described a strictureplasty technique that combines the elements of both the Heineke–Mikulicz and the Finney technique. This strictureplasty technique is best used in patients with two consecutive strictures separated by a short segment of normal bowel (5–6 cm). With this technique a longitudinal entertomy is made initially on the antimesenteric side of both strictures and through the intervening grossly normal bowel segment (Figure 4). Traction and countertraction sutures are applied at the midpoint of each stricture. Interrupted, full-thickness sutures are then applied starting at the midpoint of the enterotomy. After the posterior mesenteric wall is constructed, the suture lines are carried anteriorly to close the remaining portion of the enterotomy in a transverse fashion very much like in a Heineke–Mikulicz strictureplasty. Although it is easier to start with interrupted, full-thickness sutures on the

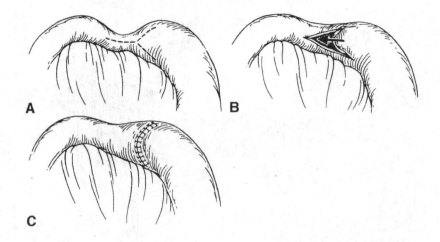

**Figure 3** Moskel–Walske–Neumayer strictureplasty. **A**: When the bowel proximal to the stricture is proportionally dilated in comparison to the distal bowel, a Y-shaped enterotomy can be performed. **B**: The proximal intestinal wall is advanced over the stricture. **C**: The enterotomy is closed in a transverse fashion, completing the V-Y advancement. (Courtesy of Mulholland MW, Zelenock GB, Michelassi F, Atlas of small intestinal surgery. In: Bell RH, Rikkers LF. Mulholland MW, editors. Digestive Tract Surgery: A Text and Atlas. Philadelphia, JB Lippincott, 1996:1263–1311. Copyright Lippincott Williams and Wilkins)

posterior wall, the anterior wall can usually be closed with a running suture. The inner layer can be reinforced with an outer layer of interrupted, seromuscular non-absorbable sutures.

## Finney strictureplasty

A Finney strictureplasty can be used for longer segments of disease, up to 15–25 cm in length. With this technique the affected bowel is folded onto itself (Figure 5), and a row of interrupted, seromuscular sutures is placed between the two limbs at the mesenteric side. A longitudinal enterotomy is then made 3–4 mm above the suture line on both intestinal limbs. A full-thickness absorbable running suture is then placed on the posterior wall of the strictureplasty and continued anteriorly to approximate the proximal and distal loop of the strictureplasty. The anterior portion of the inner suture line is then reinforced with an outer layer of interrupted, seromuscular non-absorbable sutures.

The Finney strictureplasty may not be technically feasible if the wall of the diseased intestine lacks the necessary pliability to fold on itself. Even when a Finney strictureplasty is feasible, the functional consequences may be an intestinal bypass with a large lateral diverticulum[24], resulting in bacterial overgrowth and the potential for neoplastic degeneration.

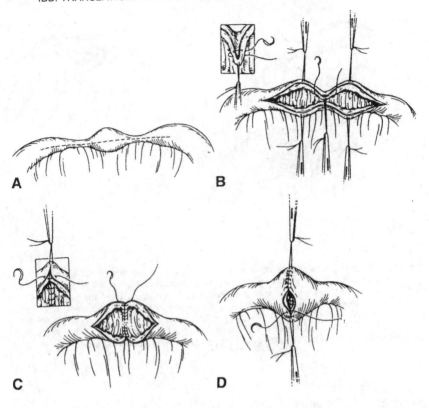

**Figure 4** Combined Heineke–Mikulicz and Finney strictureplasty. **A**: This technique is best used when two consecutive strictures are separated by a short segment of normal bowel. **B**: A longitudinal enterotomy is made on the antimesenteric side of both strictures and through the intervening grossly normal bowel segment. Traction and countertraction sutures are applied at the midpoint of each stricture. **C**: Interrupted, full-thickness sutures are then applied starting at the midpoint of the enterotomy. **D**: After the posterior mesenteric wall is constructed, the suture lines are carried anteriorly to close the remaining portion of the enterotomy in a transverse fashion. (Courtesy of Fazio VW, Tjandra JJ. Strictureplasty for Crohn's disease with multiple long strictures. Dis Colon Rectum. 1993;36:71–2. Copyright Lippincott Williams and Wilkins)

## Side-to-side isoperistaltic strictureplasty

In patients with multiple strictures located in close proximity to each other we have described the use of a side-to-side isoperistaltic strictureplasty[21]. With this technique the mesentery of the small bowel is divided at its midpoint, and the small bowel is severed between atraumatic intestinal clamps (Figure 6). The proximal intestinal loop is moved over the distal one in a side-to-side fashion ensuring that stenotic areas of one loop face dilated areas of the other loop (Figure 7). The two loops are approximated by a layer of interrupted,

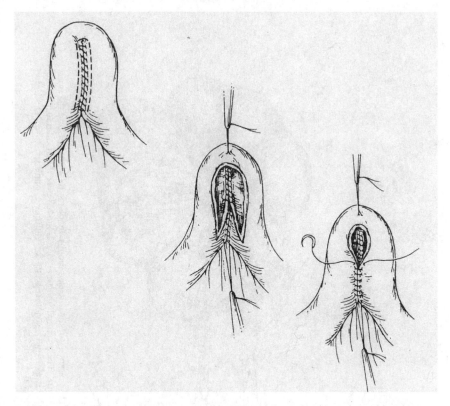

**Figure 5**   Finney strictureplasty. With this technique the affected bowel is folded onto itself and a row of interrupted, seromuscular sutures is placed between the two limbs at the mesenteric side. A longitudinal enterotomy is then made above the suture line on both intestinal limbs. A full-thickness absorbable running suture is then placed on the posterior wall of the strictureplasty and continued on the anterior wall to approximate the proximal and distal intestinal loop. The anterior portion of the inner suture line is then reinforced with an outer layer of interrupted seromuscular sutures. (Courtesy of Sharif H, Alexander-Williams J. The role of strictureplasty in Crohn's disease. Int Surg. 1992;77:15–18. Used with permission)

seromuscular non-absorbable stitches. A longitudinal enterotomy is performed on both loops, and the intestinal ends are tapered to avoid blind stumps (Figure 8). The posterior outer suture line is reinforced with an internal row of running, full-thickness absorbable sutures, continued anteriorly as a running Connell's suture. This layer is reinforced by an outer layer of interrupted, seromuscular non-absorbable stitches (Figure 9).

The side-to-side strictureplasty is particularly useful in patients with multiple strictures located in close proximity to each other. In these cases the stricture-plasty avoids an extended bowel resection with consequent sacrifice of normal small bowel between strictures, and also avoids the use of multiple Heineke–Mikulicz and Finney strictureplasties, which may result in a somewhat bulky and unyielding segment of intestine with considerable tension on each suture

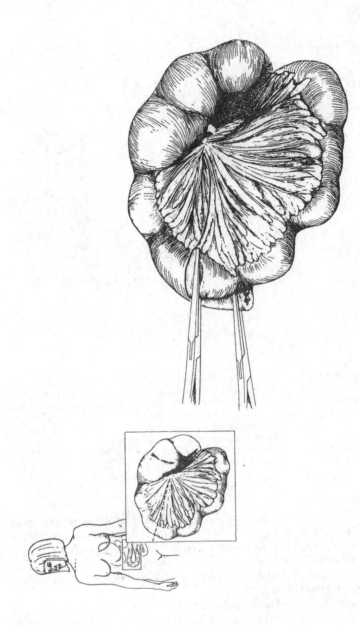

**Figure 6** Side-to-side isoperistaltic strictureplasty. The mesentery of the diseased loop is divided at its midpoint, and the small bowel is severed between atraumatic and intestinal clamps. The proximal intestinal loop is moved over the distal one in a side-to-side fashion. (Courtesy of Michelassi F. Side-to-side isoperistaltic strictureplasty for multiple Crohn's strictures. Dis Colon Rectum. 1996;39:345–9. Copyright Lippincott Williams and Wilkins)

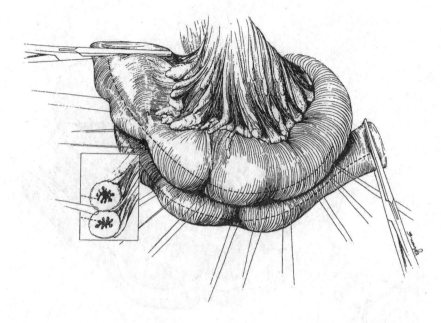

**Figure 7** Side-to-side isoperistaltic strictureplasty. The two loops are approximated by a layer of interrupted seromuscular stitches using nonabsorbable sutures. (Courtesy of Michelassi F. Side-to-side isoperistaltic strictureplasty for multiple Crohn's strictures. Dis Colon Rectum. 1996;39:345–9. Copyright Lippincott Williams and Wilkins)

line. This procedure should not be attempted in the presence of a very thickened mesentery, which makes it difficult to slide the proximal loop over the distal one. These patients are better treated with resection. Long, tight strictures with minimal intestinal lumen are also not amenable to this kind of strictureplasty. In these cases, if the proximal and distal loops have alternating strictures and skip lesions, they can be incorporated in a side-to-side strictureplasty after resection of the most severely midsegment. Finally, if the strictures are far apart from each other, patients are better treated with multiple Heineke–Mikulicz or Finney strictureplasties.

## STRATEGY ACCORDING TO SPECIFIC CLINICAL PRESENTATIONS

### Gastroduodenal CD

About 2–4% of patients with CD have involvement of the stomach or duodenum[25,26]. Extensive CD of the duodenum is usually treated with a gastrojejunal bypass. Strictureplasty has been advocated as an alternative to bypass procedures in selected patients[27–29]. If a tension-free closure is feasible, a Heineke–Mikulicz strictureplasty is used for short strictures of the first three portions of the duodenum, and a Finney for strictures of the fourth portion. A

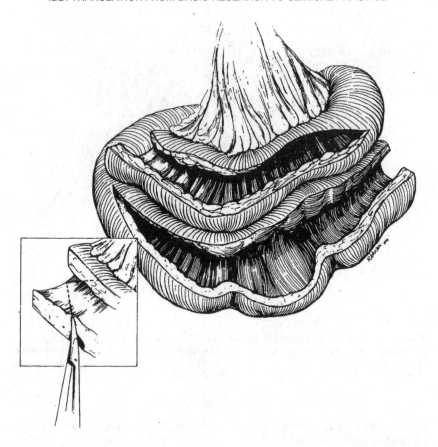

**Figure 8** Side-to-side isoperistaltic strictureplasty. A longitudinal enterotomy is performed on both loops, and the intestinal ends are spatulated to avoid blind stumps, as detailed in the inset. (Courtesy of Michelassi F. Side-to-side isoperistaltic strictureplasty for multiple Crohn's strictures. Dis Colon Rectum. 1996;39:345–9. Copyright Lippincott Williams and Wilkins)

stricture of the pylorus extending onto the duodenal bulb and first portion of the duodenum can alternatively be treated with a Jaboulay side-to-side gastroduodenostomy. This technique is preferred to a Heineke–Mikulicz strictureplasty if the chronic gastric distension caused by the stricture has approximated the greater curvature of the stomach to the second portion of the duodenum.

**Table 1** Long-term results after strictureplasty

| Study | Patients (no.) | Strictureplasties (no.) | Resections (no.) | Patients requiring reoperation for recurrence (no.) | Recurrence on previous strictureplasty (no.) | Follow-up (month) |
|---|---|---|---|---|---|---|
| Dietz et al.[15] | 314 | 1124 | 205 | 116 | n.a. | 90 |
| Yamamoto et al.[59] | 111 | 285 | 46 | 49 | n.a. | 107 |
| Alexander-Williams[60] | 80 | 309 | 7 | 26 | 4 | 42 |
| Baba and Nakai[61] | 69 | 187 | 52 | 13 | 4 | 37 |
| Hurst and Michelassi[20] | 57 | 109 | 40 | 7 | 5 | 38 |
| Serra et al.[62] | 43 | 154 | 32 | 14 | 3 | 54.4 |
| Spencer et al.[63] | 35 | 71 | 24 | 6 | 0 | 36 |
| Quandalle et al.[64] | 22 | 107 | 15 | 5 | 4 | 36 |
| Michelassi et al.[13]* | 21 | 21 | 17 | 3 | 0 | 48 |
| Pritchard et al.[65] | 13 | 62 | 4 | 4 | 1 | 24 |

*All side-to-side isoperistaltic strictureplasty

n.a., not applicable

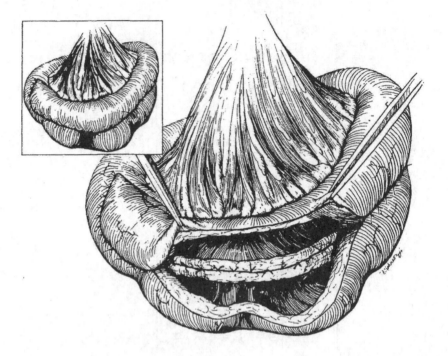

**Figure 9**  Side-to-side isoperistaltic strictureplasty. Both outer and inner suture lines are continued and finished anteriorly. A completed side-to-side isoperistaltic strictureplasty is shown in the inset. (Courtesy of Michelassi F. Side-to-side isoperistaltic strictureplasty for multiple Crohn's strictures. Dis Colon Rectum. 1996;39:345–9. Copyright Lippincott Williams and Wilkins)

## RECURRENT ANASTOMOTIC STRICTURE

CD frequently recurs at the anastomotic site. The incidence of anastomotic strictures has been reported to be between 30% and 46%[30]. Early series have reported their results with balloon dilation of the strictured anastomosis[31]. This technique can be performed only when the stricture is situated within the reach of an endoscope; it is also associated with early recurrence and high rate of perforations. More recently, strictureplasties have been used for the treatment of anastomotic strictures secondary to recurrent CD after ileocolic or entero-enteric resections, with excellent results[10,32,33]. Heineke–Mikulicz, Finney, or side-to-side isoperistaltic strictureplasties can all be used depending on the extent of the anastomotic recurrence and the presence of additional disease.

## Colonic CD

The colon is affected by CD in up to 30% of patients[34-36]. Segmental CD of the colon occurs in only 20–25% of patients with Crohn's colitis. Most commonly, the disease affects the sigmoid and is best approached with a resection and primary anastomosis. Rarely, isolated, ring-like strictures of the ascending or transverse colon are found as the only evidence of Crohn's colitis[37]. Once an occult malignancy is ruled out, these strictures may be amenable to a Heineke-Mikulicz strictureplasty. A strictureplasty is a better option than a colonic resection in patients who have already lost a sizeable amount of small bowel and may rely heavily on the colonic mucosa for absorption of water and electrolytes.

## Paediatric CD

CD is being reported with increasing frequency in the paediatric population[38]. Furthermore, in comparison with the adult population, the clinical course of the disease seems to be more aggressive in children. Therefore, the surgical principle to resect only enough intestine to relieve the complication and to preserve as much intestinal length as possible is particularly relevant in children, who have a longer life expectancy and may have a more aggressive form of the disease. As a consequence, strictureplasty has become a worthwhile surgical adjunct, especially when dealing with jejunoileitis[39]. Few series are available, but they have all shown that strictureplasty is as beneficial in children as it is in adults[39-41].

## RESULTS OF BOWEL-SPARING SURGERY IN CD

Unlike resections in which diseased tissue is removed to grossly normal margins, strictureplasties are fashioned in diseased bowel. This initially gives rise to concern about the risks of early postoperative morbidity, including anastomotic dehiscence and bleeding, recurrent symptomatic disease requiring surgery, and in the long-term, small bowel adenocarcinoma. A growing number of reports indicate that, in appropriately selected patients, perioperative morbidity after strictureplasty is similar to that reported for resection and primary anastomosis[11,42-51]. Indeed, the Cleveland Clinic group recently published the largest series with 1124 strictureplasties in 314 patients with a median follow-up of 7.5 years[15]. The overall morbidity rate was 18%, including a 7% postoperative haemorrhage rate from the strictureplasty suture line and a 2% dehiscence rate.

Haemorrhage from the strictureplasty suture line is the most common perioperative complication, occurring in 2–9% of patients[22,23]. However, gastrointestinal haemorrhage after strictureplasty is typically minor and can usually be managed conservatively with blood transfusions alone[23]. When haemorrhage persists, intra-arterial infusion of vasopressin may be successful in slowing and eventually stopping the bleeding. At the time of the selected catheterization of the superior mesenteric artery, an angiogram should be obtained to identify which strictureplasty suture line is bleeding. In case of multiple strictureplasty and anastomotic suture lines this information is particularly valuable when reoperation is required to control ongoing haemorrhage. Septic complications, including dehiscence, intra-abdominal abscess, wound infection, and enterocutaneous fistula formation, occur in only 5–6% of patients who undergo strictureplasty[52].

The validity of strictureplasties as a useful addition to the surgical armamentarium needs to stand the test of time in view of the recurrent nature of CD and the potential for malignant degeneration. So far, the long-term results after strictureplasty are encouraging. Recurrence of symptoms after strictureplasty is not more common than after resection and anastomosis. Although no randomized controlled studies directly compare recurrence rates after resection versus after strictureplasty, several authors have observed recurrence rates after strictureplasty similar to recurrence rates after resection[53,54]. In addition, it is commonly observed that postoperative recurrences in need of surgical treatment are often caused by new strictures in previously grossly normal intestine rather than recurrences at strictureplasty sites (Table 1).

Several authors have even suggested that strictureplasties induce regression of disease and possibly recovery of intestinal function. They postulate that resolution of chronic obstruction by a strictureplasty may interrupt the cascade of inflammatory events that perpetuate active disease and, in turn, lead to quiescent disease and possibly recovery of absorptive intestinal function. Stebbing et al.[11] noticed that regression of serositis and fat wrapping at strictureplasty sites in patients who underwent subsequent laparotomies. Tjandra and Fazio[10] reporting on 22 patients who had undergone strictureplasty of stenotic ileocolonic anastomosis, noted that, after surgery, none of the four patients in whom a colonoscopy was carried out showed severe and active inflammation of the strictureplasty site. Poggioli et al.[55] reported their experience with five patients who underwent a side-to-side enterocolic anastomosis for disease of the terminal ileum; in a subsequent article[12] they described two additional patients. Of the seven patients, four underwent a colonoscopy 6–12 months after surgery; the disappearance of mucosal lesions and recovery of the submucosal vascular pattern was noticed in all. Michelassi et al.[13] provided radiographic, endoscopic, and histopathological evidence that active CD regresses at the site of side-to-side isoperistaltic strictureplasties.

Patients with CD are at increased risk for small and large bowel carcinomas[56]. Although cancer may occur anywhere throughout the gastrointestinal tract, it is more common at the site of overt disease. As a consequence it has been suggested that strictureplasties may carry an increased risk for adenocarcinoma. Indeed, several cases of adenocarcinoma have been reported at strictureplasty sites. The Cleveland Clinic described one patient with coeliac sprue and CD who developed an adenocarcinoma at a strictureplasty site 7 years after the index operation[15]. Negative biopsies of the stricture had been obtained at the time of the index operation[57]. Jaskowiak and Michelassi[58] have recently reported a case of adenocarcinoma developing at a strictureplasty site 7 years after a strictureplasty in a patient whose only risk factor for intestinal cancer was CD. These reports bring undeniable evidence that malignant transformation may occur at the site of a strictureplasty, and point to the need to carefully monitor the actual occurrence of cancer at these sites. However, the available evidence justifies continuous use of bowel-sparing techniques.

## CONCLUSIONS

Current available data suggest that, with appropriate selection of patients, strictureplasty is a safe and effective technique for treating CD. Morbidity is low, and reoperative rates seem to be comparable to those obtained after resection and anastomosis. Better understanding and quantification of the neoplastic risk and of the risk for recurrent disease after strictureplasty is necessary to determine the validity of bowel-sparing surgery in patients with CD. The potential for complete reversal of Crohn's lesions and restoration of intestinal function is an interesting and stimulating concept that deserves further investigation. The surgeon treating patients with inflammatory bowel disease needs to be proficient with the different strictureplasty techniques.

## References

1. Cooke WT, Mallas E, Prior P et al. Crohn's disease: course, treatment and long term prognosis. Q J Med. 1980;49:363–84.
2. Heller G. Crohn's disease in Stockholm county 1955–1974. A study of epidemiology, results of surgical treatment and long-term prognosis. Acta Chir Scand Suppl. 1979;490:1–84.
3. Montreuil B, Bernard D, Heppell J et al. Dilatation enteroplasty for obstructive Crohn's disease. Experience of the University of Montreal [in French]. Ann Chir. 1995;49:664–8.
4. Alexander-Williams J, Haynes IG. Conservative operations for Crohn's disease of the small bowel. World J Surg. 1985;9:945–51.
5. Alexander-Williams J, Fornaro M. Strictureplasty in Crohn's disease [in German]. Chirurg. 1982;53:799–801.
6. Lee EC, Papaioannou N. Minimal surgery for chronic obstruction in patients with extensive or universal Crohn's disease. Ann R Coll Surg Engl. 1982;64:229–33.
7. Cristaldi M, Sampietro GM, Danelli PG et al. Long-term results and multivariate analysis of prognostic factors in 138 consecutive patients operated on for Crohn's disease using 'bowel-sparing' techniques. Am J Surg. 2000;179:266–70.
8. Dehn TC, Kettlewell MG, Mortensen NJ et al. Ten-year experience of strictureplasty for obstructive Crohn's disease. Br J Surg. 1989;76:339–41.
9. Broering DC, Eisenberger CF, Koch A et al. Quality of life after surgical therapy of small bowel stenosis in Crohn's disease. Dig Surg. 2001;18:124–30.
10. Tjandra JJ, Fazio VW. Strictureplasty for ileocolonic anastomosis strictures in Crohn's disease. Dis Colon Rectum. 1993;36:1099–103.
11. Stebbing JF, Jewell DP, Kettlewell MG et al. Recurrence and reoperation after stricture-plasty for obstructive Crohn's disease: long-term results. Br J Surg. 1995;82:1471–4.
12. Poggioli G, Selleri S, Stocchi L et al. Conservative surgical management of perforating Crohn's disease: side-to-side enteroenteric neoileocolic anastomosis: report of two cases. Dis Colon Rectum. 1998;41:1577–80.
13. Michelassi F, Hurst RD, Melis M et al. Side-to-side isoperistaltic strictureplasty in extensive Crohn's disease: a prospective longitudinal study. Ann Surg. 2000;232:401–8.
14. Sahai A, Belair M, Gianfelice D et al. Percutaneous drainage of intra-abdominal abscesses in Crohn's disease: short and long-term outcome. Am J Gastroenterol. 1997;92:275–8.
15. Dietz DW, Laureti S, Strong SA et al. Safety and longterm efficacy of strictureplasty in 314 patients with obstructing small bowel Crohn's disease. J Am Coll Surg. 2001;192:330–7.
16. Thompson JS. Strategies for preserving intestinal length in the short-bowel syndrome. Dis Colon Rectum. 1987;30:208–13.
17. Tonelli F, Ficari F. Strictureplasty in Crohn's disease: surgical option. Dis Colon Rectum. 2000;43:920–6.

18. Katariya RN, Sood S, Rao PG et al. Stricture-plasty for tubercular strictures of the gastro-intestinal tract. Br J Surg. 1977;64:496–8.
19. Alexander-Williams J. The technique of intestinal strictureplasty. Int J Colorectal Dis. 1986;1:54–7.
20. Hurst RD, Michelassi F. Strictureplasty for Crohn's disease: techniques and long-term results. World J Surg. 1998;22:359–63.
21. Michelassi F. Side-to-side isoperistaltic strictureplasty for multiple Crohn's strictures. Dis Colon Rectum. 1996;39:345–9.
22. Gardiner KR, Kettlewell MG, Mortensen N. Intestinal haemorrhage after strictureplasty for Crohn's disease. Int J Colorectal Dis. 1996;11:1880–2.
23. Ozuner G, Fazio VW. Management of gastrointestinal bleeding after strictureplasty for Crohn's disease. Dis Colon Rectum. 1995;38:297–300.
24. Yamamoto T, Mylonakis E, Keighley MR. Two surgical procedures for strictures at Finney strictureplasty for small bowel Crohn's disease. Dig Surg. 2000;17:451–3.
25. Murray JJ, Schoetz DJ Jr, Nugent FW et al. Surgical management of Crohn's disease involving the duodenum. Am J Surg. 1984;147:58–65.
26. Poggioli G, Stocchi L, Laureti S et al. Duodenal involvement of Crohn's disease: three different clinicopathologic patterns. Dis Colon Rectum. 1997;40:179–83.
27. Hirati K. Is strictureplasty a safe and efficacious procedure for duodenal Crohn's disease? J Gastroenterol. 2000;35:953–4.
28. Takesue Y, Yokoyama T, Akagi S et al. Strictureplasty for short duodenal stenosis in Crohn's disease. J Gastroenterol. 2000;35:929–32.
29. Worsey MJ, Hull T, Ryland L et al. Strictureplasty is an effective option in the operative management of duodenal Crohn's disease. Dis Colon Rectum. 1999;42:596–600.
30. Sharif H, Alexander-Williams J. Strictureplasty for ileo-colonic anastomotic strictures in Crohn's disease. Int J Colorectal Dis. 1991;6:214–16.
31. Alexander-Williams J, Allan A, Morel P et al. The therapeutic dilatation of enteric strictures due to Crohn's disease. Ann R Coll Surg Engl. 1986;68:95–7.
32. Yamamoto T, Keighley MR. Long-term results of strictureplasty for ileo-colonic anasto-motic recurrence in Crohn's disease. J Gastrointest Surg. 1999;3:555–60.
33. Yamamoto T, Allan RN, Keighley MR. Strategy for surgical management of ileocolonic anastomotic recurrence in Crohn's disease. World J Surg. 1999;23:1055–60.
34. Farmer RG, Hawk WA, Turnbull RB Jr. Clinical patterns in Crohn's disease: a statistical study of 615 cases. Gastroenterology. 1975;68:627–35.
35. Hurst RD, Molinari M, Chung TP et al. Prospective study of the features, indications, and surgical treatment in 513 consecutive patients affected by Crohn's disease. Surgery. 1997;122:661–7.
36. Michelassi F, Balestracci T, Chappell R et al. Primary and recurrent Crohn's disease. Experience with 1379 patients. Ann Surg. 1991;214:230–8.
37. Broering DC, Eisenberger CF, Koch A et al. Strictureplasty for large bowel stenosis in Crohn's disease: quality of life after surgical therapy. Int J Colorectal Dis. 2001;16:81–7.
38. von Allmen D, Goretsky MJ, Ziegler MM. Inflammatory bowel disease in children. Curr Opin Pediatr. 1995;7:547–52.
39. Markowitz JF. Strictureplasty in pediatric Crohn's disease. J Pediatr Gastroenterol Nutr. 1994;18:266–8.
40. Oliva L, Wyllie R, Alexander F et al. The results of strictureplasty in pediatric patients with multifocal Crohn's disease. J Pediatr Gastroenterol Nutr. 1994;18:306–10.
41. Telandor RL. Surgical management of Crohn's disease in children. Curr Opin Pediatr. 1995;7;328–34.
42. Fazio VW, Tjandra JJ, Lavery IC et al. Long-term follow-up of strictureplasty in Crohn's disease. Dis Colon Rectum. 1993;36:355–61.
43. Fazio VW, Tjandra JJ. Strictureplasty for Crohn's disease with multiple long strictures. Dis Colon Rectum. 1993;36:71–2.
44. Gaetini A, De Simone M, Resegotti A. Our experience with strictureplasty in the surgical treatment of Crohn's disease. Hepatogastroenterology. 1989;36:511–15.
45. Nivatvongs S. Strictureplasty for Crohn's disease of small intestine. Present status in Western countries. J Gastroenterol. 1995;30:139–42S.
46. Ozuner G, Fazio VW, Lavery IC et al. Reoperative rates for Crohn's disease following strictureplasty. Long-term analysis. Dis Colon Rectum. 1996;39:1199–203.

47. Ozuner G, Fazio VW, Lavery IC et al. How safe is strictureplasty in the management of Crohn's disease? Am J Surg. 1996;171:57–60.
48. USampietro GM, Cristaldi M, Porretta T et al. Early perioperative results and surgical recurrence after strictureplasty and miniresection for complicated Crohn's disease. Dig Surg. 2000;17:261–7.
49. Sayfan J, Wilson DA, Allan A et al. Recurrence after strictureplasty or resection for Crohn's disease. Br J Surg. 1989;76:335–8.
50. Tjandra JJ, Fazio VW. Strictureplasty without concomitant resection for small bowel obstruction in Crohn's disease. Br J Surg. 1994;81:561–3.
51. Tjandra JJ, Fazio VW, Lavery IC. Results of multiple strictureplasties in diffuse Crohn's disease of the small bowel. Aust NZ J Surg. 1993;63:95–9.
52. Tichansky D, Cagir B, Yoo E et al. Strictureplasty for Crohn's disease: meta-analysis. Dis Colon Rectum. 2000;43:911–19.
53. Alexander-Williams J, Haynes IG. Up-to-date management of small-bowel Crohn's disease. Adv Surg. 1987;20:245–64.
54. Laureti S, Fazio VW. Obstruction in Crohn's disease: strictureplasty versus resection. Curr Treat Options Gastroenterol. 2000;3:191–202.
55. Poggioli G, Stocchi L, Laureti S et al. Conservative surgical management of terminal ileitis: side-to-side enterocolic anastomosis. Dis Colon Rectum. 1997;40:234–7.
56. Michelassi F, Testa G, Pomidor WJ et al. Adenocarcinoma complicating Crohn's disease. Dis Colon Rectum. 1993;36:654–61.
57. Marchetti F, Fazio VW, Ozuner G. Adenocarcinoma arising from a strictureplasty site in Crohn's disease. Report of a case. Dis Colon Rectum. 1996;39:1315–21.
58. Jaskowiak NT, Michelassi F. Adenocarcinoma at a strictureplasty site in Crohn's disease. Report of a case. Dis Colon Rectum. 2001;44:284–7.
59. Yamamoto T, Bain IM, Allan RN et al. An audit of strictureplasty for small-bowel Crohn's disease. Dis Colon Rectum. 1999;42:797–803.
60. Alexander-Williams J. A life dedicated to Crohn's disease. Neth J Med. 1994;45:86–7.
61. Baba S, Nakai K. Strictureplasty for Crohn's disease in Japan. J Gastroenterol. 1995;30(Suppl. 8):135–8
62. Serra J, Cohen Z, McLeod RS. Natural history of strictureplasty in Crohn's disease: 9-year experience. Can J Surg. 1995;38:481–5.
63. Spencer MP, Nelson H, Wolff BG et al. Strictureplasty for obstructive Crohn's disease: The Mayo experience. Mayo Clin Proc. 1994;69:33–6.
64. Quandalle P, Gambiez L, Colombel JF et al. Long-term follow-up of strictureplasty in Crohn's disease. Acta Gastroenterol Belg. 1994;57:314–19.
65. Pritchard TJ, Schoetz DJ Jr, Caushaj FP et al. Strictureplasty of the small bowel in patients with Crohn's disease. An effective surgical option. Arch Surg. 1990;125:715–17.

# 24
# Mechanisms of steroid-refractory disease

## D. FRANCHIMONT

## INTRODUCTION

Corticosteroids* have been used for many years in the treatment of inflammatory bowel disease (IBD): Crohn's disease (CD) and ulcerative colitis (UC)[1–3]. Despite these 50 years of use a major question remains unanswered; that is, why do some patients respond to corticosteroids (steroid-sensitive (SS)) while others do not (steroid-resistant (SR))? Corticosteroids are indicated in the treatment of patients with moderate to severe disease and are accompanied by a significant set of short- and long-term deleterious adverse reactions. Thus, the early assessment of drug response is critical given the unfavourable natural history of these diseases.

The understanding of inter-individual variation in steroid response remains a most daunting challenge in clinical medicine today. A major bias exists in the definition of steroid response. Steroid response is often defined as changes in disease activity criteria. However, a disease may fail to respond because of a number of concomitant factors that are independent of drug efficacy. These include disease-related complications, drug interactions, environmental factors (i.e. smoking), inadequate doses, inefficient administration routes, and patient adherence. Moreover, attempts to predict steroid response based on disease type and behaviour have met with poor success. For example, many cohort studies have always demonstrated an increased prevalence of steroid resistance in UC as compared with CD. However, this has been challenged by recent population-based studies demonstrating that the frequency of steroid resistance remains similar in the two diseases[4,5]. Also, the clinical parameters that could be isolated from epidemiological studies to better predict steroid resistance are also only clinical evidences of exuberant inflammation[6–9]. Thus, a better

---

*The different terms, such as steroids, glucocorticoids, and corticosteroids, are often confusing. In this chapter, *steroid sensitivity* is used to define the sensitivity of the cell or tissue to corticosteroids; *steroid response* is used to define clinical response to corticosteroids. We will use *corticosteroids*, instead of glucocorticoids or steroids throughout the chapter, the glucocorticoid receptor (GR) being the receptor of corticosteroids.

understanding of mechanisms of steroid resistance is essential to be able to predict steroid resistance.

## THE GLUCOCORTICOID RECEPTOR AND ITS SIGNALLING PATHWAY

Corticosteroids exert their molecular and cellular actions through their glucocorticoid receptor (GR)[10,11]. Glucocorticoid receptor α shuttles between the cytoplasm and the nucleus, depending on its association with corticosteroids. In the absence of ligand, GRα resides mainly in the cytoplasm, in the form of a hetero-oligomer with several heat-shock proteins (HSP)[12]. After binding to its ligand, GRα dissociates from the HSP and enters the nucleus through nuclear pores, depending on two nuclear localization signals (NLS1 and 2)[13]. In the dimerized condition GRα binds to glucocorticoid-responsive elements (GRE) located in the promoter regions of specific genes and exerts either transactivating or transrepressing effects by respectively stimulating or inhibiting the transcription rates of its dependent genes. As a dimer and/or a monomer, GRα also interacts with other transcription factors and modulates their effects through direct protein–protein interactions[14–16]. After exerting its transcriptional effects GRα is exported to the cytoplasm via an energy-dependent transport system. Phosphorylation and dephosphorylation of the receptor may participate in the activation–inactivation, recycling and turnover of the receptor[12]. In contrast to GRα, the other isoform, GRβ, binds loosely to HSP and is predominantly nuclear. Its dominant negative activity is exerted by heterodimerization with GRα and/or by competing with GRα for GRE as a homodimer[17]. In addition to the transcriptional modulation described above, corticosteroids also influence the processing of mRNA and translation of proteins, probably through transactivation or transrepression of genes that regulate mRNA stability and translation[12].

Ligand-activated GRα regulates the transcription of responsive genes by forming complexes with the recently discovered coregulators, i.e. coactivators and/or corepressors, of transcription and several chromatin modulators[18]. These molecules not only alter chromatin structure, but also enhance or inhibit the transduction of the trancriptional signal of the ligand-activated GRα to the general transcription complex, which includes RNA polymerase II (RNPII) and its ancillary factors. There are three types of coactivator molecules involved in the action of corticosteroids: p300 and its homologous cyclic AMP-responsive element-binding protein-binding protein (CBP); the p160 type coactivators steroid receptor coactivator (SRC)-1a, SRC-2 and SRC-3; and p300/CBP-associated factor (p/CAF)[19,20]. All contain histone acetyltransferase activity through which they loosen chromatin structure by acetylating chromatin-incorporated histones[21,22]. The ligand-activated GRα also employs ATP-dependent chromatin modulators, such as the PBP/DRIP/TRAP and SWI/SNF complexes to modulate the transcription rate of target genes[23]. Collectively, coactivators/cointegrators and corepressors have been called 'coregulators' of steroid receptors. Their interaction with steroid nuclear receptors and other transcription factors is characterized by varying degrees of specificity and flexibility. Their relative abundance may also vary from tissue to tissue, and explain tissue differences in steroid sensitivity. Finally, they can be stoichiometrically limiting for some nuclear receptors and transcription factors in a particular tissue[24].

Corticosteroids exert many of their anti-inflammatory effects through protein–protein interactions with a set of transcription factors known to stimulate the inflammation, such as the nuclear factor κB (NFκB), several signal transducers and activators of transcription (STAT) and activator protein-1 (AP1). Conversely, these factors influence the effects of the GRα on its own GRE-containing responsive genes[25]. There is a very good example of how corticosteroids suppress inflammation by blocking NFκB activity. NFκB is a heterodimeric protein, composed of subunits p65 and p50, which constitutively reside in the cytoplasm in an inactive form stabilized by the inhibitory protein IκBα. Cellular activation by a multitude of inflammatory signals leads to phosphorylation and proteolytic degradation of IκBα. The liberated NFκB then translocates into the nucleus and binds to κB-responsive elements in the promoter regions of NFκB-responsive genes, such as TNF-α, IL-1 and IL-6, and stimulates transcriptional activity. Corticosteroids block NFκB activation through two different mechanisms:

1. Corticosteroid-activated GRα directly binds to p65 and suppresses NFκB activity[26].

2. Corticosteroids increase IκBα protein synthesis by stimulating its promoter through classic GRE and hence help segregate active p65/p50 heterodimer from the nucleus by forming inactive heterocomplexes with IκB in the cytoplasm[27–29].

## CORTICOSTEROIDS ON THE IMMUNE RESPONSE

The myriad of studies carried out during the past 50 years have revealed that corticosteroids are both immunosuppressive and immunostimulant. Indeed, it appears that the spectrum of corticosteroid actions is dynamic, with diametrically opposite effects on several components of the inflammatory and immune response. Corticosteroids play a major role in cell trafficking by modulating endothelial transmigration through regulation of endothelial and leucocyte membrane expression of integrins, and by influencing a wide range of CC (cystein–cystein) and CXC (cystein–X-cystein) chemokines[30–33]. At the tissue level they quickly shut down the inflammatory reaction; indeed, they are a potent immunostimulant since they restrain the inflammatory response by improving opsonization and phagocytic ability of polymorphonuclear cells and macrophages[34]. Interestingly, they regulate new families of genes such as scavengers and thrombospondins involved in the early phase of the innate immune response[35–37]. Simultaneously they are immunosuppressants and prevent noxious inflammatory reactions by blocking a number of inflammatory mediators such as cytokines, prostaglandins, leukotrienes, enzymes, NO and free radicals[38–42]. Their actions on the antigen cell presentation and adaptive immune response are rather selective[43]; indeed, corticosteroids promote humoral Th2 immune response while they suppress cellular Th1 immune response[44-46]. They strongly inhibit Th1 cytokines such as TNF-$\alpha$, IFN-$\gamma$ and IL-12, and may enhance Th2 cytokines such as IL-5 and IL-10[47-53]. Corticosteroids also differentially act on IL-12 and IL-4 signalling by selectively inhibiting IL-12-induced Stat4 phosphorylation without altering IL-4-induced Stat6 phosphorylation, or through the inhibition of IL-12 receptor $\beta_1$ and $\beta_2$ expression[54]. These two mechanisms may explain the well-described cellular immunodeficiency after corticosteroid therapy, and the progressive corticosteroid-induced shift from a Th1 to a Th2 immune response[11,55]. Finally, corticosteroids together with IL-4 may promote B cell development and isotype switching, leading to IgE and IgG$_4$ secreting plasma cells while they strongly suppress allergic inflammation[44,45,53].

Thus, the action of corticosteroids is discriminative, modulating different limbs of the immune response to help the termination of inflammation or preventing the inflammation to flare-up. Their selective actions, both in the innate and adaptive immune response, result in simultaneous and/or successive possibly positive and negative effects during the evolution of the immune response. Inversely, the different early and late stages of the immune response within the inflamed tissue (the immune phenotype) may influence the overall corticosteroid actions, and hence the response to treatment[56]. This may explain the variable steroid sensitivity observed between Th1 and Th2 inflammatory and autoimmune diseases, and during the course of a given inflammatory/autoimmune disease[57].

## SYSTEMIC OR TISSUE-SPECIFIC STEROID RESISTANCE

Mounting evidence suggests that steroid resistance is rather limited to the diseased tissue, the immune system and the digestive tract (for IBD), than systemic. Indeed, studies of SR asthma patients have revealed similar body distribution of corticosteroids in both SS and SR patients, with similar albumin and cortisol-binding globulin plasma levels[58]. More importantly, SR asthma and IBD patients have a normal dexamethasone suppression test, underscoring their normal systemic steroid sensitivity[59,60]. Most SR asthma patients develop a cushingoid appearance and metabolic dysfunction, suggesting that the systemic steroid sensitivity is well conserved; indeed, there is a tight correlation between clinical steroid resistance and the *in-vitro* lymphocyte steroid sensitivity of UC patients; this *in-vitro* steroid sensitivity being measured as the percentage of inhibition of lymphocyte proliferation with graded concentrations of dexamethasone[61]. This observation was also reported in asthma, rheumatoid arthritis and sytemic lupus erythematosus, demonstrating that the altered steroid sensitivity, and hence steroid response, takes place within the diseased tissue itself.

## PRIMARY OR SECONDARY INDUCED STEROID RESISTANCE

The recent studies carried out in SR patients with inflammatory diseases suggest that the process of inflammation (the disease itself), and the genetic and constitutional background of the patient (the host of the disease) may influence his or her variable response to corticosteroids[62,63]. To better delineate steroid resistance one should consider the pharmacology of corticosteroids: the *pharmacodynamics* of corticosteroids through which they exert their actions, the GR and its associated signalling pathway, and the *pharmacokinetics* of corticosteroids that include drug-metabolizing enzymes and drug transporters.

### Pharmacodynamics of corticosteroids

*Primary steroid resistance*

Although it is true that resistance to corticosteroids could be due to a primary defect in the GR, such as observed in familial glucocorticoid resistance (FGR), this has not yet been reported, and must be extremely rare[64–67]. These patients with FGR demonstrate very high levels of cortisol similar to those observed in Cushing syndrome, without developing a cushingoid appearance and disease complications. Lane et al. have examined some of the mutations, described in FGR kindreds, in SR asthma patients, but could not identify any of these mutations[68]. Interestingly, differential dexamethasone-dependent transcriptional activation and repression was observed among human GR variants associated with FGR[69]. The dexamethasone-induced repression of transcription from elements in the promoter of the intercellular adhesion molecule-1 via NFκB was in fact more efficient for the D641V variant GR than for the wild-type GR. Finally, five GR cDNA polymorphisms have been reported in the normal population, which were at variance with the original sequence reported by Hollenberg et al. In a population of 216 healthy subjects a reduced response in a short dexamethasone suppression test as a marker of relative systemic steroid resistance was observed in 20, otherwise healthy, persons[70]. However, this relative steroid resistance was not associated with any of these polymorphisms[71,72]. Nevertheless, this selected sample of the general population could represent the lowest section of a normal distribution of steroid sensitivity, defined by differences in the GR signal transduction system unrelated to GR polymorphisms[73]. In conclusion, it is expected that variants in the signalling molecules or transcription factors within the GR signalling pathway will account for primary steroid resistance in some IBD patients; however, these variants have not yet been characterized. Thus, there is currently very limited data that would suggest that steroid resistance in IBD is related to a polymorphism or a mutation in the GR gene.

*Secondary steroid resistance induced by inflammation*

The clinical response to corticosteroids can be correlated with *in-vitro* leucocyte sensitivity to corticosteroids. Interestingly enough, Hearing et al. showed that the steroid sensitivity (sensitivity to the inhibitory action of dexamethasone on proliferation) of lymphocytes of SR UC patients returns to normal during follow-up at 3 months, underscoring that the steroid sensitivity of these lymphocytes is reversible[61]. In asthma, steroid resistance has also been associated with a decreased inhibition of peripheral blood mononuclear cell proliferation and cytokine secretion by corticosteroids[74]. Hyperactivation of T lymphocytes, with overexpression of IL-2R and HLA-DR reported by many groups in SR patients, suggests that inflammation does indeed induce steroid resistance[75]. In these patients steroid resistance appears related to alterations of T cell glucocorticoid receptor density and binding affinity and/or to changes in post-receptor mechanisms while GR density and binding affinity in mononuclear cell subpopulations remain unaffected[74,76]. Two different populations of SR asthma patients have been described: type 1, as defined by a decreased GR binding affinity and type 2, as defined by a decreased GR number[77]. Type 1 SR asthma was shown to be reversible and secondary to inflammation, namely IL-2 and IL-4 dependent, while type 2 SR asthma appeared to be genetically determined[57,77]. Leung et al. showed that the GR binding affinity of peripheral blood mononuclear cells (PBMCs) from type 1 SR asthma returned to normal, when cultured *in vitro* overnight in medium alone, but decreased again when exposed to IL-2 and IL-4 for another 48 hours[77]. This elegant experiment underscored that steroid sensitivity was reversible and induced by inflammation.

## Mechanisms of inflammation-mediated steroid resistance

A few mechanisms have shed light on how inflammation can induce steroid resistance. Cytokines can alter tissue and cell steroid sensitivity by interfering at multiple levels with the GR signalling pathway, either at the receptor, signalling molecule or transcription factor level. Some cytokines, such as IL-1β, IL-6, TNF-α, IFN-γ, IL-2, IL-4 and IL-13, modulate GR number and binding affinity with parallel changes of steroid sensitivity[51,78–80]. It is likely that the changes in GR binding affinity and capacity after cytokine exposures indeed reflect post-translational modifications of the GR. There is some evidence to suggest that MAPK family members extracellular regulated kinases (ERK) and c-Jun N-terminal kinase (JNK) and p38 MAPK (p38) directly phosphorylate GR and inhibit GR-mediated transcriptional activation[81–83]. This could represent an early repression effect of proinflammatory signals on the expression of GR-dependent genes, and explain how cytokines induce steroid resistance during inflammation. In IBD a few studies have reported that IBD *per se* is associated with changes in GR number (not only at the leucocyte level but also at the mucosal level) and in GR binding affinity[59,84–86]. These observations emphasized that mild to moderate inflamed tissue already alters cell or tissue steroid sensitivity; yet the description of changes in GR binding affinity and capacity of PBMCs from SR IBD patients has never been reported. In truth, this remains very difficult to be investigated because of the great heterogeneity of patient populations and inter- and intra-variability of whole cell or cytosol binding assays.

The GR belongs to the nuclear receptor superfamily with members that function as ligand-activated transcription factors. There are two splicing variants of the human hGR, isoforms hGRα and hGRβ, depending on the alternative use of specific exon 9α or 9β, respectively. GRα is a 777 amino acid protein, while hGRβ consists of 742 amino acids. The first 727 amino acids from the N-terminus are identical in both isoforms. GRα possesses an additional 50 amino acids, while GRβ has an additional 15 non-homologous amino acids in the C-terminus[12]. While GRα is the classic receptor that binds to corticosteroids and transduces their biological activities, GRβ does not bind to corticosteroids and exerts weak dominant negative activity on GR-mediated genomic actions[17]. A recent study reported an increased expression of GRβ on PBMCs in SR UC patients (83%) as compared to SS UC patients (9%)[87]. However, the relatively low tissue levels of GRβ *in vivo*, as compared to the levels of GRα, question the pathophysiological role of hGRβ in this clinical condition. A markedly increased hGRβ/hGRα ratio was also reported in PBMCs and bronchial alveolar lavage (BAL) cells in SR asthma patients[88]. The relative stability of its expression over time, when following up some SR UC patients[87], suggested that GRβ could be a predictive marker for steroid resistance in UC patients. However, this has been recently challenged since hGRβ seems more related to disease activity and is cytokine inducible[89]. Importantly, cytokines such as IL-2 and IL-4 have been shown to enhance GRβ expression and this may contribute to the decreased steroid sensitivity of PBMCs (see above) following exposure to these cytokines[88]. Thus, hGRβ appears inducible and may contribute to cytokine-mediated steroid resistance[57,90].

Certain transcription factors activated by cytokine signalling molecules may also interact in the nucleus with ligand-activated GRα and either prevent ligand-activated GRα from binding to GRE or from exerting its transcriptional actions. Excessive nuclear factor (NF)-κB, activator protein (AP)-1 and signal transducers and activators of transcription (STAT) expression and activity in an inflammatory site may inhibit corticosteroid activity, contributing to a decreased steroid sensitivity[26,91,92]. NFκB is a key regulator of the transcriptional activity of many cytokines. NFκB can be viewed as an intracellular amplification mechanism that exacerbates/sustains chronic inflammatory processes. Indeed, cytokines activate NFκB leading to a positive feed-forward cycle at an inflammatory site[93]. Enhanced activation and concentration of P65 inside the nucleus inhibits the ability of GR to transactivate its GRE complementary DNA sequence[25]. Type II cytokines exert their cellular effects through the Jak/Stat signalling pathways. The transcription factor Stat5 of the type II cytokine receptors, such as IL-2Rα and IL-7Rα, directly interacts with the GR so that the Stat5/GRα complex strongly suppresses the response of a GRE-containing promoter to corticosteroids[92]. The ability of GR to bind to GRE seems impaired in SR patients[94]; indeed, increased level of AP-1 DNA binding with increased basal and PMA-stimulated protein levels of c-fos were recognized in SR compared to SS asthma patients[95]. Pretreatment of PBMCs from SR asthma patients with c-fos antisense oligonucleotides enhanced GR-DNA binding activity, suggesting that overexpression of c-fos was responsible for the decreased GR-DNA binding observed in these patients[95].

Ligand-activated glucocorticoid receptors regulate the transcription of responsive genes by forming complexes with the recently discovered coregulators (i.e. coactivators or corepressors) of transcription and several chromatin modulators (see above). A change in this flexible mix could lead to hypersensitivity and/or resistance to GR with hormonal and/or tissue predilection. There is some evidence suggesting that transcription factors such as AP-1 prevent GR transcriptional activity by quenching its coactivators. Interestingly, it should be examined whether some of these coregulators could undergo post-translational modifications and integrate multiple signals from different signalling pathways such as the MAP kinase pathway[96,97].

## Pharmacokinetics of corticosteroids

Pharmacokinetics of corticosteroids includes drug-metabolizing enzymes and drug transporters that influence body and tissue distribution of corticosteroids. Importantly two of these have been identified: MDR-1 and 11β HSD1. Recently there has been growing interest in the MDR-1 gene product, P-glycoprotein 170, which is a membrane-based drug efflux pump that actively transports multidrug resistance (MDR) substrates, such as corticosteroids, out of the cells, thereby lowering their intracellular concentration. High constitutive MDR-1 expression appears to be associated with poor medical response to corticosteroids, in both CD and UC patients, affects steroid requirements for maintenance of remission in patients with SLE, and influences disease outcome in patients with rheumatoid arthritis[98–100]. Independent of disease activity, the level of expression of MDR-1 appears constitutive and not inducible, suggesting that it could be genetically determined. Importantly, it remains to be examined whether any variants in the promoter region or the coding sequence of this gene might be associated with poor steroid sensitivity and corticosteroid response.

The key enzyme in the regulation of the intracellular active metabolite (cortisol) concentration, the 11β-hydroxysteroid dehydrogenase, may be modulated by cytokines. Tumour necrosis factor α and IL-1β have been shown to increase the expression and activity of 11β HSD1, which in turn increase the intracellular concentration of cortisol, thereby enhancing steroid sensitivity[101]. The tissue level of 11β HSD1 remains to be examined in SR IBD patients.

## PHARMACOGENETICS AND GENOMICS OF CORTICOSTEROIDS

Physicians face major dilemmas on a day-to-day basis. When treating a patient with corticosteroids they cannot predict whether or not this patient will be suffering from a steroid-dependent (SD) or SR inflammatory disease. This is an important issue because of the short- and long-term adverse effects of corticosteroids. With the completion of the human genome it is anticipated that the new and evolving discipline of pharmacogenetics will afford the opportunity to quickly predict steroid response and adverse effects based on the genotype of the patient. This focused approach relies on establishing an individual's genetic composite in order to better target patients (or patient subgroups), hence improving the efficacy and the safety of corticosteroids. An important goal of pharmacogenetics is to identify the sequence variants of each important gene (candidate gene approach) involved in the pharmacokinetics and pharmacodynamics of corticosteroids. This represents a daunting task when considering not only the pharmacology of corticosteroids, but also the genetic determinants of inflammation and the immune response, together with the specific disease variants (variants of disease *causing* or *modifying* gene) that could also potentially influence corticosteroid response. Finally, drug response is related not only to patient genotype but also to the disease itself, namely the type and stage of the disease, and environmental factors. Therefore, it is perhaps more important to first provide strong evidence of drug response before being able to predict drug response. This can only be achieved either by specifically monitoring drug response with biomarkers or by accurately phenotyping the disease itself prior to treatment.

An integrative and comprehensive approach that can resolve these issues is pharmacogenomics and proteomics, that encompass the phenotypic expression of the genomic DNA, namely the transcriptome and proteome. For example, the effect of corticosteroids on the gene expression profile of peripheral blood mononuclear cells from healthy donors with *DNA chip microarray* has been recently realized[35]. Many potential molecular markers have been identified and could be used for quick monitoring of corticosteroid response given the magnitude or pattern of their regulation. Finally, a recent study suggests that steroid resistance could be predicted on the gene profiling of the diseased tissue. A microarray exploratory analysis of renal biopsies from kidney transplant patients with acute allograft rejection demonstrated a strong B cell genomic signature in SR patients[102]. This gene profiling allowed the authors ultimately to discover that steroid resistance was associated with a significant infiltration of B cell aggregates, thus providing us with an important prognostic marker of steroid resistance, namely the presence of CD20$^+$ B cells.

## CONCLUDING REMARKS

The selective spectrum of actions of corticosteroids on a specific immune phenotype of the diseased tissue is likely to be the main cause of steroid resistance in IBD. However, it is clear that the genetic background of the patient, influencing both the magnitude of inflammatory/immune response, and the pharmacodynamics and kinetics of corticosteroids, also leads to tissue–steroid resistance. A reliable clinical definition of steroid resistance, the characterization of the specific tissue phenotype associated with a SR disease, and the identification of (or panel of) biomarkers to monitor steroid response, might help define clear patient population subgroups and help discover important pharmacogenetic markers of response that will ultimately guide our clinical decision-making.

## References

1. Shanahan F. Crohn's disease. Lancet. 2002;359:62–9.
2. Farrell RJ, Peppercorn MA. Ulcerative colitis. Lancet. 2002;359:331–40.
3. Podolsky DK. Inflammatory bowel disease. N Engl J Med. 2002;347:417–29.
4. Faubion WA Jr, Loftus EV Jr, Harmsen WS, Zinsmeister AR, Sandborn WJ. The natural history of corticosteroid therapy for inflammatory bowel disease: a population-based study. Gastroenterology. 2001;121:255–60.

5. Munkholm P, Langholz E, Davidsen M, Binder V. Frequency of glucocorticoid resistance and dependency in Crohn's disease. Gut. 1994;35:360–2.
6. Gelbmann CM. Prediction of treatment refractoriness in ulcerative colitis and Crohn's disease – do we have reliable markers? Inflamm Bowel Dis. 2000;6:123–31.
7. Gelbmann CM, Rogler G, Gross V et al. Prior bowel resections, perianal disease, and a high initial Crohn's disease activity index are associated with corticosteroid resistance in active Crohn's disease. Am J Gastroenterol. 2002;97:1438–45.
8. Daperno M, Sostegni R, Scaglione N et al. Outcome of a conservative approach in severe ulcerative colitis. Dig Liver Dis. 2004;36:21–8.
9. Lindgren SC, Flood LM, Kilander AF, Lofberg R, Persson TB, Sjodahl RI. Early predictors of glucocorticosteroid treatment failure in severe and moderately severe attacks of ulcerative colitis. Eur J Gastroenterol Hepatol. 1998;10:831–5.
10. Hollenberg SM, Weinberger C, Ong ES et al. Primary structure and expression of a functional human glucocorticoid receptor cDNA. Nature. 1985;318:635–41.
11. Franchimont D, Kino T, Galon J, Meduri GU, Chrousos G. Glucocorticoids and inflammation revisited: the state of the srt. NIH Clinical Staff Conference. Neuroimmuno-modulation. 2003;10:247–60.
12. Bamberger CM, Schulte HM, Chrousos GP. Molecular determinants of glucocorticoid receptor function and tissue sensitivity to glucocorticoids. Endocrinol Rev. 1996;17:245–61.
13. Ullman KS, Powers MA, Forbes DJ. Nuclear export receptors: from importin to exportin. Cell. 1997;90:967–70.
14. Giguere V, Hollenberg SM, Rosenfeld MG, Evans RM. Functional domains of the human glucocorticoid receptor. Cell. 1986;46:645–52.
15. Hollenberg SM, Giguere V, Segui P, Evans RM. Colocalization of DNA-binding and transcriptional activation functions in the human glucocorticoid receptor. Cell. 1987;49:39–46.
16. Hollenberg SM, Evans RM. Multiple and cooperative trans-activation domains of the human glucocorticoid receptor. Cell. 1988;55:899–906.
17. Bamberger CM, Bamberger AM, de Castro M, Chrousos GP. Glucocorticoid receptor beta, a potential endogenous inhibitor of glucocorticoid action in humans. J Clin Invest. 1995;95:2435–41.
18. Glass CK, Rose DW, Rosenfeld MG. Nuclear receptor coactivators. Curr Opin Cell. Biol. 1997;9:222–32.
19. Chakravarti D, LaMorte VJ, Nelson MC et al. Role of CBP/P300 in nuclear receptor signalling. Nature. 1996;383:99–103.
20. Leo C, Chen JD. The SRC family of nuclear receptor coactivators. Gene. 2000;245:1–11.
21. Struhl K, Moqtaderi Z. The TAFs in the HAT. Cell. 1998;94:1–4.
22. Struhl K. Histone acetylation and transcriptional regulatory mechanisms. Genes Dev. 1998;12:599–606.
23. Hsiao PW, Deroo BJ, Archer TK. Chromatin remodeling and tissue-selective responses of nuclear hormone receptors. Biochem Cell Biol. 2002;80:343–51.
24. McKenna NJ, O'Malley BW. Nuclear receptors, coregulators, ligands, and selective receptor modulators: making sense of the patchwork quilt. Ann NY Acad Sci. 2001;949:3–5.
25. McKay LI, Cidlowski JA. Molecular control of immune/inflammatory responses: interactions between nuclear factor-kappa B and steroid receptor-signaling pathways. Endocrinol Rev. 1999;20:435–59.
26. Ray A, Prefontaine KE. Physical association and functional antagonism between the p65 subunit of transcription factor NF-kappa B and the glucocorticoid receptor. Proc Natl Acad Sci USA. 1994;91:752–6.
27. Auphan N, DiDonato JA, Rosette C, Helmberg A, Karin M. Immunosuppression by glucocorticoids: inhibition of NF-kappa B activity through induction of I kappa B synthesis. Science. 1995;270:286–90.
28. Scheinman RI, Cogswell PC, Lofquist AK, Baldwin AS, Jr. Role of transcriptional activation of I kappa B alpha in mediation of immunosuppression by glucocorticoids. Science. 1995;270:283–6.
29. Wissink S, van Heerde EC, van der Burg B, van der Saag PT. A dual mechanism mediates repression of NF-kappaB activity by glucocorticoids. Mol Endocrinol. 1998;12:355–63.

30. Cronstein BN, Kimmel SC, Levin RI, Martiniuk F, Weissmann G. A mechanism for the antiinflammatory effects of corticosteroids: the glucocorticoid receptor regulates leukocyte adhesion to endothelial cells and expression of endothelial–leukocyte adhesion molecule 1 and intercellular adhesion molecule 1. Proc Natl Acad Sci USA. 1992;89:9991–5.

31. Jahnsen FL, Haye R, Gran E, Brandtzaeg P, Johansen FE. Glucocorticosteroids inhibit mRNA expression for eotaxin, eotaxin-2, and monocyte-chemotactic protein-4 in human airway inflammation with eosinophilia. J Immunol. 1999;163:1545–51.

32. Nakagawa M, Bondy GP, Waisman D, Minshall D, Hogg JC, van Eeden SF. The effect of glucocorticoids on the expression of L-selectin on polymorphonuclear leukocyte. Blood. 1999;93:2730–7.

33. Perretti M, Croxtall JD, Wheller SK, Goulding NJ, Hannon R, Flower RJ. Mobilizing lipocortin 1 in adherent human leukocytes downregulates their transmigration. Nat Med. 1996;2:1259–62.

34. Liu Y, Cousin JM, Hughes J et al. Glucocorticoids promote nonphlogistic phagocytosis of apoptotic leukocytes. J Immunol. 1999;162:3639–46.

35. Galon J, Franchimont D, Hiroi N et al. Gene profiling reveals unknown enhancing and suppressive actions of glucocorticoids on immune cells. FASEB J. 2002;16:61–71.

36. Hogger P, Dreier J, Droste A, Buck F, Sorg C. Identification of the integral membrane protein RM3/1 on human monocytes as a glucocorticoid-inducible member of the scavenger receptor cysteine-rich family (CD163). J Immunol. 1998;161:1883–90.

37. Pan LY, Mendel DB, Zurlo J, Guyre PM. Regulation of the steady state level of Fc gamma RI mRNA by IFN-gamma and dexamethasone in human monocytes, neutrophils, and U-937 cells. J Immunol. 1990;145:267–75.

38. Amezaga MA, Bazzoni F, Sorio C, Rossi F, Cassatella MA. Evidence for the involvement of distinct signal transduction pathways in the regulation of constitutive and interferon gamma-dependent gene expression of NADPH oxidase components (gp91-phox, p47-phox, and p22-phox) and high-affinity receptor for IgG (Fc gamma R-I) in human polymorphonuclear leukocytes. Blood. 1992;79:735–44.

39. Kawata R, Reddy ST, Wolner B, Herschman HR. Prostaglandin synthase 1 and prosta-glandin synthase 2 both participate in activation-induced prostaglandin D2 production in mast cells. J Immunol. 1995;155:818–25.

40. Salvemini D, Manning PT, Zweifel BS et al. Dual inhibition of nitric oxide and prostaglandin production contributes to the antiinflammatory properties of nitric oxide synthase inhibitors. J Clin Invest. 1995;96:301–8.

41. Sebaldt RJ, Sheller JR, Oates JA, Roberts LJ 2nd, FitzGerald GA. Inhibition of eicosanoid biosynthesis by glucocorticoids in humans. Proc Natl Acad Sci USA. 1990;87:6974–8.

42. Yoss EB, Spannhake EW, Flynn JT, Fish JE, Peters SP. Arachidonic acid metabolism in normal human alveolar macrophages: stimulus specificity for mediator release and phospholipid metabolism, and pharmacologic modulation in vitro and in vivo. Am J Respir Cell Mol Biol. 1990;2:69–80.

43. Piemonti L, Monti P, Allavena P et al. Glucocorticoids affect human dendritic cell differentiation and maturation. J Immunol. 1999;162:6473–81.

44. Akdis CA, Blesken T, Akdis M, Alkan SS, Heusser CH, Blaser K. Glucocorticoids inhibit human antigen-specific and enhance total IgE and IgG4 production due to differential effects on T and B cells in vitro. Eur J Immunol. 1997;27:2351–7.

45. Barnes PJ. Corticosteroids, IgE, and atopy. J Clin Invest. 2001;107:265–6.

46. Kimata H, Lindley I, Furusho K. Effect of hydrocortisone on spontaneous IgE and IgG4 production in atopic patients. J Immunol. 1995;154:3557–66.

47. Blotta MH, DeKruyff RH, Umetsu DT. Corticosteroids inhibit IL-12 production in human monocytes and enhance their capacity to induce IL-4 synthesis in CD4+ lymphocytes. J Immunol. 1997;158:5589–95.

48. DeKruyff RH, Fang Y, Umetsu DT. Corticosteroids enhance the capacity of macrophages to induce Th2 cytokine synthesis in CD4+ lymphocytes by inhibiting IL-12 production. J Immunol. 1998;160:2231–7.

49. Elenkov IJ, Papanicolaou DA, Wilder RL, Chrousos GP. Modulatory effects of glucocorti-coids and catecholamines on human interleukin-12 and interleukin-10 production: clinical implications. Proc Assoc Am Phys. 1996;108:374–81.

50. Franchimont D, Louis E, Dewe W et al. Effects of dexamethasone on the profile of cytokine secretion in human whole blood cell cultures. Regul Pept. 1998;73:59–65.

51. Franchimont D, Martens H, Hagelstein MT et al. Tumor necrosis factor alpha decreases, and interleukin-10 increases, the sensitivity of human monocytes to dexamethasone: potential regulation of the glucocorticoid receptor. J Clin Endocrinol Metab. 1999;84: 2834–9.

52. Ramierz F, Fowell DJ, Puklavec M, Simmonds S, Mason D. Glucocorticoids promote a TH2 cytokine response by CD4+ T cells *in vitro*. J Immunol. 1996;156:2406–12.

53. Wu CY, Sarfati M, Heusser C et al. Glucocorticoids increase the synthesis of immunoglobulin E by interleukin 4-stimulated human lymphocytes. J Clin Invest. 1991;87:870–7.

54. Wu CY, Wang K, McDyer JF, Seder RA. Prostaglandin E2 and dexamethasone inhibit IL-12 receptor expression and IL-12 responsiveness. J Immunol. 1998;161:2723–30.

55. Franchimont D, Galon J, Gadina M et al. Inhibition of Th1 immune response by glucocorticoids: dexamethasone selectively inhibits IL-12-induced Stat4 phosphorylation in T lymphocytes. J Immunol. 2000;164:1768–74.

56. Leung DY, Martin RJ, Szefler SJ et al. Dysregulation of interleukin 4, interleukin 5, and interferon gamma gene expression in steroid-resistant asthma. J Exp Med. 1995;181:33–40.

57. Leung DY, de Castro M, Szefler SJ, Chrousos GP. Mechanisms of glucocorticoid-resistant asthma. Ann NY Acad Sci. 1998;840:735–46.

58. Lane SJ, Palmer JB, Skidmore IF, Lee TH. Corticosteroid pharmacokinetics in asthma. Lancet. 1990;336:1265.

59. Flood L, Lofberg R, Stierna P, Wikstrom AC. Glucocorticoid receptor mRNA in patients with ulcerative colitis: a study of responders and nonresponders to glucocorticosteroid therapy. Inflamm Bowel Dis. 2001;7:202–9.

60. Lane SJ, Atkinson BA, Swaminathan R, Lee TH. Hypothalamic–pituitary–adrenal axis in corticosteroid-resistant bronchial asthma. Am J Respir Crit Care Med. 1996;153:557–60.

61. Hearing SD, Norman M, Probert CS, Haslam N, Dayan CM. Predicting therapeutic outcome in severe ulcerative colitis by measuring *in vitro* steroid sensitivity of proliferating peripheral blood lymphocytes. Gut. 1999;45:382–8.

62. Barnes PJ, Greening AP, Crompton GK. Glucocorticoid resistance in asthma. Am J Respir Crit Care Med. 1995;152:S125–40.

63. Chrousos GP, Castro M, Leung DY et al. Molecular mechanisms of glucocorticoid resistance/hypersensitivity. Potential clinical implications. Am J Respir Crit Care Med. 1996;154:S39–43; discussion S43–4.

64. Lamberts SW, Poldermans D, Zweens M, de Jong FH. Familial cortisol resistance: differential diagnostic and therapeutic aspects. J Clin Endocrinol Metab. 1986;63:1328–33.

65. Lamberts SW, Koper JW, Biemond P, den Holder FH, de Jong FH. Cortisol receptor resistance: the variability of its clinical presentation and response to treatment. J Clin Endocrinol Metab. 1992;74:313–21.

66. Lamberts SW, Koper JW, de Jong FH. Familial and iatrogenic cortisol receptor resistance. J Steroid Biochem Mol Biol. 1992;43:385–8.

67. Chrousos GP, Detera-Wadleigh SD, Karl M. Syndromes of glucocorticoid resistance. Ann Intern Med. 1993;119:1113–24.

68. Lane SJ, Arm JP, Staynov DZ, Lee TH. Chemical mutational analysis of the human glucocorticoid receptor cDNA in glucocorticoid-resistant bronchial asthma. Am J Respir Cell Mol Biol. 1994;11:42–8.

69. de Lange P, Koper JW, Huizenga NA et al. Differential hormone-dependent transcriptional activation and -repression by naturally occurring human glucocorticoid receptor variants. Mol Endocrinol. 1997;11:1156–64.

70. Huizenga NA, Koper JW, de Lange P et al. Interperson variability but intraperson stability of baseline plasma cortisol concentrations, and its relation to feedback sensitivity of the hypothalamo–pituitary–adrenal axis to a low dose of dexamethasone in elderly individuals. J Clin Endocrinol Metab. 1998;83:47–54.

71. Huizenga NA, Koper JW, De Lange P et al. A polymorphism in the glucocorticoid receptor gene may be associated with and increased sensitivity to glucocorticoids *in vivo*. J Clin Endocrinol Metab. 1998;83:144–51.

72. Koper JW, Stolk RP, de Lange P et al. Lack of association between five polymorphisms in the human glucocorticoid receptor gene and glucocorticoid resistance. Hum Genet. 1997; 99:663–8.

231

73. Huizenga NA, de Lange P, Koper JW et al. Five patients with biochemical and/or clinical generalized glucocorticoid resistance without alterations in the glucocorticoid receptor gene. J Clin Endocrinol Metab. 2000;85:2076–81.
74. Corrigan CJ. Glucocorticoid-resistant asthma. T-lymphocyte defects. Am J Respir Crit Care Med. 1996;154:S53–5; discussion S55–7.
75. Corrigan CJ, Brown PH, Barnes NC, Tsai JJ, Frew AJ, Kay AB. Glucocorticoid resistance in chronic asthma. Peripheral blood T lymphocyte activation and comparison of the T lymphocyte inhibitory effects of glucocorticoids and cyclosporin A. Am Rev Respir Dis. 1991;144:1026–32.
76. Lane SJ, Lee TH. Glucocorticoid receptor characteristics in monocytes of patients with corticosteroid-resistant bronchial asthma. Am Rev Respir Dis. 1991;143:1020–4.
77. Sher ER, Leung DY, Surs W et al. Steroid-resistant asthma. Cellular mechanisms contributing to inadequate response to glucocorticoid therapy. J Clin Invest. 1994;93:33–9.
78. Rakasz E, Gal A, Biro J, Balas G, Falus A. Modulation of glucocorticosteroid binding in human lymphoid, monocytoid and hepatoma cell lines by inflammatory cytokines interleukin (IL)-1 beta, IL-6 and tumour necrosis factor (TNF)-alpha. Scand J Immunol. 1993; 37:684–9.
79. Salkowski CA, Vogel SN. IFN-gamma mediates increased glucocorticoid receptor expression in murine macrophages. J Immunol. 1992;148:2770–7.
80. Spahn JD, Szefler SJ, Surs W, Doherty DE, Nimmagadda SR, Leung DY. A novel action of IL-13: induction of diminished monocyte glucocorticoid receptor-binding affinity. J Immunol. 1996;157:2654–9.
81. Irusen E, Matthews JG, Takahashi A, Barnes PJ, Chung KF, Adcock IM. p38 Mitogen-activated protein kinase-induced glucocorticoid receptor phosphorylation reduces its activity: role in steroid-insensitive asthma. J Allergy Clin Immunol. 2002;109:649–57.
82. Rogatsky I, Logan SK, Garabedian MJ. Antagonism of glucocorticoid receptor transcriptional activation by the c-Jun N-terminal kinase. Proc Natl Acad Sci USA. 1998;95:2050–5.
83. Tsitoura DC, Rothman PB. Enhancement of MEK/ERK signaling promotes glucocorticoid resistance in CD4+ T cells. J Clin Invest. 2004;113:619–27.
84. Rogler G, Meinel A, Lingauer A et al. Glucocorticoid receptors are down-regulated in inflamed colonic mucosa but not in peripheral blood mononuclear cells from patients with inflammatory bowel disease. Eur J Clin Invest. 1999;29:330–6.
85. Schottelius A, Wedel S, Weltrich R et al. Higher expression of glucocorticoid receptor in peripheral mononuclear cells in inflammatory bowel disease. Am J Gastroenterol. 2000;95: 1994–9.
86. Raddatz D, Middel P, Bockemuhl M et al. Glucocorticoid receptor expression in inflammatory bowel disease: evidence for a mucosal down-regulation in steroid-unresponsive ulcerative colitis. Aliment Pharmacol Ther. 2004;19:47–61.
87. Honda M, Orii F, Ayabe T et al. Expression of glucocorticoid receptor beta in lymphocytes of patients with glucocorticoid-resistant ulcerative colitis. Gastroenterology. 2000;118:859–66.
88. Leung DY, Hamid Q, Vottero A et al. Association of glucocorticoid insensitivity with increased expression of glucocorticoid receptor beta. J Exp Med. 1997;186:1567–74.
89. Orii F, Ashida T, Nomura M et al. Quantitative analysis for human glucocorticoid receptor alpha/beta mRNA in IBD. Biochem Biophys Res Commun. 2002;296:1286–94.
90. Leung DY, Chrousos GP. Is there a role for glucocorticoid receptor beta in glucocorticoid-dependent asthmatics? Am J Respir Crit Care Med. 2000;162:1–3.
91. Schule R, Rangarajan P, Kliewer S et al. Functional antagonism between oncoprotein c-Jun and the glucocorticoid receptor. Cell. 1990;62:1217–26.
92. Stocklin E, Wissler M, Gouilleux F, Groner B. Functional interactions between Stat5 and the glucocorticoid receptor. Nature. 1996;383:726–8.
93. Barnes PJ, Karin M. Nuclear factor-kappaB: a pivotal transcription factor in chronic inflammatory diseases. N Engl J Med. 1997;336:1066–71.
94. Lane SJ, Lee TH. Mechanisms of corticosteroid resistance in asthmatic patients. Int Arch Allergy Immunol. 1997;113:193–5.
95. Lane SJ, Adcock IM, Richards D, Hawrylowicz C, Barnes PJ, Lee TH. Corticosteroid-resistant bronchial asthma is associated with increased c-fos expression in monocytes and T lymphocytes. J Clin Invest. 1998;102:2156–64.

96. Yuan LW, Gambee JE. Phosphorylation of p300 at serine 89 by protein kinase C. J Biol Chem. 2000;275:40946–51.
97. Rowan BG, Weigel NL, O'Malley BW. Phosphorylation of steroid receptor coactivator-1. Identification of the phosphorylation sites and phosphorylation through the mitogen-activated protein kinase pathway. J Biol Chem. 2000;275:4475–83.
98. Farrell RJ, Murphy A, Long A et al. High multidrug resistance (P-glycoprotein 170) expression in inflammatory bowel disease patients who fail medical therapy. Gastroenterology. 2000;118:279–88.
99. Diaz-Borjon A, Richaud-Patin Y, Alvarado de la Barrera C, Jakez-Ocampo J, Ruiz-Arguelles A, Llorente L. Multidrug resistance-1 (MDR-1) in rheumatic autoimmune disorders. Part II: Increased P-glycoprotein activity in lymphocytes from systemic lupus erythematosus patients might affect steroid requirements for disease control. Joint Bone Spine. 2000;67:40–8.
100. Maillefert JF, Jorgensen C, Sany J. Multidrug resistance in rheumatoid arthritis. J Rheumatol. 1996;23:2182.
101. Escher G, Galli I, Vishwanath BS, Frey BM, Frey FJ. Tumor necrosis factor alpha and interleukin 1beta enhance the cortisone/cortisol shuttle. J Exp Med. 1997;186:189–98.
102. Sarwal M, Chua MS, Kambham N et al. Molecular heterogeneity in acute renal allograft rejection identified by DNA microarray profiling. N Engl J Med. 2003;349:125–38.

# 25
# Management of perianal Crohn's disease

## S. B. HANAUER

The evaluation and management of perianal Crohn's disease (CD) has recently been reviewed[1–3]. Perianal manifestations are diverse, including skin tags (ranging from innocuous to prolapsing 'elephant ears'), anal canal fissures and ulcers, fistulae (low→high, simple→complex), rectovaginal fistulas, perianal abscesses, anorectal strictures, and anal canal cancers. The described frequency of perianal manifestations is dependent upon the definition of findings, the intensity of examination and the duration of follow-up. The aetiology of perianal manifestations remains to be determined, although most perianal fistulas are thought to arise from inflamed anal glands or by direct extension of ulcers within the rectum. The cumulative frequency of perianal fistulas ranges between 14% and 38%, and in population series the cumulative frequency has ranged from 12% at 1 year to 26% at 20 years[4] The frequency of perianal fistulas also depends upon the lumenal location of disease and is least frequent in patients with ileal disease (12%) and higher in patients with ileocolonic disease (15%) and Crohn's colitis (41%). The highest frequency of perianal disease is seen in patients with CD involving the rectum (92%). Perianal manifestations may precede the diagnosis of CD, present simultaneously with lumenal disease, or persist as the singular manifestation of CD.

The evaluation of perianal CD is evolving along with technological advances in gut and body imaging, and will be dependent upon the best technology available at different sites. In the past, reliance on fistulography and computed tomography were of limited value due to poor diagnostic accuracy (50–60%) depending upon the nature and severity of the lesions. Most recently, surgical examination under anaesthesia, pelvic MRI and anorectal endoscopic ultra-sonography have proven to be more sensitive and are useful in defining the anatomical pathology that contributes to the combined surgical/medical approach to treatment. Most recently, a perianal CD activity index has been developed that includes anatomical and functional components in five categories (discharge, pain, restriction of sexual activities, type of perianal disease, and degree of induration), but the instrument has yet to be formally validated in clinical trials[5]. To date the only prospectively defined assessment of fistula to be tested in clinical trials was developed for use in a multicentre trial to evaluate infliximab in fistula healing[6]. The pre-specified therapeutic endpoints included 'improvement', defined as 'decrease from baseline in the number of open draining fistulas of $\geqslant 50\%$ for at least two consecutive visits (at least 4 weeks)'. 'Remission' is defined as closure (no fistula drainage despite gentle finger compression) of all fistulas that were draining at baseline for (at least) 4 weeks. Subsequently, a maintenance trial for infliximab incorporated a similar definition for 'relapse' as $< 50\%$ reduction from baseline in the number of draining fistulas over a period of $\geqslant 4$ weeks or the need for medical or surgical intervention[7]. Even these pre-specified, endpoints have a degree of subjectivity and do not account for the impact of diarrhoea on fistula drainage. Many patients will report reduction in drainage of fistula as their stools become more firm and future endpoints should require evidence of anatomical resolution of fistula tracks along with the functional assessments.

The initial approach to the management of perianal CD is to determine whether or not there is an associated abscess, and to identify the anatomical location of any associated fistula with particular respect to the anal sphincter. A primary point of emphasis is that suppuration must be drained concordant with medical therapy (e.g. antibiotics) and prior to initiation of immunomodulatory or biologic therapies that may increase the risk of systemic infection. The spectrum of surgical interventions ranges from simple fistulotomy, to placement of a cutting or draining seton, to performance of an endorectal mucosal flap, to faecal diversion with or without a permanent proctectomy. It is hopeful that modern management with immunomodulation and biologic therapy will reduce the 'historical,' 10–18% proctectomy rate for perianal CD. Further discussion regarding the specifics of perianal surgical techniques are beyond the scope of this discussion but are reviewed in the recent AGA Technical Review on Perianal Crohn's disease[3].

While the medical treatments of lumenal CD include aminosalicylates and corticosteroids, these agents have not been evaluated as therapeutic agents specifically for perianal fistulas. Nevertheless, as lumenal disease (and associated diarrhoea) improves, there is an expected reduction in fistula drainage that has not been formally evaluated. In contrast, the medical therapies that have been described to treat perianal fistulas in CD include antibiotics, immunomodulators and anti-TNF therapies. Only tacrolimus and infliximab have been evaluated in controlled clinical trials.

Despite over 20 years of empirical use of antibiotics to treat perianal fistulas in CD there have been no reported controlled clinical trials in this setting. Case series with metronidazole date back to the early 1980s, when Bernstein, Brandt and co-workers first described a series of 21 patients who were treated with metronidazole at a dose of 20 mg/kg per day. The authors described a 'clinical response' (decreased pain and tenderness) with 10/18 patients reported to have 'complete healing' of their perianal disease[8]. In a follow-up report describing 26 patients that included 17 from the previous report, they described relapse after dose-reduction or cessation of therapy and improvement with resumption of treatment[9]. However, 50% of patients developed paraesthesias with the administered dose. Metronidazole has also been combined with ciprofloxacin and azathioprine in an uncontrolled experience[10] or added to 6-mercaptopurine in a Markov model of cost-effectiveness[11]. The overall responses were described as 'improvement' in 50% and 'remission' in 25%, and were suggested in the cost-efficacy analysis to be equally efficacious and more cost-effective than infliximab therapy. Ciprofloxacin, alone[12], or in combination with metrondazole[13], has been reported to be effective for the treatment of perianal fistulas in two small, uncontrolled series totalling 13 patients. These data, while underwhelming from an evidence-based standpoint, have led to a near-ubiquitous use of antibiotics, primarily metronidazole, as a first-line approach to perianal fistula in CD[1–3]. Neither dose–response nor duration of treatment has been defined.

The purine antimetabolites have also been used to treat perianal fistulas in CD subsequent to the landmark double-blind, placebo-controlled, crossover study by Present et al.[14], in which a subgroup of 36 patients with 40 fistulas were reported to have complete closure of fistulas in 31% and partial healing in 24% after at least 3 months of 6-mercaptopurine. In a meta-analysis of five controlled trials of azathioprine and 6-mercaptopurine, that included only 70 patients with perianal disease, 22 of 41 (54%) patients treated with the immunomodulators compared with 6 of 29 (21%) placebo-treated patients had 'complete' closure of fistula, providing a pooled odds ratio of 4.44 in favour of the purine antimetabolites[15]. These data have been supported by uncontrolled series from New York[16] and Johns Hopkins[17]. To date a dose–response or optimal therapeutic metabolite level have not been established in the setting of perianal disease.

Methotrexate has also been described in uncontrolled series to be of benefit in the treatment of perianal fistula. Mahadaven and Marion reported that 4/16 (25%) patients treated with methotrexate (who had been refractory or intolerant to 6-mercaptopurine) had complete fistula closure and another five (31%) had partial closure[18]. The authors also reported a tendency of patients to relapse when parenteral therapy was changed to oral methotrexate. In a recent uncontrolled series of 12 patients who also were refractory or intolerant to azathioprine, the combination of methotrexate, 20 mg/week, and infliximab was reported to induce long-term (>6 month) 'closure' of fistula in four patients and a 'partial response' in another three patients[19]. Again, optimal dosing and/or combination approaches have yet to be established for methotrexate in the setting of perianal fistulas in CD.

The uncontrolled experience with parenteral cyclosporine, 4 mg/kg, as an effective mode to treat perianal fistulas in CD from Chicago[20] and New York[21], in which 10/12 and 14/16 patients rapidly responded to an intravenous regimen, respectively also noted the rapid recurrence with cessation of therapy. Nevertheless, Sandborn and colleagues[22] performed a double-blind, randomized, controlled trial with oral tacrolimus. Forty-eight patients were enrolled and the primary endpoint was a 50% reduction in fistula drainage. Complete closure was a secondary endpoint in this 10-week trial. Although response was seen in 43% of tacrolimus-treated patients vs 8% of placebo-treated patients ($p$ = 0.004) only 10% of tacrolimus-treated patients achieved fistula remission compared to 8% of patients receiving placebo. Despite the efficacy at inducing improvement there were a large number of side-effects reported, including increased creatinine, headaches, paraesthesias and tremors. For the most part cyclosporine and tacrolimus, despite their short-term efficacy, have been superseded by the use of infliximab for perianal fistulas.

Infliximab has become the most extensively evaluated therapy for the treatment of fistulizing CD. In the first controlled trial specifically evaluating the treatment of Crohn's fistula, 94 patients (85 with perianal fistula) with actively draining fistula despite aminosalicylates, antibiotics, corticosteroids, or immunomodulators were randomized to receive three infusions of infliximab, at 5 or 10 mg/kg, or placebo at 0, 2 and 6 weeks. The primary endpoint, 'closure of at least 50% of fistula' for at least 4 weeks, was achieved in 68% of patients receiving 5 mg/kg infusions, 56% of patients receiving 10 mg/kg, and 26% of placebo-treated patients. 'Closure of all fistulas' was observed in 55%, 38% and 13% of patients receiving 5 mg/kg, 10 mg/kg, or placebo, respectively. However, the median duration of closure after cessation of therapy was only 12 weeks. Subsequently, the ACCENT II trial was designed to determine the durability of fistula response with maintenance therapy following the 5 mg/kg, three-dose induction regimen[7]. This trial enrolled 306 patients with refractory, actively draining fistula (83% with perianal fistula); all received an induction regimen and were then randomized to receive 8-week maintenance doses of infliximab 5 mg/kg or placebo over 54 weeks. Patients who relapsed were able to be escalated from 5 mg/kg to 10 mg/kg dosing or from placebo to 5 mg/kg dosing. Similar to the previous trial, 69% of patients had a fistula 'response' by week 14. The median time to loss of response through week 54 was greater than 40 weeks for patients receiving maintenance infliximab compared to 14 weeks for patients randomized to placebo maintenance. In the intent-to-treat population 39% of patients receiving infliximab maintenance had complete closure of all fistulas at week 54 vs 19% who received placebo 'maintenance'. Of note, approximately 60% of both groups who 'lost their response' improved, again, with crossover therapy to 10 mg/kg infliximab or 5 mg/kg infliximab infusions. In addition, quality-of-life scores, need for surgery, and need for hospitalization were secondary endpoints that were benefited by infliximab maintenance. These data are impressive because of the lack of response to prior therapies (30% of patients had been on antibiotics, 28% on corticosteroids and 34% on immunomodulators) and may be improved by combining surgical drainage and seton placement, or combining an endoscopic or MRI evaluation to rule out subclinical abscess formation[3]. Optimal combination therapy with infliximab (i.e. azathioprine or methotrexate) has not yet been established, although these immunomodulators have been effective at reducing formation of antibodies to infliximab[23].

In summary, the optimal approach to perianal fistulas in CD involves a cooperative effort by physicians and surgeons to identify the location of fistula origins and complications, drain suppuration, place setons when appropriate (active rectal disease or with inter-sphincteric or supra-sphincteric fistula), and to induce and maintain cessation of fistula drainage. Throughout the course, treatment of fistulas will be within the context of treating the entire CD scenario (lumenal disease and other complications). The medical regimen of choice will be dependent upon the severity and chronicity of fistula drainage. For mildly affected patients most clinicians will begin with antibiotics and/or immunomodulators and reserve infliximab for severely affected patients or those refractory to non-biologic approaches. Although complete 'healing' (permanent closure of fistula tracks) may not be achieved, improvement in the patients' quality of life and overall morbidity can often be established and maintained. It is hopeful that, with the advent of anti-TNF and combination therapies, that the devastating impact of faecal diversion and/or permanent proctectomy will be avoided in even the most troublesome settings of perianal fistula.

## References

1. Hanauer SB, Sandborn W. Management of Crohn's disease in adults. Am J Gastroenterol. 2001;96:635–43.
2. Lichtenstein GR. Treatment of fistulizing Crohn's disease. Gastroenterology. 2000;119: 1132–47.
3. Sandborn WJ, Fazio VW, Feagan BG, Hanauer SB. AGA technical review on perianal Crohn's disease. Gastroenterology. 2003;125:1508–30.
4. Schwartz DA, Loftus EV Jr, Tremaine WJ et al. The natural history of fistulizing Crohn's disease in Olmsted County, Minnesota. Gastroenterology. 2002;122:875–80.
5. Irvine EJ. Usual therapy improves perianal Crohn's disease as measured by a new disease activity index. McMaster IBD Study Group. J Clin Gastroenterol. 1995;20:27–32.
6. Present DH, Rutgeerts P, Targan S et al. Infliximab for the treatment of fistulas in patients with Crohn's disease. N Engl J Med. 1999;340:1398–405.
7. Sands BE, Anderson FH, Bernstein CN et al. Infliximab maintenance therapy for fistulizing Crohn's disease. N Engl J Med. 2004;350:876–85.
8. Bernstein LH, Frank MS, Brandt LJ, Boley SJ. Healing of perineal Crohn's disease with metronidazole. Gastroenterology. 1980;79:357–65.
9. Brandt LJ, Bernstein LH, Boley SJ, Frank MS. Metronidazole therapy for perineal Crohn's disease: a follow-up study. Gastroenterology. 1982;83:383–7.
10. Dejaco C, Harrer M, Waldhoer T, Miehsler W, Vogelsang H, Reinisch W. Antibiotics and azathioprine for the treatment of perianal fistulas in Crohn's disease. Aliment Pharmacol Ther. 2003;18:1113–20.
11. Arseneau KO, Cohn SM, Cominelli F, Connors AF Jr. Cost-utility of initial medical management for Crohn's disease perianal fistulae. Gastroenterology. 2001;120:1640–56.
12. Turunen U, Farkkila M, Seppala K. Long-term treatment of perianal or fistulous Crohn's disease with ciprofloxacin. Scand J Gastroenterol. 1989;24:144.
13. Solomon M, McLeod RS, O'Connor B, Steinhart AH, Greenberg GR, Cohen Z. Combination ciprofloxacin and metronidazole in severe perianal Crohn's disease. Clin Invest Med. 1992;15(Suppl.):A41.
14. Present DH, Korelitz BI, Wisch N, Glass JL, Sachar DB, Pasternack BS. Treatment of Crohn's disease with 6-mercaptopurine. A long-term, randomized, double-blind study. N Engl J Med. 1980;302:981–7.
15. Pearson DC, May GR, Fick GH, Sutherland LR. Azathioprine and 6-mercaptopurine in Crohn disease. A meta-analysis. Ann Intern Med. 1995;123:132–42.
16. Korelitz BI, Adler DJ, Mendelsohn RA, Sacknoff AL. Long-term experience with 6-mercaptopurine in the treatment of Crohn's disease. Am J Gastroenterol. 1993;88:1198–205.
17. O'Brien JJ, Bayless TM, Bayless JA. Use of azathioprine or 6-mercaptopurine in the treatment of Crohn's disease. Gastroenterology. 1991;101:39–46.
18. Mahadevan U, Marion JF, Present DH. Fistula response to methotrexate in Crohn's disease: a case series. Aliment Pharmacol Ther. 2003;18:1003–8.
19. Schroder O, Blumenstein I, Schulte-Bockholt A, Stein J. Combining infliximab and methotrexate in fistulizing Crohn's disease resistant or intolerant to azathioprine. Aliment Pharmacol Ther. 2004;19:295–301.
20. Hanauer SB, Smith MB. Rapid closure of Crohn's disease fistulas with continuous intravenous cyclosporin A [see comments]. Am J Gastroenterol. 1993;88:646–9.
21. Present DH, Lichtiger S. Efficacy of cyclosporine in treatment of fistula of Crohn's disease. Dig Dis Sci. 1994;39:374–80.

22. Sandborn WJ, Present DH, Isaacs KL et al. Tacrolimus for the treatment of fistulas in patients with Crohn's disease: a randomized, placebo-controlled trial. Gastroenterology. 2003;125:380–8.

23. Sandborn WJ. Optimizing anti-tumor necrosis factor strategies in inflammatory bowel disease. Curr Gastroenterol Rep. 2003;5:501–5.

# Section VIII
# The difficult cases of ulcerative colitis

**Chair: C. GASCHÉ and S. MARKOVIC**

# 26
# A case of refractory proctitis

## W. MIEHSLER

## CASE REPORT

### History

In 1982 a policeman, born in 1930, started to suffer from rectal tenesmus, urgently defecating small portions with bloody admixtures. His preceding medical history was uneventful. The patient was initially treated at a primary-health facility where he underwent sigmoidoscopy, which showed mucosal inflammation with ulceration and friability of the rectum with normal-appearing mucosa proximal to the rectosigmoid junction (15 cm). Histologically biopsies of the sigmoid colon showed regular mucosa, whereas rectal biopsies showed a continuous inflammation with crypt distortion and crypt abscesses. Based on these results the diagnosis of ulcerative proctitis was established and the patient received sulphasalazine suppositories. Within a few weeks, he went into remission and remained free of symptoms until 1993.

### 'Refractory proctitis'

In November 1993 symptoms of proctitis reoccurred with urgent rectal tenesmus, and defecation of six to eight small portions with bloody admixtures. Additionally the patient had dyspnoea, palpitation and fatigue. He was admitted to a secondary health facility. Laboratory tests showed a hypochromic, microcytic anaemia compatible with iron deficiency (haemoglobin 8.1 g/dl, haematocrit 26%, MCV 66 fl, MCH 21 pg, serum iron 12 µg/dl, ferritin 16 µg/L, transferrin 336 mg/dl). The patient underwent colonoscopy, again showing inflammatory, ulcerative lesions in the rectum and a normal-appearing mucosa beyond the rectosigmoid junction consistent with ulcerative proctitis. As stated in the report of the colonoscopy, the investigation was stopped at the left flexure due to pain and the 'clear diagnosis'. Dyspnoea, palpitation and fatigue were considered to be due to the severe iron-deficiency anaemia, secondary to ulcerative proctitis. Therefore the patient was treated with mesalamine enemas and oral iron substitution, but symptoms and anaemia persisted. A subsequent course of oral prednisolone starting at an initial dose of 50 mg/day led to little improvement and a repeated colonoscopy in June 1994, again just reaching the hepatic flexure, still showed moderately active proctitis. Now the patient was referred to our centre as 'refractory proctitis'.

## Course

At the time the patient was referred to our centre (tertiary referral centre) he had five or six urgent defecations and sometimes visible blood, no extraintestinal manifestations and suffered dyspnoea and fatigue. He had mildly elevated inflammatory parameters (C-reactive protein 3.4 mg/dl, blood sedimentation rate 41 mm/h) and a microcytic, hypochromic anaemia (haemoglobin 8.1 g/dl, MCV 66 fl, MCH 21 pg) as dominant signs of refractoriness. Therapy was extended by oral mesalamine (4 g/day) and iron substitution with intravenous iron sucrose. After 3 months the patient was not yet in remission, though some improvement was achieved. He had three of four pulpy stools with little blood, little urgency, no dyspnoea and less fatigue (haemoglobin 11.7 g/dl, MCV 79 fl, MCH 27 pg, C-reactive protein 1.0 mg/dl, blood sedimentation rate 19 mm/h). However, the patient now presented a new symptom: right-sided abdominal pain and bloating. He underwent abdominal sonography with transabdominal bowel sonography (TABS) which showed wall thickening of the ascending colon (so-called 'target lesion') suggestive of advanced Crohn's disease. It was assumed that 'refractoriness' was due to a wrong diagnosis, which prompted re-colonoscopy.

Complete colonoscopy, done during conscious sedation, showed a healing rectal mucosa but a polypoid and ulcerating lesion of the ascending colon, histologically being an invasive adenocarcinoma. The patient underwent subtotal colectomy and ileorectostomy (G2, pT3, pN0, M0). An ileoanal pouch was not installed, since the patient feared functional limitations (continence was little impaired), most of the 'refractory' symptoms (especially anaemia) seemed to be tumour-related and colonoscopy showed a healing rectal mucosa. In 1995 the occurrence of metastases of the liver and peritoneal carcinosis prompted chemotherapy with 5-fluorouracil, leucovorin, mitomycin and methotrexate. The patient finally died in 1997 of his terminal malignant disease.

## COMMENTS AND DISCUSSION

This case deals with a patient with known ulcerative proctitis, which was initially managed simply and successfully, who presented with a relapse refractory to standard therapy. The patient presented with misleading symptoms that were interpreted as 'refractory proctitis', which led to delay of the diagnosis of another, in this case, malignant disease.

Ulcerative colitis (UC) is defined as chronic inflammatory bowel disease at least involving the rectum but possibly affecting the entire colon in a continuous way[1]. In about 20–30% of cases the disease is confined to the rectum not exceeding the rectosigmoid junction which is referred to as ulcerative proctitis[2,3]. Usually, ulcerative proctitis is thought to take a benign course, not being a challenge for the treating physician. However, if no complete remission is achieved after treatment with topical steroids or mesalamine enemas after 6–8 weeks, and no response is seen even after addition of oral mesalamine, the condition may be referred to as 'refractory distal UC' according to a definition of an international working team report[4].

How often does refractory procitits occur, and what are possible reasons for refractoriness? In a recent follow-up study of 273 patients with ulcerative proctitis it was shown that 14% of patients were considered to have a refractory course of disease which was mainly due to more proximal extension of UC, thus being in fact a left-sided colitis or pancolitis[2]. Additionally, it has been demonstrated that patients initially diagnosed as having ulcerative proctitis who developed a refractory course were later diagnosed as having Crohn's disease[5]. In the present patient Crohn's disease was also one of the differential diagnoses for refractoriness due to the 'target sign' seen in transabdominal bowel sonography which is suggestive for Crohn's disease[6]. Other reasons for refractoriness are certainly infections with e.g. *Clostridium difficile* or cytomegalovirus, the latter receiving increasing attention in ulcerative colitis refractory to medical therapy[7–9]. In the aforementioned follow-up study only 6% of patients were finally considered to have refractory ulcerative proctitis which was defined as more than three relapses per year, persisting disease activity despite continuous therapy or the need for steroids or immunosuppressants[2]. To sum up, a refractory course in ulcerative proctitis is a rare condition and often mimicked by another problem, e.g. left-sided colitis or pancolitis, CD, infectious diseases or, as in this case, something completely different.

There have been countless trials evaluating different therapeutic options for ulcerative colitis; however, few of them exclusively addressed or reported results in patients with ulcerative proctitis. The therapy of ulcerative colitis and proctitis is reviewed in detail in the chapter by Rachmilewitz in this book. Briefly, it has clearly been demonstrated that the topical therapy with mesalamine is the backbone of therapy in this entity with enemas as galenic version for left-sided colitis and suppositories for proctitis[10]. Additionally, the combination of mesalamine enemas with oral mesalamine was shown to have advantages over rectal therapy alone, a therapeutic option for patients with proven proctitis refractory to local therapy[11]. Budesonide enemas (2 mg) were also studied in distal ulcerative colitis vs. mesalamine enemas (1 g). Although showing equal efficacy in terms of endoscopic and histological outcome parameters the rate of clinical remission was significantly higher in the mesalamine group[12]. If patients remain refractory to these agents most clinicians would favour a course of systemic steroids, although such a strategy is not supported by firm data from randomized controlled trials[10,13].

The present patient had a history of ulcerative proctitis and presented with symptoms compatible with relapse. Colonoscopy and histology were also compatible with this diagnosis; however, there are two points that need to be discussed: two colonoscopies were incompletely performed during the course of 'refractory proctitis' that would have achieved an earlier diagnosis of right-sided colonic cancer if they were done completely. The main reason for breaking off the colonoscopies was pain, which again indicates the need for patient-tailored analgosedation during endoscopy, particularly to allow complete investigation which is the basis for high-quality results that answer important questions[14]. Complete colonoscopy is mandatory in such a patient for three reasons. First the patient has exceeded the age of 50 years, nowadays entailing the recommendation for a complete screening colonoscopy[15] (albeit the patient was treated in the 1990s when a simple sigmoidoscopy was recommended for screening purpose[16]). Second, the patient had an IBD refractory to therapy, raising the question of whether the classification of IBD (UC vs. CD) was correct. An ileocolonoscopy with segment biopsies is extremely helpful for differentiation between these two entities[17]. Third, a surveillance colonoscopy is recommended in IBD patients after 8–10 years of disease duration which was the case in the present patient[18].

The other relevant item is the severe anaemia which was a dominating symptom of refractoriness. Anaemia is a common finding in IBD but it is certainly also seen in colorectal carcinoma (CRC)[19,20]. It is important to stress that anaemia also has a high impact on disease activity scores with a haemoglobin below 10 g/dl loading four points into the colitis activity index by Rachmilewitz, which by itself already reaches the cut-off for active disease[21]. Since the rectal mucosa was almost healed at the final colonoscopy it is obvious that the dominating 'refractory' symptoms, i.e. severe anaemia as well as visible blood in stools, were in fact symptoms of colonic cancer rather than IBD, thus representing a misleading pitfall. Anaemia is refractory to oral iron substitution in 26% of patients with UC, but its degree usually parallels the severity of disease and should therefore cause doubt if severity of anaemia does not match severity of IBD[22].

Another point of interest is the question of whether colonic cancer in this patient was a sporadic carcinoma or an IBD-associated cancer. There are some arguments that contradict the thesis that this carcinoma was an IBD-associated CRC. On the one hand it has been shown that IBD-associated CRC is related to inflammation and extent of disease, ulcerative proctitis surprisingly not being associated with an increased risk of CRC[23]. Thus, the present patient was not at increased risk for developing IBD-associated CRC. On the other hand the risk for CRC in IBD is related to inflammation, and no colonoscopy in the present patient ever showed an inflammation exceeding the rectosigmoid junction[24]. Additionally, anaplastic and mucinous cancers are more common in IBD-associated CRC, which was not the histological type of CRC in the present patient[25].

The patient underwent subtotal colectomy and ileorectostomy although an ileal pouch–anal anastomosis (IPAA) usually produces a high quality of life in UC patients and is usually favoured[26]. From the oncological point of view the role of IPAA as a resective method for CRC is less clear. There are some data in patients with familial polyposis coli and associated CRC that suggest that IPAA might also be safe for this indication[27]. However, an IPAA was not performed in the present patient, since the patient feared functional limitations (his continence was little impaired), most of the 'refractory' symptoms seemed to be tumour-related and colonoscopy showed that proctitis was controlled by medical means. Incontinence is an important functional limitation which a patient with IPAA can have, and the frequency of this problem increases with the age of the patient, being highest over 65 years of age[28].

In conclusion, in my view refractory proctitis is a rare condition, and any refractory course of ulcerative proctitis should raise the suspicion that there is another problem lurking behind the diagnosis of refractory proctitis.

## References

1.  Lennard-Jones J, Shivananda S. Clinical uniformity of inflammatory bowel disease at presentation and during the first year of disease in the north and south of Europe. Eur J Gastroenterol Hepatol. 1997;9:353–9.
2.  Meucci G, Vecchi M, Astegiano M et al. The natural history of ulcerative proctitis: a multicenter, retrospective study. Gruppo di Studio per le Malattie Infiammatorie Intestinali (GSMII). Am J Gastroenterol. 2000;95:469–73.
3.  Moum B, Vatn M, Ekbom A et al. Incidence of ulcerative colitis and indeterminate colitis in four counties of southeastern Norway, 1990–93. A prospective population-based study. The Inflammatory Bowel South-Eastern Norway (IBSEN) Study Group of Gastroenterologists. Scand J Gastroenterol. 1996;31:362–6.
4.  Järnerot G, Lennard-Jones J, Bianchi Porro G, Brynskov J, Campieri M, Present D. Working team report: medical treatment of refractory distal ulcerative colitis. Gastroenterol Int. 1991;4:93–8.
5.  Langevin S, Menard D, Haddad H, Beaudry R, Poisson J, Devroede G. Idiopathic ulcerative proctitis may be the initial manifestation of Crohn's disease. J Clin Gastroenterol. 1992;15:199–204.
6.  Worlicek H, Lutz H, Heyder N, Matek W. Ultrasound findings in Crohn's disease and ulcerative colitis: a prospective study. J Clin Ultrasound. 1987;15:153–63.
7.  Kochhar R, Ayyagari A, Goenka M, Dhali G, Aggarwal R, Mehta S. Role of infectious agents in exacerbation of ulcerative colitis in India. A study of *Clostridium difficile*. J Clin Gastroenterol. 1993;16:26–30.
8.  Kambham N, Vij R, Cartwright CA, Longacre T. Cytomegalovirus infection in steroid-refractory ulcerative colitis: a case–control study. Am J Surg Pathol. 2004;28:365–73.
9.  Itzkowitz S. Conditions that mimic inflammatory bowel disease. Diagnostic clues and potential pitfalls. Postgrad Med. 1986;80:219–31.
10. Cohen R, Woseth D, Thisted R, Hanauer S. A meta-analysis and overview of the literature on treatment options for left-sided ulcerative colitis and ulcerative proctitis. Am J Gastroenterol. 2000;95:1263–76.
11. Safdi M, DeMicco M, Sninsky C et al. A double-blind comparison of oral versus rectal mesalamine versus combination therapy in the treatment of distal ulcerative colitis. Am J Gastroenterol. 1997;92:1867–71.
12. Lemann M, Galian A, Rutgeerts P et al. Comparison of budesonide and 5-aminosalicylic acid enemas in active distal ulcerative colitis. Aliment Pharmacol Ther. 1995;9:557–62.
13. Kjeldsen J. Treatment of ulcerative colitis with high doses of oral prednisolone. The rate of remission, the need for surgery, and effect of prolonging the treatment. Scand J Gastroenterol. 1993;28:821–6.
14. Bell G. Premedication, preparation and surveillance. Endoscopy. 2002;34:2–12.
15. Helm J, Choi J, Sutphen R, Barthel J, Albrecht T, Chirikos T. Current and evolving strategies for colorectal cancer screening. Cancer Control. 2003;10:193–204.
16. Hart A, Wicks A, Mayberry J. Colorectal cancer screening in asymptomatic populations. Gut. 1995;36:590–8.
17. Dejaco C, Oesterreicher C, Angelberger S et al. Diagnosing colitis: a prospective study on essential parameters for reaching a diagnosis. Endoscopy. 2003;35:1004–8.
18. Eaden J, Mayberry J. Gudelines for screening and surveillance of asymptomatic colorectal cancer in patients with inflammatory bowel disease. Gut. 2002;51(Suppl. V):V10–12.
19. Wilson A, Reyes E, Ofman J. Prevalence and outcomes of anemia in inflammatory bowel disease: a systematic review of the literature. Am J Med. 2004;116(Suppl. 7A):44–9S.
20. Tas F, Eralp Y, Basaran M et al. Anemia in oncology practice: relation to diseases and their therapies. Am J Clin Oncol. 2002;25:371–9.
21. Rachmilewitz D. Coated mesalazine (5-aminosalicylic acid) versus sulphasalazine in the treatment of active ulcerative colitis: a randomised trial. Br Med J. 1989;298:82–6.

22. Schreiber S, Howaldt S, Schnoor M et al. Recombinant erythropoietin for the treatment of anemia in inflammatory bowel disease. N Engl J Med. 1996;334:619–23.
23. Ekbom A, Helmick C, Zack M, Adami H. Ulcerative colitis and colorectal cancer. A population-based study. N Engl J Med. 1990;323:1228–33.
24. Rhodes J, Campbell B. Inflammation and colorectal cancer: IBD-associated and sporadic cancer compared. Trends Mol Med. 2002;8:10–16.
25. Choi P, Zelig M. Similarity of colorectal cancer in Crohn's disease and ulcerative colitis: implications for carcinogenesis and prevention. Gut. 1994;35:950–4.
26. Robb B, Pritts T, Gang G et al. Quality of life in patients undergoing ileal pouch–anal anastomosis at the University of Cincinnati. Am J Surg. 2002;183:353–60.
27. Ziv Y, Church JM, Oakley JR, McGannon E, Schroeder TK, Fazio VF. Results after restorative proctocolectomy and ileal pouch–anal anastomosis in patients with familial adenomatous polyposis and coexisting colorectal cancer. Br J Surg. 1996;83:1578–80.
28. Delaney C, Fazio V, Remzi F et al. Prospective, age-related analysis of surgical results, functional outcome, and quality of life after ileal pouch–anal anastomosis. Ann Surg. 2003; 238:221–8.

# 27
# Inflammatory bowel diseases and innate immunity

## J. WEHKAMP, K. FELLERMANN and E. F. STANGE

---

## INTRODUCTION

Despite active research for many decades the aetiology of Crohn's disease (CD) is still enigmatic. Most research has focused on a potential dysregulation of specific mucosal immunology. These investigations have elegantly described the mucosal cellular populations and cytokine profiles associated with inflammatory bowel disease (IBD) but have not succeeded in finding the aetiological culprit. In this overview we will outline a novel concept of how epidemiological, pathophysiological, genetic, molecular, clinical and pharmacological sets of data may be synthesized into a unifying hypothesis compatible with many features of this disease.

Several years ago we became interested in the innate mucosal system of antibiotic peptides contributing to the defensive array of substances and structures opposed to the invasion of lumenal bacteria and other potential invaders. This system of antibiotic peptides is apparently synthesized and secreted by the intestinal mucosa as part of innate immunity, but has received little attention. In this overview we will outline some recent findings regarding innate immunity which might be crucial for the pathophysiology of IBD.

## EPIDEMIOLOGY: THE ROLE OF HYGIENE

There is a clearcut north to south gradient of IBD incidences worldwide, including Europe[1]. In developing countries infectious intestinal diseases represent the rule, and idiopathic IBD, especially CD, the rare exception[1]. Since migration in many instances is associated with adaptation to the incidence rates in the immigrant country there is little doubt that environmental factors are involved[2]. This is supported by the finding that good domestic hygiene in infancy has been shown to be a risk factor for CD but not for ulcerative colitis (UC), even within a country[3]. Similarly, *Helicobacter pylori* seroprevalence was substantially reduced in CD (odds ratio 0.18) but not in UC. In addition, CD occurs more often in members of small families as

**Table 1** Some features of innate and adaptive immunity that contribute together to mucosal host defence. The arrows demonstrates the crosstalk and linkage between these systems

| *Innate immunity* | ⇄ | *Adaptive immunity* |
|---|---|---|
| Hard-wired | | Require priming |
| Immediate | | Immediate or late |
| Cell-bound pattern recognition receptors | | Soluble + cellular receptors/antibodies |
| Macrophages, Paneth + epithelial cells, fibroblasts, neutrophils | | B and T cells, professional antigen-presenting cells |
| cd1, TLR, NOD | | T cell receptors, antibodies |

opposed to those with many children. Since intrafamilial transmission of common pathogens is frequent, the single child is particularly prone to be raised under more hygienic conditions, with less risk of acquiring gut infections[4,5]. Most probably these various factors associated with the incidence of CD serve as indicators of a relatively clean environment leading to a diminished confrontation with pathogenic or non-pathogenic microorganisms. As a result the intestinal innate immune system is probably not 'trained' to confront minor infections without recruiting the full array of specific immune functions which act only at the expense of a relevant inflammation.

Another important aspect in this regard is the apparently frequent association of a recent intestinal infection with the first appearance of CD and the prevalence of superinfection in pre-established IBD[6]. Although these relationships have not been fully understood the interpretation has been made that an infection in some way triggers a relapse of the idiopathic bowel disease by breaking mucosal tolerance. Despite their self-limited character these infections may initiate a cascade of inflammatory events leading to chronic relapsing disease in genetically susceptible hosts ('hit-and-run' hypothesis). Alternatively, the host with IBD may be more likely to contract an intestinal infection because of a defective innate defence system

## PATHOPHYSIOLOGY: THE ROLE OF LUMENAL BACTERIA

It has always been an intriguing hypothesis that IBD are caused by a specific, hitherto unrecognized infection. For example, *Mycobacterium paratuberculosis* has been considered by various groups to cause not only Johne's disease in cattle but also CDe in humans[7,8]. The debate has been ongoing for many years and is beyond the scope of the present considerations. It should be noted, however, that even very recently, using novel techniques such as granuloma isolation with laser-capture microdissection many more Crohn's samples were shown to be positive for mycobacteria than were controls[9]. Thus, although mycobacteria are far from proven to be causative agents, it is apparent that the mucosa in CD frequently harbours unusual and potentially pathogenic bacteria. In some instances *Listeria* have been isolated[10] or specific mucosal adherent *Escherichia coli*[11]. Interestingly, there is a tremendous increase in the mucosal-associated bacterial counts in the neoterminal ileum after ileocaecal resection for CD, and this colonization may be related to postoperative relapse[12]. Measles infection in CD is also a very controversial issue, but it adds to the list of transmissible agents recovered from Crohn's mucosa[13]. Taken together, these findings indicate that Crohn's mucosa is often the target of various infections, but positive proof that the disease is caused by these agents is missing. Most importantly, the immune response in the gut mucosa is not specific to any of these suspicious agents, but rather unspecific to a multitude of organisms.

However, it has only recently been appreciated that the mucosal immune response in IBD is directed towards a multitude of common lumenal bacteria. The most convincing evidence for a break in mucosal tolerance in intestinal inflammation stems from the observation that knockout mice lacking several relevant genes, including interleukins 2 or 10, develop experimental colitis only when raised in contaminated but not in sterile conditions[14]. This fits well with the consistent finding of a break in mucosal tolerance towards various luminal bacteria in inflammatory bowel diseases[15,16]. It may be concluded that these diseases are not autoimmune diseases in the strict sense, i.e. reactivity against autologous tissues, but only in a more general sense, i.e. immune response towards commensal bacteria. The permeable mucosal barrier may also explain the development of anti-*Saccharomyces cerevisiae* antibodies, especially in familial CD as well as antibodies to various other microbes, including *E. coli.*

The most surprising finding in this regard is the demonstration by Swidsinski et al. that the mucosa in IBD is heavily contaminated by adherent and sometimes invading bacteria entering from the lumen[17]. In contrast, normal mucosa is virtually sterile when washed a few times in saline. These findings are difficult to reconcile with an immunological dysregulation as the sole basis of intestinal inflammation in these diseases. Rather, it is conceivable that there may be a primary defect in the chemical barrier of intestinal antibiotic defensins which protect the normal mucosa extremely efficiently against adherent or entering microbes. It is conceivable that a thorough understanding of these functionally relevant peptides is paramount to understanding the true pathogenesis of IBD. Indirectly, a change in the expression or function of this chemical defence may indeed explain the changes in bacterial flora in IBD reviewed by Linskens et al.[18].

## DEFENSINS: GENETICS, EXPRESSION AND REGULATION

Probably the most important peptide family of endogenous antibiotics is the still-growing number of defensins[19-21]. They comprise a class of cationic antimicrobial peptides with a molecular weight of 3–5 kDa conserved throughout phylogeny. All defensins have been mapped to chromosome 8 in humans[22-24]. Six α-defensins and 4 β-defensins have been identified in humans so far. The α-defensins comprise human neutrophil peptide 1–4, abundant in granulocytes, and human defensin 5 and 6 synthesized in Paneth cells[25]. The β-defensins are of epithelial origin and abundant in skin, intestine and lung. The concept of a certain defensin exclusively formed by specialized tissues or cells needs revision, as inflammation induces epithelial expression of human neutrophil peptides[26] and β-defensins in monocytes and lymphocytes[27].

Defensins can be divided into constitutive forms, e.g. HBD-1 with its widespread stable distribution[28] and inducible peptides such as HBD-2[29]. The mechanisms of activation are currently under investigation. A cytokine-driven induction, e.g. by IL-1β and TNF-α, has been shown besides a direct response to bacterial components such as lipopolysaccharides and lipoproteins. Possible signalling pathways involve Toll-like receptors, especially TLR2 and 4, eventually leading to NFκB-mediated activation of transcription. NOD2/CARD15 as an intracellular LPS receptor induces NFκB[30], which in turn is known to trigger HBD-2 transcription. Interestingly, this NFκB response is impaired in the NOD2 insertion mutation associated with CD[31,32], suggesting a diminished innate response to bacterial components. Human defensin 5 is released as a propeptide from Paneth cells and activated by trypsinogen in the lumen of the intestinal crypts[33]. The functional significance in bacterial infection has recently been shown in HD-5 transgenic mice, which are protected from lethal *Salmonella* infection[34]. On the other hand, matrilysin-deficient mice fail to process defensins efficiently and exhibit higher bacterial counts[35].

## DEFENSINS AND IBD

Some defensins appear to be induced in both CD and UC. Human neutrophil peptides 1–3, as well as lysozyme, are expressed in surface enterocytes of mucosa with active IBD but surprisingly not in controls[26]. HD-5 is stored in a precursor form in normal Paneth cells and is expressed by metaplastic colonic Paneth cells[36]. Notably, both α-defensins HD-5 and HD-6 are induced in the colonic mucosa of IBD patients[36–38]. The alterations in β-defensins are more intriguing because there is a conspicuous difference between CD and UC. It has been suggested that HBD-1 is constitutively expressed in the intestinal epithelium[39] and qualitative investigations indeed showed constitutive expression in normal tissue and IBD mucosa[40]. With the quantitative approach a decrease of HBD-1 was found in inflamed mucosa of both CD and UC. However, it remains to be shown that such a decrease actually translates into a diminished mucosal antibacterial activity.

The inducible HBD-2, which has been described originally in skin[29], is also expressed in the colon during inflammation[39], particularly in UC[40]. It has now been shown by three different independent studies that HBD-2 is highly induced in inflamed mucosa of UC patients. As compared to UC this induction is missing in CD[40–42]. Most probably there is a lack of β-defensin induction in CD contributing to a defective antimicrobial barrier or, alternatively, there is an excessive induction in UC.

The third defensin studied was HBD-3, which was reported by Harder et al. as a novel inducible β-defensin in skin[43]. Another group described HBD-3 based on genomic analysis[44]. Our recent study first described HBD-3 in the human colon. Although HBD-3 was also slightly induced in inflamed Crohn's mucosa, its expression was preferentially enhanced in inflamed and non-inflamed UC[41]. A deficiency in the antimicrobial defence systems of defensins may be a reasonable and plausible explanation for the break of the antibacterial barrier function in IBD.

In conclusion, the decrease of HBD-1 in both IBD, and the lack of induction of both inducible β-defensins HBD-2 and HBD-3 in CD, suggest a deficient mucosal barrier function. This may in part be compensated by the induction of the α-defensins. A lack in the innate defence system of antimicrobial peptides may lead to a permanent but slow bacterial invasion triggering the inflammatory process, but further direct studies on antimicrobial peptide activity in IBD mucosa are required to validate this hypothesis.

## NOD2, A PETIDOGLYCAN RECEPTOR AND DEFENSIN EXPRESSION

Although the aetiology of CD is still enigmatic, the recent finding in about one-third of CD patients of a loss of function mutation in the putative intracellular peptidoglycan receptor NOD2 represents a major advance[45,46]. The pathophysiology of NOD2 in CD was proposed to link to immunological dysregulation in monocytes. Alternatively, intestinal epithelial cells[47,48] and Paneth cells[49] which have also been demonstrated to express this putative receptor might be compromised in their antibacterial response. It has been demonstrated that NOD2-mutated epithelial cells display a disturbed response against *Salmonella*[50], supporting the hypothesis of a defensine deficiency. Unpublished data from our laboratory show that mutation in the NOD2 gene leads to a disturbed expression of antimicrobial Paneth cell defensins. This disturbed Paneth cell defensin expression might explain the ileal preference in NOD2-mutated CD patients.

## TOLL-LIKE RECEPTORS AND THEIR EXPRESSION IN IBD

Intestinal epithelial cells express various pattern-recognition receptors recognizing microbial 'pathogen-associated molecular patterns' as 'non-self' and to rapidly initiate innate immune responses of survival and active defence strategies against lumenal pathogens.

This system of several functional Toll-like receptors (TLR) appears to be a key regulator of the innate response system. Different TLR are responding to different pathogens and bacterial components, including lipopolysaccharide, flagellin and others[51-52]. In active IBD the expression of TLR3 and TLR4 is differentially modulated in the intestinal epithelium. TLR3 is significantly down-regulated in active CD but not in UC. In contrast, TLR4 is up-regulated in both UC and CD. TLR2 and TLR5 expression remained unchanged in IBD[53]. These data suggest that IBD may be associated with distinctive changes in selective TLR expression in the intestinal epithelium, implying that alterations in the innate response system may contribute to the pathogenesis of these disorders.

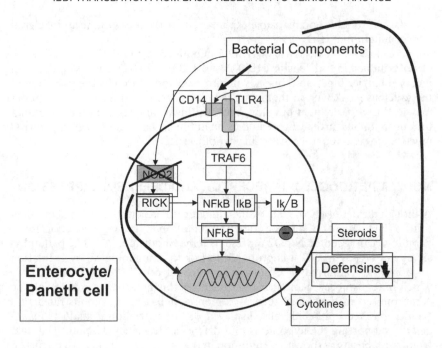

**Figure 1** The putative pathomechanism of the NOD2 mutations in CD. NOD2 is also expressed in epithelial cells and a loss of function mutation may lead to disturbed Paneth cell defensin expression as well as diminished barrier

## THERAPY: THE ROLE OF ANTIBIOTICS AND PROBIOTICS

If a deficiency of these endogenous antibiotics was triggering relapse one would expect exogenous antibiotics to be an efficacious treatment option. Indeed, antibiotics appear to have a limited effect in CD and probiotics in UC. In CD the exogenous antibiotics may compensate for the deficient endogenous antibiotic response to infection or commensal bacterial invasion. In UC the pattern is different, with low basal activity but normal induction during inflammation. Therefore antibiotics may not work, and the benefit of probiotics may be due to the induction of β-defensins, as demonstrated recently *in vitro*[54]. In contrast to the majority of tested *Escherichia coli* the Nissle 1917 strain potently up-regulated HBD-2 expression in colonic cell culture. In pouchitis both approaches work for induction and maintenance, respectively.

## CONCLUDING REMARKS

Seventy years after Crohn's description of the disease named after him it becomes apparent that CD is not a disease but a syndrome. It is not surprising that the diverse facets of genetic predisposition, where only a minority of patients display a defective NOD2 gene, modified by environmental factors such as childhood hygiene and others, may lead to very different forms of disease with respect to localization, natural course and therapeutic response. Although in no way perfect, the present hypothesis appears to be plausible for the reasons presented above, but particularly since the multitude of defensins, other antibiotic peptides and related transcription factors or transporters leaves enough room for clinical diversity. The link between mutations of the NOD2 gene in CD patients related to a diminished defensin expression is further evidence for our hypothesis that CD may be a defensin deficiency syndrome[55]. In contrast, UC appears not to be related to a defensin induction problem. However, the fact that ANCA are directed against the endogenous antibiotic BPI (bactericidal permeability-increasing protein)[56–58], or that the disease is related to certain MDR-1 polymorphisms[59] which may alter defensin export leaves enough room for speculation and, more importantly, hypothesis-driven future work.

### Acknowledgements

Helpful discussions and joint investigations with JM Schröder and J Harder are gratefully acknowledged. The work was supported generously by the Robert Bosch Foundation, Stuttgart, Germany.

### References

1.  Shivananda S, Lennard-Jones J, Logan R et al. Incidence of inflammatory bowel disease across Europe: is there a difference between north and south? Results of the European Collaborative Study on Inflammatory Bowel Disease (EC-IBD). Gut. 1996;39:690–7.
2.  Probert CS, Jayanthi V, Rampton DS, Mayberry JF. Epidemiology of inflammatory bowel disease in different ethnic and religious groups: limitations and aetiological clues. Int J Colorectal Dis. 1996;11:25–8.
3.  Gent AE, Hellier MD, Grace RH, Swarbrick ET, Coggon D. Inflammatory bowel disease and domestic hygiene in infancy. Lancet. 1994;343:766–7.
4.  Feeney MA, Murphy F, Clegg AJ, Trebble TM, Sharer NM, Snook JA. A case–control study of childhood environmental risk factors for the development of inflammatory bowel disease. Eur J Gastroenterol Hepatol. 2002;14:529–34.
5.  Gilat T, Hacohen D, Lilos P, Langman MJ. Childhood factors in ulcerative colitis and Crohn's disease. An international cooperative study. Scand J Gastroenterol. 1987;22:1009–24.
6.  Stallmach A, Carstens O. Role of infections in the manifestation or reactivation of inflammatory bowel diseases. Inflamm Bowel Dis. 2002;8:213–18.
7.  Moss MT, Sanderson JD, Tizard ML et al. Polymerase chain reaction detection of *Mycobacterium paratuberculosis* and *Mycobacterium avium* subsp *silvaticum* in long term cultures from Crohn's disease and control tissues. Gut. 1992;33:1209–13.
8.  McFadden JJ, Butcher PD, Chiodini R, Hermon-Taylor J. Crohn's disease-isolated mycobacteria are identical to *Mycobacterium paratuberculosis*, as determined by DNA probes that distinguish between mycobacterial species. J Clin Microbiol. 1987;25:796–801.

9.  Ryan P, Bennett MW, Aarons S et al. PCR detection of *Mycobacterium paratuberculosis* in Crohn's disease granulomas isolated by laser capture microdissection. Gut. 2002;51:665–70.
10. Liu Y, van Kruiningen HJ, West AB, Cartun RW, Cortot A, Colombel JF. Immunocyto-chemical evidence of *Listeria*, *Escherichia coli*, and *Streptococcus* antigens in Crohn's disease. Gastroenterology. 1995;108:1396–404.
11. Darfeuille-Michaud A, Neut C, Barnich N et al. Presence of adherent *Escherichia coli* strains in ileal mucosa of patients with Crohn's disease. Gastroenterology. 1998;115:1405–13.
12. Neut C, Bulois P, Desreumaux P et al. Changes in the bacterial flora of the neoterminal ileum after ileocolonic resection for Crohn's disease. Am J Gastroenterol. 2002;97:939–46.
13. Wakefield AJ, Ekbom A, Dhillon AP, Pittilo RM, Pounder RE. Crohn's disease: pathogenesis and persistent measles virus infection. Gastroenterology. 1995;108:911–16.
14. Sellon RK, Tonkonogy S, Schultz M et al. Resident enteric bacteria are necessary for development of spontaneous colitis and immune system activation in interleukin-10-deficient mice. Infect Immun. 1998;66:5224–31.
15. Duchmann R, May E, Heike M, Knolle P, Neurath M, Meyer zum Büschenfelde K-H. T cell specificity and cross reactivity towards enterobacteria, *Bacteroides*, *Bifidobacterium*, and antigens from resident intestinal flora in humans. Gut. 1999;44:812–8.
16. Landers CJ, Cohavy O, Misra R et al. Selected loss of tolerance evidenced by Crohn's disease-associated immune responses to auto- and microbial antigens. Gastroenterology. 2002;123:689–99.
17. Swidsinski A, Ladhof A, Pernthaler A et al. Mucosal flora in inflammatory bowel disease. Gastroenterology. 2002;122:44–54.
18. Linskens RK, Huijsdens XW, Savelkoul PH, Vandenbroucke-Grauls CM, Meuwissen SG. The bacterial flora in inflammatory bowel disease: current insights in pathogenesis and the influence of antibiotics and probiotics. Scand J Gastroenterol Suppl. 2001;29–40.
19. Fellermann K, Stange EF. Defensins – innate immunity at the epithelial frontier. Eur J Gastroenterol Hepatol. 2001;13:771–6.
20. Cunliffe RN, Mahida YR. Antimicrobial peptides in innate intestinal host defence. Gut. 2000;47:16–7.
21. Lehrer RI, Ganz T. Defensins of vertebrate animals. Curr Opin Immunol. 2002;14:96–102.
22. Harder J, Siebert R, Zhang Y et al. Mapping of the gene encoding human beta-defensin-2 (DEFB2) to chromosome region 8p22-p23.1. Genomics. 1997;46:472–5.
23. Liu L, Zhao C, Heng HH, Ganz T. The human beta-defensin-1 and alpha-defensins are encoded by adjacent genes: two peptide families with differing disulfide topology share a common ancestry. Genomics. 1997;43:316–20.
24. Liu L, Wang L, Jia HP et al. Structure and mapping of the human beta-defensin HBD-2 gene and its expression at sites of inflammation. Gene. 1998;222:237–44.
25. Bevins CL, Jones DE, Dutra A, Schaffzin J, Muenke M. Human enteric defensin genes: chromosomal map position and a model for possible evolutionary relationships. Genomics. 1996;31:95–106.
26. Cunliffe RN, Kamal M, Rose FR, James PD, Mahida YR. Expression of antimicrobial neutrophil defensins in epithelial cells of active inflammatory bowel disease mucosa. J Clin Pathol. 2002;55:298–304.
27. Duits LA, Ravensbergen B, Rademaker M, Hiemstra PS, Nibbering PH. Expression of beta-defensin 1 and 2 mRNA by human monocytes, macrophages and dendritic cells. Immunology. 2002;106:517–25.
28. Zhao C, Wang I, Lehrer RI. Widespread expression of beta-defensin hBD-1 in human secretory glands and epithelial cells. FEBS Lett. 1996;396:319–22.
29. Harder J, Bartels J, Christophers E, Schröder JM. A peptide antibiotic from human skin. Nature. 1997;387:861.
30. Ogura Y, Bonen DK, Inohara N et al. A frameshift mutation in NOD2 associated with susceptibility to Crohn's disease. Nature. 2001;411:603–6.
31. Hugot JP, Chamaillard M, Zouali H et al. Association of NOD2 leucine-rich repeat variants with susceptibility to Crohn's disease. Nature. 2001;411:599–603.
32. Ogura Y, Inohara N, Benito A, Chen FF, Yamaoka S, Nunez G. Nod2, a Nod1/Apaf-1 family member that is restricted to monocytes and activates NF-kappaB. J Biol Chem. 2001;276:4812–8.

33. Ghosh D, Porter E, Shen B et al. Paneth cell trypsin is the processing enzyme for human defensin-5. Nat Immunol. 2002;3:583–90.
34. Salzman NH, Gottsch JD, Huttner KM, Paterson Y, Bevins CL. Protection against enteric salmonellosis in transgenic mice expressing a human intestinal defensin. Nature advance online publication, 19 March 2003.
35. Wilson CL, Ouellette AJ, Satchell DP et al. Regulation of intestinal α-defensin activation by the metalloproteinase matrilysin in innate host defense. Science. 1999;286:113–7.
36. Cunliffe RN, Rose FRAJ, Keyte J, Abberley L, Chan WC, Mahida YR. Human defensin 5 is stored in precursor form in normal Paneth cells and is expressed by some villous epithelial cells and by metaplastic Paneth cells in the colon in inflammatory bowel disease. Gut. 2001;48:176–85.
37. Wehkamp J, Schwind B, Herrlinger KR et al. Innate immunity and colonic inflammation: enhanced expression of epithelial alpha-defensins. Dig Dis Sci. 2002;47:1349–55.
38. Lawrance IC, Fiocchi C, Chakravarti S. Ulcerative colitis and Crohn's disease: distinctive gene expression profiles and novel susceptibility candidate genes. Hum Mol Genet. 2001;10:445–56.
39. O'Neil DA, Porter EM, Elewaut D et al. Expression and regulation of the human b-defensins hBD-1 and hBD-2 in intestinal epithelium. J Immunol. 1999;163:6718–24.
40. Wehkamp J, Fellermann K, Herrlinger KR et al. Human beta-defensin 2 but not beta-defensin 1 is expressed preferentially in colonic mucosa of inflammatory bowel disease. Eur J Gastroenterol Hepatol. 2002;14:745–52.
41. Wehkamp J, Harder J, Weichenthal M et al. Inducible and constitutive beta-defensins are differentially expressed in Crohn's disease and ulcerative colitis. Inflamm Bowel Dis. 2003; 9:215–23.
42. Fahlgren A, Hammarstrom S, Danielsson A, Hammarstrom ML. Increased expression of antimicrobial peptides and lysozyme in colonic epithelial cells of patients with ulcerative colitis. Clin Exp Immunol. 2003;131:90–101.
43. Harder J, Bartels J, Christophers E, Schröder JM. Isolation and characterization of human b-defensin-3, a novel human inducible peptide antibiotic. J Biol Chem. 2001;276:5707–13.
44. Garcia JR, Jaumann F, Schulz S et al. Identification of a novel, multifunctional beta-defensin (human beta-defensin 3) with specific antimicrobial activity. Its interaction with plasma membranes of Xenopus oocytes and the induction of macrophage chemoattraction. Cell Tissue Res. 2001;306:257–64.
45. Ogura Y, Bonen DK, Inohara N et al. A frameshift mutation in NOD2 associated with susceptibility to Crohn's disease. Nature. 2001;411:603–6.
46. Hugot J-P, Chamaillard C, Zouali H et al. Association of NOD2 leucine-rich repeat variants with susceptibility to Crohn's disease. Nature. 2001;411:599–603.
47. Berrebi D, Maudinas R, Hugot JP et al. Card15 gene overexpression in mononuclear and epithelial cells of the inflamed Crohn's disease colon. Gut. 2003;52:840–6.
48. Rosenstiel P, Fantini M, Brautigam K et al. TNF-alpha and IFN-gamma regulate the expression of the NOD2 (CARD15) gene in human intestinal epithelial cells. Gastroenterology. 2003;124:1001–9.
49. Lala S, Ogura Y, Osborne C et al. Crohn's disease and the NOD2 gene: a role for paneth cells. Gastroenterology. 2003;125:47–57.
50. Hisamatsu T, Suzuki M, Reinecker HC, Nadeau WJ, McCormick BA, Podolsky DK. CARD15/NOD2 functions as an antibacterial factor in human intestinal epithelial cells. Gastroenterology. 2003;124:993–1000.
51. Cario E, Brown D, McKee M, Lynch-Devaney K, Gerken G, Podolsky DK. Commensal-associated molecular patterns induce selective Toll-like receptor-trafficking from apical membrane to cytoplasmic compartments in polarized intestinal epithelium. Am J Pathol. 2002;160:165–73.
52. Cario E, Rosenberg IM, Brandwein SL, Beck PL, Reinecker HC, Podolsky DK. Lipopolysaccharide activates distinct signaling pathways in intest epithelial cell lines expressing Toll-like receptors. J Immunol. 2000;164:966–72.
53. Cario E, Podolsky DK. Differential alteration in intestinal epithelial cell expression of toll-like receptor 3 (TLR3) and TLR4 in inflammatory bowel disease. Infect Immun. 2000;68: 7010–17.

54.  Wehkamp J, Harder J, Wehkamp-von Meissner B et al. NFκB- and AP-1-mediated induction of human beta defensin-2 in intestinal epithelial cells by Escherichia coli Nissle 1917: a novel effect of a probiotic bacterium. Infect Immun. 2004;72:5750–8.
55.  Fellermann K, Wehkamp J, Herrlinger KR, Stange EF. Crohn's disease: a defensin deficiency syndrome? Eur J Gastroenterol Hepatol. 2003;15:627–34.
56.  Vecchi M, Sinico A, Bianchi MB et al. Recognition of bactericidal/permeability-increasing protein by perinuclear anti-neutrophil cytoplasmic antibody-positive sera from ulcerative colitis patients: prevalence and clinical significance. Scand J Gastroenterol. 1998;33:1284–8.
57.  Elzouki AN, Eriksson S, Lofberg R, Nassberger L, Wieslander J, Lindgren S. The prevalence and clinical significance of alpha 1-antitrypsin deficiency (PiZ) and ANCA specificities (proteinase 3, BPI) in patients with ulcerative colitis. Inflamm Bowel Dis. 1999; 5:246–52.
58.  Schultz H, Weiss J, Carroll S, Gross WL. The endotoxin-binding bactericidal/permeability-increasing protein (BPI): a target antigen of autoantibodies. J Leukoc Biol 2001;69:505–12
59.  Schwab M, Schaeffeler E, Marx C et al. Association between the C3435T *MDR1* gene polymorphism and susceptibility for ulcerative colitis. Gastroenterology. 2003;124:26–33.

# 28
# Biology of cancer in ulcerative colitis and consequences for chemoprevention

C. R. BOLAND, C. GASCHÉ, G. MARRA and C. L. CHANG

## INTRODUCTION

### Chronic inflammation is associated with cancer

It has long been appreciated that many cancers occur more frequently in the setting of chronic inflammation. The mechanism responsible for inflammation-associatd carcinogenesis, and rational attempts to intervene therapeutically in the process, remain enigmatic. Gastrointestinal cancers represent prime examples of this problem. Adenocarcinoma of the oesophagus and stomach occur almost exclusively in the presence of chronic (acid reflux-induced) oesophagitis and chronic (*Helicobacter pylori*-associated) gastritis, respectively. Chronic pancreatitis and inflammatory bowel disease of the small and large intestine are also associated with increased incidences of cancer. Unfortunately, gastrointestinal cancers are particularly lethal, and there is essentially no good treatment for advanced, metastatic disease. Only early detection of colon cancer has had any impact on the morbidity and mortality from this disease. The theme of work from our laboratory is that we need to better understand how these tumours develop, so that we can develop effective preventive strategies, and attack the problem expectantly.

### Multiple distinct mechanistic pathways have been identified for colorectal carcinogenesis

It is currently appreciated that there are at least three distinct pathways of colorectal tumour development: chromosomal instability (CIN), microsatellite instability (MSI), and the CpG island methylator phenotype (CIMP). We are still learning how these types of genomic instability pathways are distinct, as there are distinct examples where they overlap[1]. This chapter will present evidence that carcinogenesis in the setting of chronic inflammation may develop through yet another molecular pathway, which may be a disorder of the DNA mismatch repair (MMR) system, leading to a 'low level' of MSI (i.e. MSI-L) produced without inactivating genetic mutations in the MMR system.

## Inflammation-associated cancers may develop through a novel pathway

Since chronic inflammation generates free radicals, which might cause mutations, it is seemingly apparent that cancer is an unavoidable consequence. However, the human genome is well protected by several homeostatic systems that repair mutations, including base excision repair, nucleotide excision repair, and DNA mismatch repair (MMR). Loeb has long proposed that a hypermutable state is required to account for all of the mutations found in cancer[2]. The data we have obtained suggest that there is an inappropriate down-regulation of DNA MMR in some cells in response to oxidative stress.

## Some historical perspective on MSI

Some background on MSI is required to understand the link between abnormal DNA MMR, inflammation and cancer. MSI is the result of a type of genomic instability that results from loss of the DNA MMR system[3,4]. Investigators were developing methods to find chromosomal losses known as 'loss of heterozygosity' (LOH) events, and eventually used microsatellite analysis, which was a more powerful tool for this purpose. Serendipitously, MSI was recognized by the appearance of novel – rather than missing – bands on autoradiograms, as illustrated in Figure 1. Sequencing the DNA from the novel bands revealed that these novel bands were the result of insertion or deletion mutations at mononucleotide (e.g. $A_n$) and dinucleotide repeat (e.g. $[CA]_n$) sequences[5-8]. This enigmatic phenotype was shortly thereafter linked to mutations in the major genes of the DNA MMR system, hMSH2 and hMLH1[9].

It was recognized that some colorectal cancers had insertion/deletion mutations at most (>40%) of the loci (which is called MSL-H for 'high'), and others had mutations at fewer loci (<10%, called MSL-L, for 'low')[6]. MSI-H tumours are a clinically distinct group; they tend to occur in the proximal colon, are diploid with only rare LOH events, have a different spectrum of somatic mutations, are associated with a better prognosis, have unique pathological features, and are almost always found in Lynch syndrome colorectal cancers[9]. MSI-L tumours are clinically more like tumours without MSI[3,6].

## Microsatellite instability in inflammation-associated gastrointestinal cancers

Our laboratory and others found that ulcerative colitis-associated cancers have a higher frequency of MSI than seen in sporadic colorectal cancers[10]. This difference is perhaps even greater than currently appreciated because the results were reported before the standardization of MSI nomenclature[3]. For example, we found MSI in 27% of the colitic cancers, using a panel consisting of five

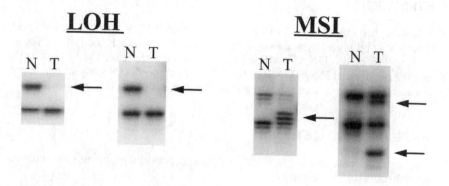

**Figure 1** A short, polymorphic DNA sequence (i.e. a microsatellite) is amplified by PCR, and the normal tissue (N), and if informative, provides two PCR products which are separated on a sequencing gel. On the left are two examples of loss of heterozygosity (LOH), in which one of the PCR products has been deleted from the genome of the tumour tissue (T), and no gene product is present on the gel, as indicated by the arrow on the autoradiogram. On the right are two examples of microsatellite instability (MSI), in which the length of the PCR product has been altered by an insertion or deletion mutation in a repetitive sequence, leading to the appearance of novel, shorter bands from the tumour tissue, as indicated by the arrows on the extreme right gel. Characteristically, five microsatellite loci are amplified, and the PCR product of the DNA from N is compared with that from T. If no mutations (novel bands) are observed, this is microsatellite stable (MSS). If two or more of five are mutated, this is MSI-H, and indicates complete loss of DNA MMR. If only one of the five loci shows a mutation in T, this is MSI-L

dinucleotide repeat markers (and no mononucleotide repeats), which would be a relatively insensitive and minimal estimate (unpublished data); this is MSI-L. Moreover, we had previously reported that the chronically inflamed, non-neoplastic colitic mucosa also has MSI[11], which has been confirmed by others[12,13].

## A mechanism for MSI-L

There is no consensus on the mechanism responsible for MSI-L in tumours. Loeb's group has shown that MSI can be induced in an *in-vitro* model of oxidative stress[14]. Another group has reported that an MSI-like mutational pattern is the 'signature' of oxidative damage[15]. Meltzer's group has demonstrated that 17% of ulcerative colitis-associated cancers had mutations in the $A_{10}$ microsatellite sequence of the TGF-β RII gene[16], suggesting a possible role for MSI in the pathogenesis of these tumours. However, there has never been evidence that the DNA MMR genes are mutated or inactivated in the inflamed tissues.

## HYPOTHESIS

Based on the above observations we hypothesized that chronically inflamed mucosae might undergo down-regulation of DNA MMR activity without inactivation of the DNA MMR genes. It this were the case the tumour cells might have normal DNA MMR activity, but the tumour DNA would still carry the mutational signature indicating the pathogenesis of the tumour.

## MATERIALS AND METHODS

The methods have been described in detail in previous publications, but are summarized here for clarity.

### Cell culture and $H_2O_2$ treatment

Our first cell model used consisted of human erythroleukaemia (HEL) cells that were treated with various concentrations of $H_2O_2$ for 1 h, in the absence of serum. At the end of treatment, the cells were resuspended in serum to recover from oxidative stress[17]. The other models were derived from this[18,19].

### Recombinant hMutSα and hMutLα

Baculovirus vectors carrying cDNA inserts encoding the hMSH6, hMSH2, hMLH1, and hPMS2 proteins were constructed to express DNA MMR proteins. In order to maintain protein stability, hMutSα and hMutLα recombinant protein complexes were purified from Sf9 cells after co-infecting hMSH2 and hMSH6 (i.e. hMutSα) or hMLH1 and hPMS2 (i.e. hMutLα) expression constructs, respectively.

### In-vitro DNA MMR assay

Twenty-four hours after HEL cells were exposed to a specified concentration of $H_2O_2$ for 1 h, cytoplasmic extracts were prepared as described previously. The extracts were used to repair a plasmid-based DNA heteroduplex containing a single G:T mismatch. In the reconstitution studies the repair assays were carried out as described above, except that the extracts were supplemented with recombinant hMutSα (hMSH2 + hMSH6) or hMutLα (hMLH1 + hPMS2). The repaired heteroduplex DNA was subsequently purified and electroporated into *Escherichia coli* and plated on soft agar[17]. When no DNA repair occurs, a high percentage of mixed plaques containing both blue and colourless progeny are observed. A reduced percentage of mixed plaques and a concomitant increase in blue plaques are indicative of DNA repair, which is quantitated by counting blue and white bacterial colonies[20].

## RESULTS

### Relaxation of DNA MMR activity in response to $H_2O_2$

In the first series of experiments we cultured HEL cells in the presence of $H_2O_2$, as a model of oxidative stress. First we determined that $H_2O_2$ doses from 0.001 mM to 1.0 mM for 1 h were tolerated by HEL cells. Using the MTT assay as a measure of metabolic activity and cell number, the cells rapidly recovered from this stress. We then performed an assay of the DNA MMR activity, as previously reported[20]. This permitted the most direct functional analysis of the intact system. We found significant reductions in DNA MMR activity after exposures to between 0.1 mM and 1.0 mM $H_2O_2$ (Figure 2).

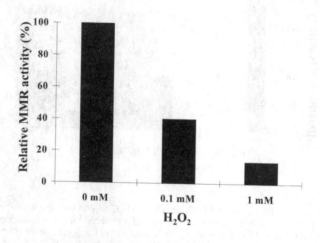

**Figure 2** DNA MMR activity of HEL cells 24 h after removal of $H_2O_2$. Administration of hydrogen peroxide at either non-lethal dose (0.1 mM or 1.0 mM) reduces DNA MMR activity, using a direct enzymatic assay for complete MMR activity[17,20]

### Reconstitution of DNA MMR activity with recombinant DNA MMR proteins

There were several possible explanations for this, including the fact that global metabolic activity of the cells had been inhibited. Furthermore, MMR activity could have been diminished by a loss of any one of the multiple proteins required in the system, including those involved in DNA excision and resynthesis. We hypothesized that the DNA MMR proteins had been down-regulated, so we restored the DNA MMR system using recombinant proteins generated in a baculovirus system. These proteins are unstable except as a heteroduplex, which is how they were prepared and used. We found that the MMR activity could be restored by adding back recombinant hMutSα (hMSH2 + hMSH6), hMutSβ (hMSH2 + hMSH3), and hMutLα (hMLH1 +

263

hPMS2). hMutSα and hMutSβ have complementary activities in recognizing different types of DNA errors[9]. hMutLα does not directly interact with DNA, but interacts with either hMutSα or hMutSβ to complete DNA repair. By adding back hMutSα together with hMutLα, complete restoration of DNA MMR activity was achieved (Figure 3). Completing the story, hMutSβ and hMutLα reconstitution restored the ability to repair 'loop-out' DNA mismatches[17].

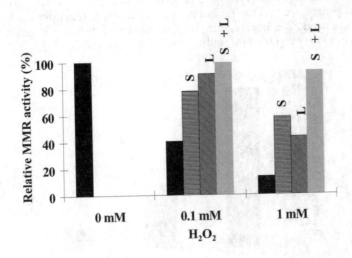

**Figure 3** Reconstitution of DNA MMR activity in HEL cells 24 h after removal of $H_2O_2$ with recombinant hMutSα and/or hMutLα. Addition of the recombinant protein heteroduplexes MutSα ('S' in the figure = hMSH2 + hMSH6), MutLα ('L' in the figure = hMLH1 + hPMS2), or all four proteins restored the relaxed DNA MMR activity induced by $H_2O_2$

## Reduced expression of DNA MMR proteins

We then performed Northern blots to look for differences in mRNA expression of the DNA MMR proteins, but could find no reproducible differences in response to $H_2O_2$ (data not shown). However, by doing Western blots we found significant reductions in hPMS2 at the 1.0 mM dose of $H_2O_2$, and significant reduction in the expression of hMSH6 at either the 0.1 mM or 1.0 mM concentrations. Thus, it appears that the relaxation of DNA MMR activity occurred by a reduction in the DNA MMR proteins, principally the two minor proteins hPMS2 and hMSH6. There were no apparent changes in the expression of hMSH2 or hMLH1[17].

## Rates of mutation with and without DNA MMR activity

We then developed a model that would permit direct measurement of mutation rates in cells with or without DNA MMR activity in response to $H_2O_2$. A DNA MMR-sensitive plasmid was created in which a microsatellite sequence was placed within the reading frame of the enhanced green fluorescent protein (EGFP) protein (Figure 4). The plasmid was constructed so that the reading frame of EGFP was shifted and no protein was expressed. However, an insertion/deletion mutation in the $(CA)_{13}$ would create a frame shift that would lead to expression of the protein, which could then be detected by fluorescent microscopy or flow cytometry.

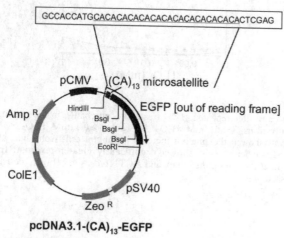

GCCACCATGCACACACACACACACACACACACACACTCGAG

pCMV    $(CA)_{13}$ microsatellite

Amp $^R$    HindIII    EGFP [out of reading frame]
BsgI
BsgI
BsgI
EcoRI

ColE1

pSV40

Zeo $^R$

**pcDNA3.1-$(CA)_{13}$-EGFP**

**Figure 4** The plasmid construct encodes an enhanced green fluorescent protein (EGFP) gene, with a dinucleotide repeat ($[CA]_{13}$) inserted between the start codon and the rest of the open reading frame in such a way that the new reading frame produces no protein. However, a mutation that deletes or adds a [CA] element from the dinucleotide repeat creates a frameshift that re-establishes the proper reading frame, and permits expression of EGFP. The appearance of 'revertants' (or fluorescent cells) permits an estimate of the mutation rates in response to oxidative stress. Dinucleotide repeat microsatellites are sensitive to the presence of intact DNA MMR activity, and can be used to estimate mutation rates in the presence or absence of DNA MMR

The plasmid was transfected into DNA MMR-deficient (HCT116 and HCT116+chr3-M2) and MMR-proficient (HCT116+chr3 cells). Under control conditions (i.e. no $H_2O_2$), a small mutant fraction ranging from 0.3% to 0.7% of the cells was observed. At 0.1 mM $H_2O_2$, only HCT 116 (DNA MMR-deficient) cells showed an increase in the mutant fraction, but at the 1.0 mM $H_2O_2$ level all lines accumulated mutations, as demonstrated by the appearance of fluorescence. HCT116+chr3-M2 cells (MMR-deficient) were most susceptible to this, being 16.4 times more hypermutable compared to

HCT116+chr3. HCT116 (DNA MMR-deficient) cells had a mutation rate that was only 4.1-fold greater than mock-treated control (Figure 5). Thus, this confirmed the utility of $H_2O_2$ as a model of oxidative stress, and demonstrated the role of an intact DNA MMR system in limiting mutagenesis[18].

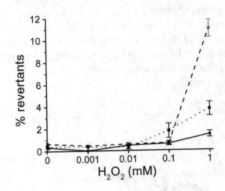

**Figure 5** Three days after repeated $H_2O_2$ treatment, no change in the mutant fraction was found at concentrations $\leqslant 0.01$ mM $H_2O_2$. However, $\geqslant 0.1$ mM $H_2O_2$, MMR-proficient HCT116+chr3 cells showed the smallest increase in mutant cells (solid line, at the bottom), whereas the MMR-deficient cells showed significantly greater revertant fractions, with HCT116+chr3-M2 at the top (dashed line), and HCT116 cells in between (dotted line between the solid and dashed lines)

## Models of chemoprevention

Finally, we developed a model of chemoprevention in which we added aspirin to HCT116+chr3 cells to determine its role in inducing apoptosis. Increasing the dose of aspirin from 0 to 10 mM reduced the growth response, and increased the level of apoptosis, as determined by Annexin V-Cy5 staining and flow cytometry. Perhaps more importantly, we found significant increases in the expression of several of the DNA MMR proteins by quantitative Western blot. Specifically, in HCT116+chr3 cells, treatment with 1 mM aspirin increased the expression of hMLH1 and hPMS2 proteins by 2.5-fold and 2.0-fold respectively, and increased expression of hMSH2 and hMSH6 proteins by 2–3-fold. For the SW480 cell line (also DNA MMR-proficient), treatment with 1–5 mM aspirin increased the expression of hMLH1 and hPMS2 proteins by 2–4-fold and 3–5-fold respectively, and increased the expression of the hMSH2 and hMSH6 proteins by 3–7-fold[19].

## DISCUSSION

This series of experiments has established new models that yielded several important new findings. First, we found that oxidative stress down-regulates the DNA MMR system*, and this appears to be regulated at the level of protein expression. Interestingly, it was the *minor* DNA MMR proteins (i.e. hMSH6 and hPMS2) that were affected, whereas the major DNA MMR proteins hMSH2 and hMLH1 did not appear to be altered in response to $H_2O_2$. The restoration of the DNA MMR enzymatic activity by replacement of the appropriate DNA MMR proteins demonstrates conclusively that it is the loss of these proteins that mediates this change in behaviour. There are many other enzymes involved in the DNA MMR repair process, but these were not required to restore full activity.

A similar model was established in collaboration with Firestein's group, which studies rheumatoid arthritis. Synovial tissues in this disease have an excess of p53 mutations, which is not well understood[21-23]. On Western blot analysis a significant decrease in the expression of hMSH6 and hMSH3 was seen in synovium obtained from patients with rheumatoid arthritis. In cultured synoviocytes, Western analysis demonstrated constitutive expression of all of the DNA MMR proteins. However, in the presence of oxidative stress generated by an NO donor, a reduction of hMSH6 was seen[24]. Thus, oxidative stress also relaxes the DNA MMR system in rheumatoid arthritis, which may account for the high incidence of mutations in the p53 gene in this setting.

In summary, the pathogenesis of cancer in the setting of inflammatory bowel disease appears to be complex. We reported in the past that inflamed, but non-neoplastic, colonic tissues in the setting of IBD contain a mutational signature called MSI-L. A mutational signature compatible with MSI has been reported in the neoplastic lesions[10,25]. Hypermethylation of the hMLH1 gene has also been reported in this setting[26,27]. A variety of other mutations have been reported, including point mutations at p53, inactivation of APC, frameshift mutations in the TGF-β receptor type 2, and others[28-32]. The pathophysiology of tumour development in IBD may be as complicated as carcinogenesis is for patients without this additional feature.

Most importantly, our new findings regarding the up-regulation of the DNA MMR system by aspirin deserve particular emphasis, because of its mechanism. It will be of considerable interest to see whether similar responses are seen with other drugs used to limit inflammation in the setting of IBD, such as sulphasalazine[33].

---

*The DNA MMR genes and proteins have been named after the bacterial genes which, when mutated, give rise to a 'mutator' phenotype. The genes responsible were then called MutS, MutL, etc. The yeast homologues of these were called the MutS and MutL homologues (MSH and MLH), and multiple homologues were found for each. The human versions of these have been called the human MutS homologues (hMSH), and humans have more homologues than yeast, which is why we have hMSH2 (and hMSH1-8, etc.). The human forms are sometimes called MSH or MLH for convenience, rather than hMSH or hMLH, which would be more correct. Finally, the PMS genes are actually homologous to the MLH family, but have completely different names, for unfortunate historial reasons.

## Acknowledgement

This work was supported in part by NIH Grant R-01 CA72851 (to C.R.B.), and by the Research Service of the Department of Veterans Affairs, USA (C.R.B.), the UCSD Comprehensive Cancer Center (C.R.B. and C.L.C.), and the Austrian Scientific Funds Grant J1702-MED (to C.G.).

## References

1. Goel A, Arnold CN, Niedzwiecki D et al. Characterization of sporadic colon cancer by patterns of genomic instability. Cancer Res. 2003;63:1608–14.
2. Loeb LA. A mutator phenotype in cancer. Cancer Res. 2001;61:3230–9.
3. Boland CR, Thibodeau SN, Hamilton SR et al. A National Cancer Institute Workshop on microsatellite instability for cancer detection and familial predisposition: development of international criteria for the determination of microsatellite instability in colorectal cancer. Cancer Res. 1998;58:5248–57.
4. Umar A, Boland CR, Terdiman JP et al. Revised Bethesda guidelines for hereditary nonpolyposis colorectal cancer (Lynch syndrome) and microsatellite instability. J Natl Cancer Inst. 2004;96:261–8.
5. Ionov Y, Peinado MA, Malkhosyan S, Shibata D, Perucho M. Ubiquitous somatic mutations in simple sequences reveal a new mechanism for colonic carcinogenesis. Nature. 1993;363:558–61.
6. Thibodeau SN, Bren G, Schaid D. Microsatellite instability in cancer of the proximal colon. Science. 1993;260:816–19.
7. Aaltonen LA, Peltomaki P, Leach FS et al. Clues to the pathogenesis of familial colorectal cancer. Science. 1993;260:812–16.
8. Peltomaki P, Lothe RA, Aaltonen LA et al. Microsatellite instability is associated with tumors that characterize the hereditary non-polyposis colorectal carcinoma syndrome. Cancer Res. 1993;53:5853–5.
9. Marra G, Boland CR. Hereditary nonpolyposis colorectal cancer: the syndrome, the genes, and historical perspectives. J Natl Cancer Inst. 1995;87:1114–25.
10. Suzuki H, Harpaz N, Tarmin L et al. Microsatellite instability in ulcerative colitis-associated colorectal dysplasias and cancers. Cancer Res. 1994;54:4841–4.
11. Brentnall TA, Crispin DA, Bronner MP et al. Microsatellite instability in nonneoplastic mucosa from patients with chronic ulcerative colitis. Cancer Res. 1996;56:1237–40.
12. Heinen CD, Noffsinger AE, Belli J et al. Regenerative lesions in ulcerative colitis are characterized by microsatellite mutation. Genes Chrom Cancer. 1997;19:170–5.
13. Lyda MH, Noffsinger A, Belli J, Fenoglio-Preiser CM. Microsatellite instability and K-ras mutations in patients with ulcerative colitis. Hum Pathol. 2000;31:665–71.
14. Jackson AL, Loeb LA. The contribution of enogenous sources of DNA damage to the multiple mutations in cancer. Mutat Res. 2001;477:7–21.
15. Turker MS, Gage BM, Rose JA et al. A novel signature mutation for oxidative damage resembles a mutational pattern found commonly in human cancers. Cancer Res. 1999;59: 1837–9.
16. Souza RF, Garrigue-Antar L, Lei J et al. Alterations of transforming growth factor-beta 1 receptor type II occur in ulcerative colitis-associated carcinomas, sporadic colorectal neoplasms, and esophageal carcinomas, but not in gastric neoplasms. Hum Cell. 1996;9: 229–36.
17. Chang CL, Marra G, Chauhan DP et al. Oxidative stress inactivates the human DNA mismatch repair system. Am J Physiol Cell Physiol. 2002;283:C148–54.
18. Gasche C, Chang CL, Rhees J, Goel A, Boland CR. Oxidative stress increases frameshift mutations in human colorectal cancer cells. Cancer Res. 2001;61:7444–8.
19. Goel A, Chang DK, Ricciardiello L, Gasche C, Boland CR. A novel mechanism for aspirin-mediated growth inhibition of human colon cancer cells. Clin Cancer Res. 2003;9: 383–90.
20. Koi M, Umar A, Chauhan DP et al. Human chromosome 3 corrects mismatch repair deficiency and microsatellite instability and reduces N-methyl-N'-nitro-N-nitrosoguanidine

tolerance in colon tumor cells with homozygous hMLH1 mutation. Cancer Res. 1994;54:4308–12.

21. Tak PP, Zvaifler NJ, Green DR, Firestein, GS. Rheumatoid arthritis and p53: how oxidative stress might alter the course of inflammatory diseases. Immunol Today. 2000;21:78–82.

22. Firestein GS, Echeverri F, Yeo M, Zvaifler NJ, Green DR. Somatic mutations in the p53 tumor suppressor gene in rheumatoid arthritis synovium. Proc Natl Acad Sci USA. 1997; 94:10895–900.

23. Seemayer CA, Kuchen S, Neidhart M et al. p53 in rheumatoid arthritis synovial fibroblasts at sites of invasion. Ann Rheum Dis. 2003;62:1139–44.

24. Lee SH, Chang DK, Goel A et al. Microsatellite instability and suppressed DNA repair enzyme expression in rheumatoid arthritis. J Immunol. 2003;170:2214–20.

25. Fleisher AS, Esteller M, Harpaz N et al. Microsatellite instability in inflammatory bowel disease-associated neoplastic lesions is associated with hypermethylation and diminished expression of the DNA mismatch repair gene, hMLH1. Cancer Res. 2000;60:4864–8.

26. Sato F, Shibata D, Harpaz N et al. Aberrant methylation of the HPP1 gene in ulcerative colitis-associated colorectal carcinoma. Cancer Res. 2002;62:6820–2.

27. Sato F, Harpaz N, Shibata D et al. Hypermethylation of the p14(ARF) gene in ulcerative colitis-associated colorectal carcinogenesis. Cancer Res. 2002;62:1148–51.

28. Greenwald BD, Harpaz N, Yin J et al. Loss of heterozygosity affecting the p53, Rb, and mcc/apc tumor suppressor gene loci in dysplastic and cancerous ulcerative colitis. Cancer Res. 1992;52:741–5.

29. Yin J, Harpaz N, Tong Y et al. p53 point mutations in dysplastic and cancerous ulcerative colitis lesions. Gastroenterology. 1993;104:1633–9.

30. Tarmin L, Yin J, Harpaz N et al. Adenomatous polyposis coli gene mutations in ulcerative colitis-associated dysplasias and cancers versus sporadic colon neoplasms. Cancer Res. 1995;55:2035–8.

31. Souza RF, Lei J, Yin J et al. A transforming growth factor beta 1 receptor type II mutation in ulcerative colitis-associated neoplasms. Gastroenterology. 1997;112:40–5.

32. Redston MS, Papadopoulos N, Caldas C, Kinzler KW, Kern SE. Common occurence of APC and K-ras gene mutations in the spectrum of colitis-associated neoplasias. Gastroenterology. 1995;108:383–92.

33. Bernstein CN, Eaden J, Steinhart AH, Munkholm P, Gordon PH. Cancer prevention in inflammatory bowel disease and the chemoprophylactic potential of 5-aminosalicylic acid. Inflamm Bowel Dis. 2002;8:356–61.

# 29
# Therapeutic strategy in ulcerative colitis – 2004

## D. RACHMILEWITZ

## INTRODUCTION

In view of the fact that the aetiology of ulcerative colitis is not known at present, successful therapy for inflammatory bowel disease (IBD) must demonstrate the ability to modify the disease process rather than exclusively treat symptoms. It has to induce rapid symptom control, to be safe, and its administration should be simple. Last but not least the ideal therapy should not only induce but also maintain the remission. Addressing the issue of the state-of-the-art evidence-based treatment of ulcerative colitis (UC) is based on published controlled studies rather than on open studies or case reports.

The treatment of active UC should differentiate between the treatment of distal and extensive colitis and between induction and maintenance of remission.

## ACUTE MILD–MODERATE COLITIS

For the acute mild–moderate active proctitis or distal colitis initial therapy includes oral or topical aminosalicylates with or without the addition of topical or oral corticosteroids. Second-line therapy includes oral corticosteroids and immunosuppressive drugs. For mild–moderate extensive UC oral with or without topical aminosalicylates with or without oral corticosteroids are the treatments of choice. Oral corticosteroids and immunosuppressants are the second-line therapy for extensive involvement.

After the first report of Svartz[1] in 1948, it took 12 years until the first controlled trial with sulphasalazine for active distal UC was published.

In 1960, in a placebo-controlled trial conducted in 50 patients, oral sulphasalazine (4 g) was shown to induce remission in 35% of the patients as opposed to 5% remission rate in the placebo-treated group[2]. Eight years later suppositories of sulphasalazine (1 g) were shown to induce remission in 83% of patients, whereas in placebo-treated subjects the remission rate was 28%[3] Moller et al. demonstrated a similar efficacy of sulphasalazine enemas (3 g), again in a placebo-controlled trial[4].

Several studies conducted in the late 1950s and early 1960s demonstrated the usefulness of local steroids for the treatment of distal active UC. In two placebo-controlled trials hydrocortisone enemas (100 mg) were shown to be five times more effective than placebo in inducing remission[5,6]. Hydrocortisone enema (100 mg) was also more effective than oral prednisolone (20 mg)[7].

However, it is of interest to note that a careful meta-analysis[8] revealed that rectal 5-ASA preparations are better than oral 5-ASA, and only if this approach is not satisfactory is the use of rectal corticosteroids advised. For distal active UC, rectal 5-ASA is better than oral 5-ASA, but when necessary a combination of rectal and oral 5-ASA is even better[9].

In mild–moderate active UC it seems that oral 5-ASA is better than oral sulphasalazine, as shown in the meta-analysis performed by Sutherland et al. in 1993[10]. There is still debate regarding the recommended 5-ASA dose to be used in active UC. Many studies have yielded different recommendations. In a recent study[11] 3.0 g/day was found to be better that 4.5 g/day or 1.5 g/day. It is quite clear that the minimal dose should be 3.0 g/day and that a higher dose may not necessarily be better.

When 5-ASA is not sufficient, systemic corticosteroids should be used. The double-blind controlled trials were conducted in the late 1950s and better ones did not follow. Oral cortisone (100 mg) over a 6-week period induced remission in 41% of patients as opposed to 16% remission in a placebo-treated group[12]; 40–60 mg of prednisone induced remission after 3–4 weeks in 68% of patients as opposed to a 17% remission rate in the placebo-treated group[13]. When oral corticosteroids fail, it is still possible to benefit from intravenous administration. An important trial conducted in the USA revealed that, in patients who had never been treated before with steroids, adrenocorticotrophic hormone (ACTH) is significantly better than intravenous administration of hydrocortisone[14]. Intravenous ACTH is not so popular nowadays, but should be considered as an alternative to hydrocortisone when oral corticosteroids do not induce remission.

Obviously the side-effects of corticosteroids are their main drawback. The synthetic steroid preparation, budesonide, was proven to be statistically as effective as oral prednisolone with fewer side-effects[15]. However, in clinical practice on many occasions the impressive statistically significant results in trials cannot be reproduced. Moreover, even budesonide usage is involved with side-effects, and long-term treatment is definitely not advised. As a rule, when systemic corticosteroids are used it is important to give them for a limited period, to induce remission and not to aim to maintain the remission with corticosteroids.

In patients with refractory disease unresponsive to the 5-ASA drugs, in those with frequent flare-ups requiring cortiticosteroids, in patients with corticosteroid dependence or complications immunomodulators should be used. Up to now azathioprine or 6-mercaptopurine (6-MP) are the drugs of choice. Therapy with azathioprine 2.0–2.5 mg/kg or with 6-MP 1.0–1.5 mg/kg should be recommended, with the expectation that its effects will be noticed after 8–10 weeks of treatment[16].

In patients with refractory distal or extensive colitis not responding to all the other modalities, intravenous cyclosporine may be used as the last possible modality that may avoid colectomy. This approach, reported for the first time in 1994, was found to induce remission in 80% of subjects[17]. Greater success in avoiding colectomy was reported when cyclosporine therapy was followed by azathioprine or 6-MP[18]. The original dosage of 4 mg/g per day administered for 7–10 days was recently reported to be exaggerated, since 2 mg/kg per day seems to induce the same remission rate.

## MAINTENANCE OF REMISSION

For the maintenance of remission of uncomplicated UC at present the modalities to be used are 5-ASA preparations. When the anatomical extent is limited to the distal colon topical preparations may be used, whereas in extensive involvement oral preparations are the only choice.

Proctitis can be maintained in remission with suppositories of mesalazine. Relapse rate during 1 year is dose-dependent. A dose of 1 g is more effective than 0.5 g, and both doses are better than placebo[19].

When the involvement is extensive, oral sulphasalazine was shown to be significantly better than placebo. A dose of 2 g during 1 year maintained the remissions in 71% of patients as opposed to 25% in the placebo-treated group[20]. Similar results were obtained in another study conducted for 6 months[21]. In an effort to find the recommended dose of sulphasalazine to be used in order to maintain remission 2 g/day was found to do better than 1 g/day, and a dose of 4 g/day was not significantly better than 2 g/day[22].

A big meta-analysis[10] revealed that, for the maintenance of remission, sulphasalazine is better than 5-ASA; therefore sulphasalazine should be preferred unless patients cannot tolerate the drug because of untoward side-effects.

Under no circumstance should systemic corticosteroids be used for maintenance of UC in remission. The argument against using them for maintenance treatment is not their side-effects; several clinical trials failed to demonstrate that corticosteroids are effective for maintenance of the disease in remission. One year treatment with 50 mg/day of hydrocortisone was as effective as placebo[23]. Six months treatment with prednisone 15 mg/day resulted in a 60% relapse rate, similar to the 58% relapse rate in a placebo-treated group[13]. In another study, in which prednisone was added to sulphasalazine during a 3-month period, the relapse rate was 17%, whereas when prednisone was added to placebo the relapse rate was significantly higher at 58%[24].

## PROBIOTICS FOR THE MAINTENANCE OF UC IN REMISSION

Probiotic preparations are now in fashion for many indications. During 1 year of treatment the probiotic preparation E. coli (Nissle 1917) was shown to be as effective as mesalazine in the maintenance of UC in remission in a large number of patients[25]. Other probiotic preparations such as VSL-3, shown to be effective in the maintenance of pouchitis in remission, await clinical trials in UC patients.

It would not be a surprise if other probiotic preparations besides *E. coli* (Nissle 1917) were found to be a potent modality to maintain UC in remission. Their potency is probably due to the recent findings concerning the immunostimulatory properties of the CpG motif in the bacterial genome. This motif was recently reported to prevent experimental colitis[26] and therefore should also be beneficial for the maintenance of UC in remission.

## MISCELLANEOUS THERAPIES FOR UC

In addition to the drugs commonly used, due to the fact that no modality is a satisfactory solution to the problem, there are permanent trials to evaluate other therapeutic options. All these trials are preliminary and no definite convincing results have been obtained so far. These trials include: nicotine[27]; short-chain fatty acids[28]; heparin[29]; and leukocytopheresis[29].

## BIOLOGIC THERAPIES

It is expected that biologic therapies, that were evaluated and used in Crohn's disease, will also be effective in UC. They may not necessarily be the first drug of choice to be used in uncomplicated UC, but deserve objective evaluation. At present a controlled trial with Remicade is in process, and other approaches are also being considered.

## References

1. Svartz N. The treatment of 124 cases of ulcerative colitis with salazopyrine and attempts of desensitization in cases of hypersensitiveness to sulfa. Acta Med Scand. 1948;139(Suppl. 206):465–72.
2. Baron JH, Connell AM, Lennard-Jones JE, Jones FA. Sulphasalazine and salicylazosulphadimidine in ulcerative colitis. Lancet. 1962;1:1094–6.
3. Watkinson, cited in Goligher JC, de Dombal FT, Watts J McK et al., editors. Ulcerative Colitis. Baltimore: Baillière, Tindall & Cassell, 1968:207–8.
4. Moller C, Kiviluoto O. Santavirta S et al. Local treatment of ulcerative proctitis with salicylazosulphapyridine (salazopyrine) enema. Clin Trials J. 1978;15:199–203.
5. Truelove SC. Treatment of ulcerative colitis with local cortisone hemisuccinate sodium. Report on a controlled therapeutic trial. Br Med J. 1958;2:1072–7.
6. Watkinson G. Treatment of ulcerative colitis with topical hydrocortisone hemisuccinate sodium: a controlled trial exemplifying restricted sequential analysis. Br Med J. 1958;2: 1077–82.
7. Truelove SC. Systemic and local corticosteroid therapy in ulcerative colitis. Br Med J. 1960; 5171:464–7.
8. Marshall JK, Irvine EJ. Rectal corticosteroids versus alternative treatments in ulcerative colitis: a meta-analysis. Gut. 1997;40:775–81.
9. Safdi M, DeMicco M, Sninsky C et al. A double-blind comparison of oral versus rectal mesalamine versus combination therapy in the treatment of distal ulcerative colitis. Am J Gastroenterol. 1997;92:1867.
10. Southerland LR, May GR, Shaffer EA. Sulfasalazine revisited: a meta-analysis of 5-aminosalicylic acid in the treatment of ulcerative colitis. Ann Intern Med. 1993;118:540.
11. Kruis W, Bar-Meir S, Feher J et al., for the International Salofalk Pellets Study Group. The optimal dose of 5-aminosalicylic acid in active ulcerative colitis: a dose-finding study with newly developed mesalamine. Clin Gastroenterol Hepatol. 2003;1:36.

12. Truelove SC, Witts LJ. Cortisone in ulcerative colitis. Final report in a Therapeutic trial. Br Med J. 1955;2:104–8.
13. Lennard-Jones JE, Longmore AJ, Newell AC et al. An assessment of prednisone, salazopyrine and topical hydrocortisone hemisuccinate used as outpatient treatment for ulcerative colitis. Gut. 1960;1:217–22.
14. Meyers S, Sachar DB, Goldberg J et al. Corticotropin versus hydrocortisone in the intravenous treatment of ulcerative colitis. A prospective, randomized double-blind clinical trial. Gastroenterology. 1983;85:351–7.
15. Lofberg R, Danielsson A, Suhr O et al. Oral budesonide versus prednisolone in patients with active extensive and left sided ulcerative colitis. Gastroenterology. 1996;110:1713–18.
16. George J, Present DH, Pou R et al. The long term outcome of ulcerative colitis treated with 6-mercaptopurine. Am J Gastroenterol. 1992;305:20–2.
17. Lichtiger S, Present DH, Kornbluth A, Gelernt I. Cyclosporine in severe ulcerative colitis refractory to steroid therapy. N Engl J Med. 1994;330:1841–5.
18. Cohen RD, Stein D Hanauer SB. Intravenous cyclosporine in ulcerative colitis; a five year experience. Am J Gastroenterol. 1999;94:1587–92.
19. D'Albasio G, Paoluzi P, Campieri M et al and the Italian IBD Study Group. Maintenance treatment of ulcerative proctitis with mesalazine suppositories: a double blind placebo-controlled trial. Am J Gastroenterol. 1998;93:799.
20. Misiewicz JJ, Lennard-Jones JE, Connell AM et al. Controlled trial of sulfasalazine in maintenance therapy for ulcerative colitis. Lancet. 1965;1:185–8.
21. Dissanayake AS, Truelove SC. A controlled trial of long-term maintenance treatment of ulcerative colitis with sulphasalazine (Salazopyrine). Gut. 1973;14:923–6.
22. Azad Khan AK, Howes DT, Piris J et al. Optimum dose of sulphasalazine for maintenance treatment of ulcerative colitis. Gut. 1908;21:232–40.
23. Truelove SC, Witts LJ. Cortisone and corticotropin in acute ulcerative colitis. Br Med J. 1959:2:387–94.
24. Powell-Tuck J, Brown RL, Chamber TJ et al. A controlled trial of alternative day prednisolone as a maintenance treatment for ulcerative colitis in remission. Digestion. 1981;22:263.
25. Rembacken BJ, Snelling AM, Hawkey, Chalmers DM, Axon ATR. Non pathogenic Escherichia coli versus mesalazine for the treatment of ulcerative colitis: a randomized trial. Lancet. 1999;354:635.
26. Rachmilewitz D, Karmeli F, Takabayashi K et al. Immunostimulatory DNA ameliorates experimental and spontaneous murine colitis. Gastroenterology. 2002;122:1428–41.
27. Pullan RD, Rhodes J, Ganesh S et al. Transdermal nicotine for active ulcerative colitis. N Engl J Med. 1994;330:811.
28. Scheppach W. Treatment of distal ulcerative colitis with short chain fatty acid enemas. A placebo controlled trail German–Austrian SCFA Study Group. Dig Dis Sci. 1996;41:2254.
29. Folwaczny C, Fricke H, Endres S et al. Anti-inflammatory properties of unfractioned heparin in patients with highly active ulcerative colitis: a pilot study. Am J Gastroenterol. 1997;92:911.

# Index

# Falk Symposium Series

*These titles were published under the MTP Press imprint.

# Falk Symposium Series

65. Hadziselimovic F, Herzog B, eds.: *Inflammatory Bowel Diseases and Morbus Hirschprung*. Falk Symposium No. 65. 1992      ISBN: 0-7923-8995-6
66. Martin F, McLeod RS, Sutherland LR, Williams CN, eds.: *Trends in Inflammatory Bowel Disease Therapy*. Falk Symposium No. 66. 1993      ISBN: 0-7923-8827-5
67. Schölmerich J, Kruis W, Goebell H, Hohenberger W, Gross V, eds.: *Inflammatory Bowel Diseases – Pathophysiology as Basis of Treatment*. Falk Symposium No. 67. 1993      ISBN: 0-7923-8996-4
68. Paumgartner G, Stiehl A, Gerok W, eds.: *Bile Acids and The Hepatobiliary System: From Basic Science to Clinical Practice*. Falk Symposium No. 68. 1993      ISBN: 0-7923-8829-1
69. Schmid R, Bianchi L, Gerok W, Maier K-P, eds.: *Extrahepatic Manifestations in Liver Diseases*. Falk Symposium No. 69. 1993      ISBN: 0-7923-8821-6
70. Meyer zum Büschenfelde K-H, Hoofnagle J, Manns M, eds.: *Immunology and Liver*. Falk Symposium No. 70. 1993      ISBN: 0-7923-8830-5
71. Surrenti C, Casini A, Milani S, Pinzani M , eds.: *Fat-Storing Cells and Liver Fibrosis*. Falk Symposium No. 71. 1994      ISBN: 0-7923-8842-9
72. Rachmilewitz D, ed.: *Inflammatory Bowel Diseases – 1994*. Falk Symposium No. 72. 1994      ISBN: 0-7923-8845-3
73. Binder HJ, Cummings J, Soergel KH, eds.: *Short Chain Fatty Acids*. Falk Symposium No. 73. 1994      ISBN: 0-7923-8849-6
73B. Möllmann HW, May B, eds.: *Glucocorticoid Therapy in Chronic Inflammatory Bowel Disease: from basic principles to rational therapy*. International Falk Workshop. 1996      ISBN 0-7923-8708-2
74. Keppler D, Jungermann K, eds.: *Transport in the Liver*. Falk Symposium No. 74. 1994      ISBN: 0-7923-8858-5
74B. Stange EF, ed.: *Chronic Inflammatory Bowel Disease*. Falk Symposium. 1995      ISBN: 0-7923-8876-3
75. van Berge Henegouwen GP, van Hoek B, De Groote J, Matern S, Stockbrügger RW, eds.: *Cholestatic Liver Diseases: New Strategies for Prevention and Treatment of Hepatobiliary and Cholestatic Liver Diseases*. Falk Symposium 75. 1994.      ISBN: 0-7923-8867-4
76. Monteiro E, Tavarela Veloso F, eds.: *Inflammatory Bowel Diseases: New Insights into Mechanisms of Inflammation and Challenges in Diagnosis and Treatment*. Falk Symposium 76. 1995.      ISBN 0-7923-8884-4
77. Singer MV, Ziegler R, Rohr G, eds.: *Gastrointestinal Tract and Endocrine System*. Falk Symposium 77. 1995.      ISBN 0-7923-8877-1
78. Decker K, Gerok W, Andus T, Gross V, eds.: *Cytokines and the Liver*. Falk Symposium 78. 1995.      ISBN 0-7923-8878-X
79. Holstege A, Schölmerich J, Hahn EG, eds.: *Portal Hypertension*. Falk Symposium 79. 1995.      ISBN 0-7923-8879-8
80. Hofmann AF, Paumgartner G, Stiehl A, eds.: *Bile Acids in Gastroenterology: Basic and Clinical Aspects*. Falk Symposium 80. 1995      ISBN 0-7923-8880-1
81. Riecken EO, Stallmach A, Zeitz M, Heise W, eds.: *Malignancy and Chronic Inflammation in the Gastrointestinal Tract – New Concepts*. Falk Symposium 81. 1995 ISBN 0-7923-8889-5
82. Fleig WE, ed.: *Inflammatory Bowel Diseases: New Developments and Standards*. Falk Symposium 82. 1995      ISBN 0-7923-8890-6
82B. Paumgartner G, Beuers U, eds.: *Bile Acids in Liver Diseases*. International Falk Workshop. 1995      ISBN 0-7923-8891-7

# Falk Symposium Series

83. Dobrilla G, Felder M, de Pretis G, eds.: *Advances in Hepatobiliary and Pancreatic Diseases: Special Clinical Topics.* Falk Symposium 83. 1995.    ISBN 0-7923-8892-5
84. Fromm H, Leuschner U, eds.: *Bile Acids – Cholestasis – Gallstones: Advances in Basic and Clinical Bile Acid Research.* Falk Symposium 84. 1995    ISBN 0-7923-8893-3
85. Tytgat GNJ, Bartelsman JFWM, van Deventer SJH, eds.: *Inflammatory Bowel Diseases.* Falk Symposium 85. 1995    ISBN 0-7923-8894-1
86. Berg PA, Leuschner U, eds.: *Bile Acids and Immunology.* Falk Symposium 86. 1996    ISBN 0-7923-8700-7
87. Schmid R, Bianchi L, Blum HE, Gerok W, Maier KP, Stalder GA, eds.: *Acute and Chronic Liver Diseases: Molecular Biology and Clinics.* Falk Symposium 87. 1996    ISBN 0-7923-8701-5
88. Blum HE, Wu GY, Wu CH, eds.: *Molecular Diagnosis and Gene Therapy.* Falk Symposium 88. 1996    ISBN 0-7923-8702-3
88B. Poupon RE, Reichen J, eds.: *Surrogate Markers to Assess Efficacy of TReatment in Chronic Liver Diseases.* International Falk Workshop. 1996    ISBN 0-7923-8705-8
89. Reyes HB, Leuschner U, Arias IM, eds.: *Pregnancy, Sex Hormones and the Liver.* Falk Symposium 89. 1996    ISBN 0-7923-8704-X
89B. Broelsch CE, Burdelski M, Rogiers X, eds.: *Cholestatic Liver Diseases in Children and Adults.* International Falk Workshop. 1996    ISBN 0-7923-8710-4
90. Lam S-K, Paumgartner P, Wang B, eds.: *Update on Hepatobiliary Diseases 1996.* Falk Symposium 90. 1996    ISBN 0-7923-8715-5
91. Hadziselimovic F, Herzog B, eds.: *Inflammatory Bowel Diseases and Chronic Recurrent Abdominal Pain.* Falk Symposium 91. 1996    ISBN 0-7923-8722-8
91B. Alvaro D, Benedetti A, Strazzabosco M, eds.: *Vanishing Bile Duct Syndrome – Pathophysiology and Treatment.* International Falk Workshop. 1996    ISBN 0-7923-8721-X
92. Gerok W, Loginov AS, Pokrowskij VI, eds.: *New Trends in Hepatology 1996.* Falk Symposium 92. 1997    ISBN 0-7923-8723-6
93. Paumgartner G, Stiehl A, Gerok W, eds.: *Bile Acids in Hepatobiliary Diseases – Basic Research and Clinical Application.* Falk Symposium 93. 1997    ISBN 0-7923-8725-2
94. Halter F, Winton D, Wright NA, eds.: *The Gut as a Model in Cell and Molecular Biology.* Falk Symposium 94. 1997    ISBN 0-7923-8726-0
94B. Kruse-Jarres JD, Schölmerich J, eds.: *Zinc and Diseases of the Digestive Tract.* International Falk Workshop. 1997    ISBN 0-7923-8724-4
95. Ewe K, Eckardt VF, Enck P, eds.: *Constipation and Anorectal Insufficiency.* Falk Symposium 95. 1997    ISBN 0-7923-8727-9
96. Andus T, Goebell H, Layer P, Schölmerich J, eds.: *Inflammatory Bowel Disease – from Bench to Bedside.* Falk Symposium 96. 1997    ISBN 0-7923-8728-7
97. Campieri M, Bianchi-Porro G, Fiocchi C, Schölmerich J, eds. *Clinical Challenges in Inflammatory Bowel Diseases: Diagnosis, Prognosis and Treatment.* Falk Symposium 97. 1998    ISBN 0-7923-8733-3
98. Lembcke B, Kruis W, Sartor RB, eds. *Systemic Manifestations of IBD: The Pending Challenge for Subtle Diagnosis and Treatment.* Falk Symposium 98. 1998    ISBN 0-7923-8734-1
99. Goebell H, Holtmann G, Talley NJ, eds. *Functional Dyspepsia and Irritable Bowel Syndrome: Concepts and Controversies.* Falk Symposium 99. 1998    ISBN 0-7923-8735-X
100. Blum HE, Bode Ch, Bode JCh, Sartor RB, eds. *Gut and the Liver.* Falk Symposium 100. 1998    ISBN 0-7923-8736-8

# Falk Symposium Series

101. Rachmilewitz D, ed. *V International Symposium on Inflammatory Bowel Diseases.* Falk Symposium 101. 1998   ISBN 0-7923-8743-0
102. Manns MP, Boyer JL, Jansen PLM, Reichen J, eds. *Cholestatic Liver Diseases.* Falk Symposium 102. 1998   ISBN 0-7923-8746-5
102B. Manns MP, Chapman RW, Stiehl A, Wiesner R, eds. *Primary Sclerosing Cholangitis.* International Falk Workshop. 1998.   ISBN 0-7923-8745-7
103. Häussinger D, Jungermann K, eds. *Liver and Nervous System.* Falk Symposium 102. 1998   ISBN 0-7924-8742-2
103B. Häussinger D, Heinrich PC, eds. *Signalling in the Liver.* International Falk Workshop. 1998   ISBN 0-7923-8744-9
103C. Fleig W, ed. *Normal and Malignant Liver Cell Growth.* International Falk Workshop. 1998   ISBN 0-7923-8748-1
104. Stallmach A, Zeitz M, Strober W, MacDonald TT, Lochs H, eds. *Induction and Modulation of Gastrointestinal Inflammation.* Falk Symposium 104. 1998   ISBN 0-7923-8747-3
105. Emmrich J, Liebe S, Stange EF, eds. *Innovative Concepts in Inflammatory Bowel Diseases.* Falk Symposium 105. 1999   ISBN 0-7923-8749-X
106. Rutgeerts P, Colombel J-F, Hanauer SB, Schölmerich J, Tytgat GNJ, van Gossum A, eds. *Advances in Inflammatory Bowel Diseases.* Falk Symposium 106. 1999   ISBN 0-7923-8750-3
107. Špičák J, Boyer J, Gilat T, Kotrlik K, Mareček Z, Paumgartner G, eds. *Diseases of the Liver and the Bile Ducts – New Aspects and Clinical Implications.* Falk Symposium 107. 1999   ISBN 0-7923-8751-1
108. Paumgartner G, Stiehl A, Gerok W, Keppler D, Leuschner U, eds. *Bile Acids and Cholestasis.* Falk Symposium 108. 1999   ISBN 0-7923-8752-X
109. Schmiegel W, Schölmerich J, eds. *Colorectal Cancer – Molecular Mechanisms, Premalignant State and its Prevention.* Falk Symposium 109. 1999   ISBN 0-7923-8753-8
110. Domschke W, Stoll R, Brasitus TA, Kagnoff MF, eds. *Intestinal Mucosa and its Diseases – Pathophysiology and Clinics.* Falk Symposium 110. 1999   ISBN 0-7923-8754-6
110B. Northfield TC, Ahmed HA, Jazwari RP, Zentler-Munro PL, eds. *Bile Acids in Hepatobiliary Disease.* Falk Workshop. 2000   ISBN 0-7923-8755-4
111. Rogler G, Kullmann F, Rutgeerts P, Sartor RB, Schölmerich J, eds. *IBD at the End of its First Century.* Falk Symposium 111. 2000   ISBN 0-7923-8756-2
112. Krammer HJ, Singer MV, eds. *Neurogastroenterology: From the Basics to the Clinics.* Falk Symposium 112. 2000   ISBN 0-7923-8757-0
113. Andus T, Rogler G, Schlottmann K, Frick E, Adler G, Schmiegel W, Zeitz M, Schölmerich J, eds. *Cytokines and Cell Homeostasis in the Gastrointestinal Tract.* Falk Symposium 113. 2000   ISBN 0-7923-8758-9
114. Manns MP, Paumgartner G, Leuschner U, eds. *Immunology and Liver.* Falk Symposium 114. 2000   ISBN 0-7923-8759-7
115. Boyer JL, Blum HE, Maier K-P, Sauerbruch T, Stalder GA, eds. *Liver Cirrhosis and its Development.* Falk Symposium 115. 2000   ISBN 0-7923-8760-0
116. Riemann JF, Neuhaus H, eds. *Interventional Endoscopy in Hepatology.* Falk Symposium 116. 2000   ISBN 0-7923-8761-9
116A. Dienes HP, Schirmacher P, Brechot C, Okuda K, eds. *Chronic Hepatitis: New Concepts of Pathogenesis, Diagnosis and Treatment.* Falk Workshop. 2000   ISBN 0-7923-8763-5

# Falk Symposium Series

# Falk Symposium Series

132A. Staritz M, Adler G, Knuth A, Schmiegel W, Schmoll H-J, eds. *Side-effects of Chemotherapy on the Gastrointestinal Tract.* Falk Workshop. 2003
ISBN 0-7923-8791-0

132B. Reutter W, Schuppan D, Tauber R, Zeitz M, eds. *Cell Adhesion Molecules in Health and Disease.* Falk Workshop. 2003          ISBN 0-7923-8786-4

133. Duchmann R, Blumberg R, Neurath M, Schölmerich J, Strober W, Zeitz M. *Mechanisms of Intestinal Inflammation: Implications for Therapeutic Intervention in IBD.* Falk Symposium 133. 2004          ISBN 0-7923-8787-2

134. Dignass A, Lochs H, Stange E. *Trends and Controversies in IBD – Evidence-Based Approach or Individual Management?* Falk Symposium 134. 2004
ISBN 0-7923-8788-0

134A. Dignass A, Gross HJ, Buhr V, James OFW. *Topical Steroids in Gastroenterology and Hepatology.* Falk Workshop. 2004          ISBN 0-7923-8789-9

135. Lukáš M, Manns MP, Špičák J, Stange EF, eds. *Immunological Diseases of Liver and Gut.* Falk Symposium 135. 2004          ISBN 0-7923-8792-9

136. Leuschner U, Broomé U, Stiehl A, eds. *Cholestatic Liver Diseases: Therapeutic Options and Perspectives.* Falk Symposium 136. 2004          ISBN 0-7923-8793-7

137. Blum HE, Maier KP, Rodés J, Sauerbruch T, eds. *Liver Diseases: Advances in Treatment and Prevention.* Falk Symposium 137. 2004          ISBN 0-7923-8794-5

138. Blum HE, Manns MP, eds. *State of the Art of Hepatology: Molecular and Cell Biology.* Falk Symposium 138. 2004          ISBN 0-7923-8795-3

138A. Hayashi N, Manns MP, eds. *Prevention of Progression in Chronic Liver Disease: An Update on SNMC (Stronger Neo-Minophagen C).* Falk Workshop. 2004
ISBN 0-7923-8796-1

139. Adler G, Blum HE, Fuchs M, Stange EF, eds. *Gallstones: Pathogenesis and Treatment.* Falk Symposium 139. 2004          ISBN 0-7923-8798-8

140. Colombel J-F, Gasché C, Schölmerich J, Vucelic C, eds. *Inflammatory Bowel Disease: Translation from Basic Research to Clinical Practice.* Falk Symposium 140. 2005.          ISBN 1-4020-2847-4